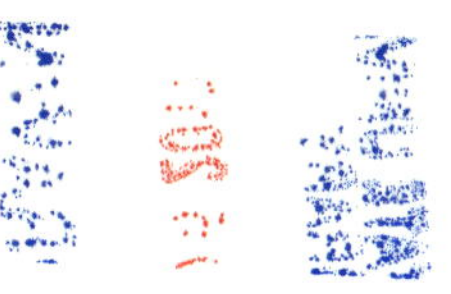

PLACE IN R   nove   be k
TO VO

Ocean Colour: Theory and Applications in a Decade of CZCS Experience

EURO

COURSES

A series devoted to the publication of courses and educational seminars organized by the Joint Research Centre Ispra, as part of its education and training program.
Published for the Commission of the European Communities, Directorate-General Telecommunications, Information Industries and Innovation, Scientific and Technical Communications Service.

The EUROCOURSES consist of the following subseries:

- Advanced Scientific Techniques

- Chemical and Environmental Science

- Energy Systems and Technology

- Environmental Impact Assessment

- Environmental Management

- Health Physics and Radiation Protection

- Computer and Information Science

- Mechanical and Materials Science

- Nuclear Science and Technology

- Reliability and Risk Analysis

- Remote Sensing

- Technological Innovation

REMOTE SENSING

Volume 3

Ocean Colour:
Theory and Applications in a Decade of CZCS Experience

Edited by

Vittorio Barale

and

Peter M. Schlittenhardt

Commission of the European Communities,
Joint Research Centre,
Institute for Remote Sensing Applications,
Ispra, Italy

KLUWER ACADEMIC PUBLISHERS
DORDRECHT / BOSTON / LONDON

Based on the lectures given during the Eurocourse on
'Ocean Colour: Theory and Applications in a Decade of CZCS Experience'
held at the Joint Research Centre, Ispra, Italy,
October 21–25, 1991

Library of Congress Cataloging-in-Publication Data

```
Ocean colour / edited by Vittorio Barale and Peter M. Schlittenhardt.
      p.   cm. -- (Euro courses.  Remote sensing ; v. 3)
   Includes index.
   ISBN 0-7923-1586-3
   1. Optical oceanography--Remote sensing.  2. Colors--Analysis.
 I. Barale, Vittorio, 1954-   . II. Schlittenhardt, Peter M., 1939-
 . III. Series.
 GC180.O24  1993
 551.46'01--dc20                                        92-46130
```


ISBN 0-7923-1586-3

Publication arrangements by
Commission of the European Communities
Directorate-General Telecommunications, Information Industries and Innovation,
Scientific and Technical Communication Unit, Luxembourg

EUR 14888
© 1993 ECSC, EEC, EAEC, Brussels and Luxembourg

Published by Kluwer Academic Publishers,
P.O. Box 17, 3300 AA Dordrecht, The Netherlands.

Kluwer Academic Publishers incorporates the publishing programmes of
D. Reidel, Martinus Nijhoff, Dr W. Junk and MTP Press.

Sold and distributed in the U.S.A. and Canada
by Kluwer Academic Publishers,
101 Philip Drive, Norwell, MA 02061, U.S.A.

In all other countries, sold and distributed
by Kluwer Academic Publishers Group,
P.O. Box 322, 3300 AH Dordrecht, The Netherlands.

Printed on acid-free paper

Printed in the Netherlands

Table of Contents

PREFACE

Remote sensing of ocean colour has been shown to be of invaluable help in monitoring the marine environment and its bio-geo-chemical and physical processes, providing information on a score of key parameters for the assessment of ecological mechanisms on our planet. Amongst the numerous remote sensors placed in Earth orbit to perform environmental observations, the Coastal Zone Color Scanner (CZCS), which functioned on board the satellite Nimbus-7 from late 1978 to early 1986, has been the main source of the early ocean colour data available for most of the marine regions of the world. Much work has been devoted to CZCS data processing and analysis techniques throughout the 1980's. After a decade of experience, various programmes for the complete archival and utilization of CZCS-derived data have been undertaken in recent times by the international scientific community in North America, Europe and Japan.

Following the succes of such initiatives, plans are being made by the major Space Agencies of the world for the carrying out of new ocean colour missions: i.e. the Sea-viewing Wide-Field-of-view Sensor (SeaWiFS), planned for 1993 by the National Aeronautics and Space Administration of the USA; the Ocean Colour and Temperature Scanner (OCTS), to be placed in orbit in 1995 by the space agency of Japan; the Medium Resolution Imaging Spectrometer (MERIS), foreseen for deployment towards the end of this decade by the European Space Agency. Setting the stage to utilize the results obtained so far in the field of ocean colour applications, and to prepare for the advent of new scientific opportunities in the coming years, is a task that was taken up by the International Space Year 1992 (ISY '92).

The material presented in this Volume is drawn from presentations given at a Workshop on OCEAN COLOUR: THEORY AND APPLICATIONS IN A DECADE OF CZCS EXPERIENCE, held in the framework of the ISY '92 Activity devoted to the topic 'Productivity of the Global Ocean' (PGO). The ISY '92 PGO Activity was established by the Space Agency Forum of the International Space Year (SAFISY), with the aim of fostering the full exploitation of remote sensing techniques for an improved understanding of primary productivity in the sea. Therefore, the Activity has been centered, from its very beginning, on the topics of (i) ocean colour data, in particular the existing CZCS archives, (ii) related algorithms and models, as well as (iii) application demonstration programmes.

The ISY '92 PGO Activity has been coordinated, in the 1990/1992 period, by two leading Institutions: the Institute for Remote Sensing Applications (IRSA), Joint Research Centre (JRC) of the Commission of the European Communities (CEC), located in Ispra, Italy; and the Biological Oceanography Division (BOD) of the Bedford Institute of Oceanography (BIO), located in Dartmouth, NS, Canada. One of the main objectives of the Activity has been that of acting as a forum for the coordination of a number of initiatives aimed at the exploitation of the historical CZCS data set - and, consequently, to the preparation of new ocean colour missions in the coming years. Such initiaves involved the attempt, by groups taking part in the Activity, to estimate annual marine primary production at the global scale, using the available CZCS data set and the results of earlier work on the estimation of marine primary production from ocean colour data. Other objectives included the realization of workshops and documents addressing the experience gained in the last decade with CZCS data, as well as the development of demonstrations concerning ocean colour-derived estimates of primary production in selected oceanic regimes.

Various application and/or demonstration programmes have been endorsed or established within the Activity, through the initiative of the PGO Working Group. Two major lines of action have emerged from the meetings of the PGO Working Group, and have been pursued by the Activity's leading Institutions. The first line is that of the Ocean Colour European Archive Network (OCEAN) Project, carried out in Europe by the IRSA, JRC CEC, and the European Space Agency (ESA), for the exploitation of the historical CZCS archives. This Project is now being followed by a proposal for an Ocean Colour Techniques for Observation, Processing and Utilization Systems (OCTOPUS) Programme, devoted to the exploitation in Europe of SeaWiFS and other future ocean colour missions. The second line concerns an attempt of estimating annual marine primary production at the global scale, using both ocean colour and complementary in situ data. This contribution to the PGO endeavour, supported also by the European Space Agency, is being carried out by BIO BOD, on an ocean by ocean basis, using monthly averaged historical CZCS data. It is expected that such activities will constitute the first steps of an effort, to be continued beyond ISY '92 under the auspices of the International Geosphere Biosphere Programme (IGBP), and in concert with its main relevant core project, the Joint Global Ocean Flux Study (JGOFS).

The Workshop, from which the present volume originates, was organized by the IRSA, JRC CEC, in the frame of the ISY '92 PGO Activity, and was held in Ispra, Italy, on 21-25 October 1991, as part of the EUROCOURSES series. Its aim was that of providing a reference in ocean colour science and of promoting the full exploitattion of CZCS data in the field of biological oceanography. The Workshop offered a series of state-of-the-art lectures, by a group of scientists prominent in these fields, on theory, applications and future perspectives of ocean colour.

In particular, after an introduction on the historical perspective of ocean colour and the CZCS mission, the first section of the Workshop was devoted to CZCS theoretical background, radiative transfer and under-water optics, as well as calibration, atmospheric correction and pigment concentration retrieval algorithms developed for the CZCS. A critical review of residual problems in CZCS algorithms was also undertaken. Further, major applications of CZCS data around the world, carried out in the past decade, were reviewed. The second part of the Workshop was centered on the application of ocean colour to the assessment of marine biological information, with particular regard to plankton biomass, primary productivity and the coupling of physical/biological models. The links between global oceanic production and climate dynamics were also addressed. The third section was devoted to future approaches and goals of ocean colour science, to scientific and technological perspectives, as well as to future plans for new sensors and systems devoted to marine optical remote sensing. Finally, the Workshop included discussions on ISY '92 PGO initiatives, current CZCS archives and plans for the full exploitation of CZCS data in experimental regional and global marine productivity assessments.

The meeting provided both an overview of the current status in the field of optical remote sensing of the sea, as well as a forum to debate new ideas and developments for the upcoming ocean colour missions. Several recommendations pertaining to these topics emerged from the debate which was conducted during the Workshop. A major requirement of the ocean colour scientific community appeared to be that of coordinating instrument development, including remote sensors and shipborne or in situ systems (such as, e.g., permanent moored-buoy systems to be used for the calibration of orbital instruments), and of optical research at sea, including activites devoted to a standard calibration of the systems employed. Other requirements concern the need for improved bio-optical algorithms (an issue which involves also the new opportunities offered by upcoming sensors with upgraded spectral capabilities). In particular, research efforts should be directed toward the development of algorithms capable of extracting information on more parameters than just chlorophyll-like pigments; and possibly toward the establishment of regional algorithms, for the retrieval of such parameters. These algorithms should take into account the special water optical properties prevailing in certain areas (e.g. those basins dominated by waters of coastal type), or the peculiar bio-geo-chemical -and environmental at large- setting of entire oceanic regions (e.g. the so-called biogeographical provinces of the world's oceans, or the high latitude regions).

The information derived from the new generation of ocean colour algorithms will undoubtably constitute a major input for primary production models. There is currently a need to discuss and clarify, within the scientific community, the relative merits and demerits, differences and similarities, among the various models proposed for application with ocean colour data. Only after such a clarification, in fact, will it be possible to establish firm guidelines necessary for the operational derivation of data products on marine primary productivity. Similarly, a complementary requirement exists for the availability and use of auxiliary parameters, such as sea surface temperature, winds, and irradiance. Future remote sensing systems devoted to the marine environment should take into account the great potential benefits arising from the availability of concurrent, co-registred data on ocean colour and on these parameters.

Finally, it is obvious that the foreseen availability of improved data sets, algorithms and models places new emphasis on the issues of data processing capabilities and of data archiving and distribution. The shear amount of data to be generated by upcoming remote sensing systems, and the complexity of their operational manipulation, will require the use of improved distributed data bases, networks, and communication links in general, in order to allow for an effective distribution of value-added data and derived information on marine environmental topics.

The Workshop was attended by about 40 scientists - from Australia, Canada, China, France, Germany, Italy, Japan, Mexico, Netherlands, Norway, Portugal, Spain, Switzerland, United Kingdom, USA, as well as from the former USSR and former Yugoslavia - involved in ocean colour and related disciplines. Their contribution to the success of the Workshop, and the compilation of the present Volume, is gratefully acknowledged. Special thanks are due to the Authors of the various Chapters of this Volume, for the time and effort they have put into their presentations and papers, and to the EUROCOURSES staff, for their collaboration in the organization of the Workshop and of the present editorial opportunity. Last, but not least, the invaluable work of Nadia Noui as Editorial Assistant, without whom this Volume would have never reached the printing press, is gratefully acknowledged.

V. Barale

P. M. Schlittenhardt

Ispra, January 1992

List of contributors

R. W. Austin, Center for Hydro-Optics and Remote Sensing, San Diego State University, 6505 Alvarado Road, Suite 206, San Diego, California 92120-5005 USA (Tel. +1/619/5942244, Fax +1/619/5944570)

V. Barale, Institute for Remote Sensing Applications , Joint Research Centre, Commission of the European Communities , 21020 Ispra (VA), Italy (Tel. +39/332/789274, Fax +39/332/789034)

R. Doerffer, Institute of Physics , GKSS Forschungszentrum Geesthacht , Postfach 1160 , 2054 Geesthacht, Germany (Tel. +49/4152/872480, Fax +49/4152/872444)

G. C. Feldman, Goddard Space Flight Centre, National Aeronautics and Space Administration , Greenbelt, MD 20771, USA (Tel. +1/301/2869428, Fax +1/301/2863221)

H. Fukushima, School of High-Technology for Human Welfare, Tokai University, 317 Nishino, Numazu, 424 Japan (Tel. +81/559/681111, Fax +81/559/681155)

F. B. Griffiths, CSIRO Division of Fisheries Research, GPO Box 1538 , Hobart, Tasmania 7001 Australia (Tel. +61/2/206538, Fax +61/2/240530)

H. R. Gordon, Department of Physics, University of Miami , Box 248046, Coral Gables, FL 33124 , USA (Tel. +1/305/2842323, Fax +1/305/2844222)

G. P. Harris, CSIRO Office of Space Science and Applications, Cnr North & Daley Rds, ANU Campus Acton ACT, GPO Box 3023, Canberra 2601, Australia (Tel. +61/6/2790811, Fax +61/6/2790812)

E. E. Hofmann, Center for Coastal Physical Oceanography, Crittenton Hall, Old Dominion University, Norfolk, VA 23529, USA (Tel. +1/804/6834945, Fax +1/804/6835550)

J. Ishizaka, National Institute for Resources and Environment, 16-3 Onogawa, Tsukuba, Ibaraki , 305 Japan (Tel. +81/298/588379, Fax +81/298/588357)

C. R. McClain, Oceans and Ice Branch (code 971), Goddard Space Flight Center, National Aeronautics and Space Administration , Greenbelt, MD 21114 , USA (Tel. +1/301/2865377, Fax +1/301/2862717)

J. Peláez-Hudlet, Instituto de Oceanografía Satelital, ICML - UNAM, A.P. 811, Mazatlán, Sinaloa 82000, México (Tel. +52/69/825546, Fax +52/69/826133)

T. Platt, Biological Oceanography Division, Bedford Institute of Oceanography , P.O. Box 1006, Dartmouth, Nova Scotia, B2Y 4A2 Canada (Tel. +1/902/4268044, Fax +1/902/4269388)

S. Sathyendranath, Department of Oceanography, Dalhousie University, Halifax, Nova Scotia, B3H 4J1 Canada (Tel. +1/902/4263739, Fax +1/902/4267827)

J. J. Simpson, Satellite Oceanography Center, Scripps Institution of Oceanography, University of California at San Diego, La Jolla, CA 92093-0237, USA (Tel. +1/619/5345426, Fax +1/619/5345602)

V. Strass, Alfred Wegener Institut für Polar- and Meeresforschung, Columbusstrasse P.O. Box 120161, 2850 Bremerhaven, Germany (Tel. +49/471/4831494, Fax +49/471/4831425)

B. Sturm, Institute for Remote Sensing Applications , Joint Research Center, Commission of the European Communities , 21020 Ispra (VA), Italy (Tel. +39/332/789934, Fax +39/332/789034)

H. Van Der Piepen, Institute of Optoelectronics, DLR,, Oberpfaffenhofen , 8031 Wessling, Germany (Tel. +49/8153/281127, Fax +49/8153/281349)

K.U. Wolf, Institut für Meereskunde an der Universität Kiel , Düsternbrooker Weg 20, 2300 Kiel, Germany (Tel. +49/431/565876, Fax +49/431/5973867)

C. S. Yentsch, Bigelow Laboratory for Ocean Sciences, P.O. Box 475, McKown Point, West Boothbay Harbor, ME 04575 USA (Tel. +1/207/6332173, Fax +1/207/6336584)

OPTICAL REMOTE SENSING OF THE OCEANS:
BC (BEFORE CZCS) AND AC (AFTER CZCS).

R.W. AUSTIN
Center for Hydro-Optics and Remote Sensing
San Diego State University
6505 Alvarado Road, Suite 206
San Diego, California 92120-5005 USA

ABSTRACT. Following the release of some spectacular over-water photography taken by the astronauts on the earth-orbital Gemini and Apollo missions in the 1960's, researchers from academia, government laboratories and industry proposed a number of potential applications for optical remote sensing of the oceans. Subsequently, various studies and experimental efforts were undertaken to better understand and exploit these applications of the color patterns in the sea. By the early 70's, the requirements for an ocean color sensor operating at satellite altitudes had become sufficiently well understood for NASA to prepare specifications for and to initiate the procurement of the Coastal Zone Color Scanner (CZCS) which was launched in October 1978. A NASA Experiment Team (NET) was formed in 1976. They were charged with the task of advising NASA on the type of output products that could be derived from the sensor, for developing the algorithms necessary to obtain those products, and for conducting the surface experiments necessary to validate the satellite derived output products. A brief account is provided of the development of our knowledge of ocean optics and ocean color over the past 100 years, and of the major influence of satellite remote sensing on our understanding and acceptance of ocean optical phenomena. The ocean science community has applied CZCS data to a variety of ocean research problems with eminent success.

1. Introduction

The Coastal Zone Color Scanner (CZCS) was the first satellite-borne sensor designed to remotely assess the biological productivity of ocean waters by means of the apparent color of the water. By ocean color, we mean the relative amounts of water-leaving radiance in the various portions of the visible spectrum. The apparent color of the water at the satellite is that water leaving color (or those relative radiances) after propagation upward through the atmosphere. We will first turn our attention to the nature of the ocean color signal and to a bit of the history of our understanding of marine optics and ocean color.

Standing on the deck of a ship or looking down at the ocean from an aircraft, one becomes aware of the changing color of the ocean. If the water is sufficiently deep that the color is not materially affected by the reflected light from the ocean floor, then the observed color is due to the sum of the reflected light from the ocean surface and the subsurface light reflected from the water and the suspended particulate material it contains.

1

V. Barale and P.M. Schlittenhardt (eds.),
Ocean Colour: Theory and Applications in a Decade of CZCS Experience, 1–15.
© 1993 *ECSC, EEC, EAEC, Brussels and Luxembourg. Printed in the Netherlands.*

The surface reflected light is an unwanted masking signal and is a consequence of the differing indices of refraction of the water and the air. This surface reflectance varies only slightly with wavelength. Thus the apparent color that one observes due to this component is attributable to the color of the sky light being reflected by the sea surface. The magnitude of this surface reflectance is dependent on the angle of observations, however, and is between 2 and 3% for angles from 0 to 45° from the nadir, increasing to 6% at 60°, 13.5% at 70°, 35% at 80° and 100% at 90°. These values are for a calm sea and are modified, particularly at angles greater than 70° by wind roughening of the surface (Austin, 1974).

The light propagating up from beneath the surface contains information about the material dissolved and suspended in the water. This component of ocean color varies markedly with wavelength and in a very different fashion for the blue unproductive waters typical of the oligotrophic central oceans and for the green productive waters found in coastal and upwelling regions. The irradiance reflectance for central oceans might be as high as 10% in the blue, decreasing rapidly to a few tenths of a percent in the yellow and longer wavelengths. For the greener waters, the reflectance might range from 1% in the blue to 2 or 3% in the green and again dropping rapidly at the longer wavelengths.

Thus we see that in order for an observer to sense the color of the water and not the reflected skylight, the angle of observation should be steeply downward (less than roughly 45° from the nadir) in order to minimize the masking effects of the reflected skylight. Furthermore, the above comments assume clear cloudless skies. Reflected radiance from a cloudy sky greatly decreases the ability of an observer or sensor to assess the true color of the ocean water.

In common with many areas of study, the progress in both marine optics and ocean color remote sensing has not been a continuum, but rather progress has been episodic and has followed certain events, discoveries, people and groups. Major advances have thus been superimposed on a background or continuum of progress. We will describe the history of marine optics, and the much briefer history of ocean color remote sensing, as epochal in nature and present some of the significant factors in the development of both of these disciplines.

The two areas obviously have many common aspects, although remote sensing introduces a number of additional complications (as for example, the atmospheric degradation of the signal, glint or sunlight reflected from the sea surface, polarization of light by the atmosphere, etc). We shall first very briefly discuss the development of marine optics and the people and events that have helped to shape it ,and then turn our attention to ocean color remote sensing.

2. Marine Optics

The first interest in what we now call marine optics was probably by mariners who wanted to know how deep the waters were beneath their ships. It was for that reason that P.A. Secchi in 1885 studied the disappearance of circular disks in the Tyrrhenian Sea. His "Reports on Experiments Made on Board the Papal Steam Sloop L'Immacolata Concezione to Determine the Transparency of the Sea" makes fascinating reading, and I recommend it to anyone interested in seeing how much information can be deduced from a simple but carefully planned (truly low budget) set of experiments (Cialdi, 1886). Although he may not have invented the device that bears his name, Secchi appears to have been one of the first physical scientists to try to understand the factors that affected the disappearance depth and to codify the techniques for its proper use. This very unsophisticated, and perhaps first, instrument for marine optics is still in use all over the world and the data base of Secchi depths was probably the largest in number and geographical coverage of any body of marine optical data until the era of the CZCS. Unfortunately, some have imputed greater significance to properties inferred from this simple measurement than it is capable of providing. Preisendorfer (1986) has written a comprehensive and definitive paper on the subject on the 100th anniversary of Secchi's original report.

The next item of interest to us was the invention of a color scale by F.A. Forel (1890) which originally was used to assess the color of Swiss lakes. The scale consists of sealed vials containing mixtures of two aqueous solutions: one of copper sulfate, the other of potassium chromate. The colors of the vials range from blue through yellow as the relative amounts of the two solutions are changed. The vials are compared by an observer to the water color when viewing a Secchi disk near the surface. This scale, or variants of it, is still in use today.

Kalle (1938) reviewed and made important contributions to the theory of the causes of the color of ocean water. He correctly described the blue of clear open ocean water as the result of molecular scattering and concluded that the green color of coastal waters was due to the addition of yellow substance or "gelbstoff" (which absorbs the blue light). He also suggested that the color of larger particles (such as phytoplankton), if present in sufficient concentration, could also impart color to the water.

In the late twenties and during the thirties, a number of investigators made measurements of the absorption coefficient of pure water and of pure sea water and concluded that they were essentially the same. It was at this time that George Clarke made his first of many important contributions to the fields of ocean optics, ocean color and finally to ocean color and remote sensing. He measured the penetration of daylight into ocean waters (Clarke 1933, 1936), and together with James (Clarke and James, 1939) measured the spectral absorption of seawater. In the same period, Utterback and his colleagues made a number of similar determinations of spectral attenuation of various bodies of ocean water.

Following World War II, there was a rapid expansion in activity in marine optics in general and in our knowledge of the theory and measurement of ocean color. There appeared several research groups that were very effective in advancing the "state of the art". In France, there were Le Grand, Lenoble, Ivanoff, and presently Morel, all of whom have made significant contributions. In the Soviet Union, there has been a very large effort in marine optics. It is unfortunate, however, that in the past the exchange of information has been so difficult. This situation now appears to be swiftly changing for the better. In Japan, the work of Sasaki, Okami, Oshiba, Watanabe and others has resulted in many additions to the marine optics data base including a large and valuable body of in situ spectral irradiance data.

One of the most important individuals in the post-war arena for marine optics was N. G. Jerlov. He is best known for his two books, "Optical Oceanography" (1968) and its later revision "Marine Optics" (1976), and for his classification scale for oceanic and coastal waters. He made many other important contributions to the literature and headed the Institute of Physical Oceanography at the University of Copenhagen, which has been the source of much excellent research in the field.

In the United States, the Visibility Laboratory was formed by S.Q. Duntley at M.I.T. in the early 1950's. It was moved to the Scripps Institution of Oceanography in 1952. This group consisted of physicists, mathematicians, psychologists, and engineers with the general charter to study the factors affecting visibility and detection of objects in the ocean or in the atmosphere, either by human observers or by physical devices. Among the studies undertaken were the optical properties of the marine and atmospheric environment. New methods for the measurement of those properties were devised and the necessary instruments were developed. R.W. Preisendorfer undertook to place a rigorous mathematical foundation beneath the various efforts of the Laboratory. This resulted in many contributions to the literature of marine optics including a six volume treatise entitled "Hydrologic Optics" (Preisendorfer, 1976). Duntley together with J.E. Tyler, and others made many contributions to both the theoretical and experimental sides of marine optics including instruments for measuring volume scattering function, radiance distribution, and spectral irradiance in the ocean. The measurements made by Tyler and R.C. Smith (1970) using the submersible spectroradiometer in a wide variety of ocean areas and fresh water lakes have provided a widely used and referenced set of spectral irradiance data. The same instrument was modified to measure upwelling spectral radiance and found wide use by Smith, Austin and others for providing data for algorithm development and surface truth for the CZCS program.

There are, of course, other groups and individuals that have contributed significantly to ocean color and marine optics. Beardsley, Zaneveld, and Mueller at Oregon State University; Howard Gordon and his colleagues and students at the University of Miami; and Charles Yentsch, the founder and first director of the Bigelow Laboratory for Ocean Sciences, to name a few in the U.S. In Australia, John Kirk at CSIRO has made important contributions to hyrologic optics.

The above listing of individuals is not intended to be complete, but only to mention some of those that have participated in the episodic nature of the advances in marine optics, and in particular to those aspects which have contributed to our knowledge of ocean color. Figure 1 provides a summary of the chronology of some of the significant people and events in the study of ocean color.

One last example of an event that had a major impact on ocean optics research in the 1960's. An *ad hoc* group of the Scientific Committee for Ocean Research (SCOR) was formed in November 1963 under the auspices of UNESCO and IAPSO as the SCOR Working Group 15. The group consisted of eight researchers from seven countries who were specialists in marine productivity studies and in marine optics. The charge of the Working Group was:

1. to identify exactly what measurement of irradiance is required by biological oceanographers, and

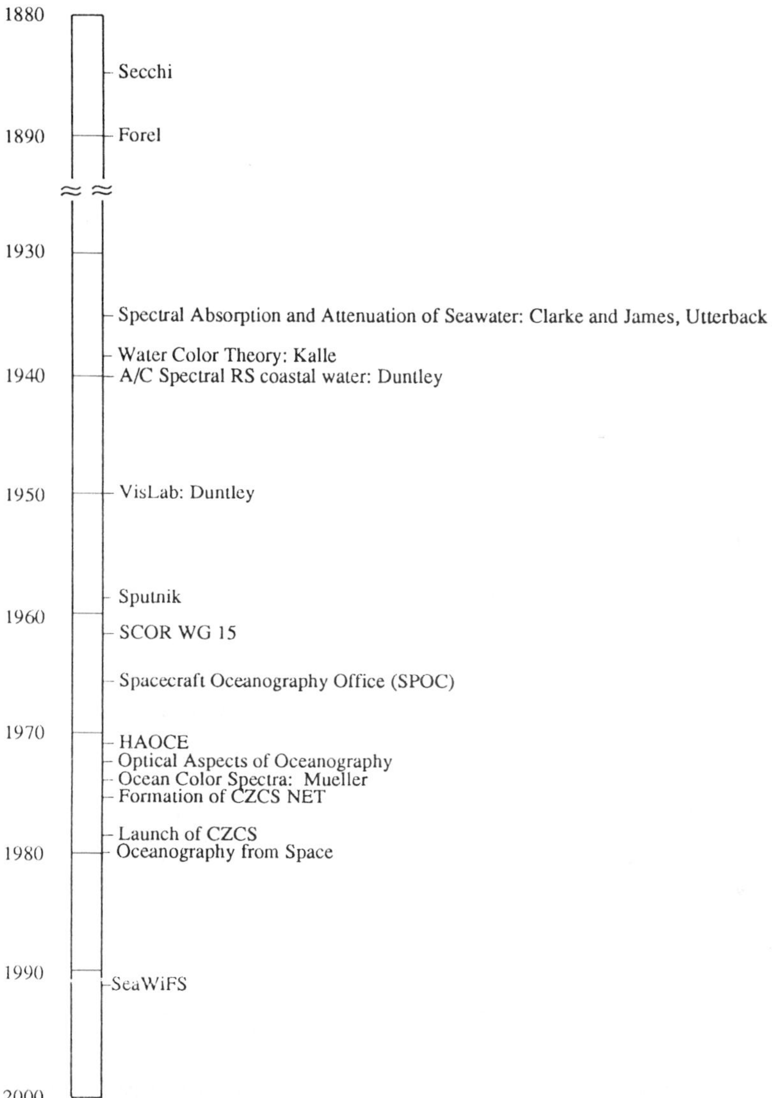

Figure 1 Chronology of Ocean Color.

2. to recommend apparatus and procedures for measuring the variable defined above.

Among its activities, the group sponsored two cruises in May of 1968 and May 1970. The second cruise on the R/V Discoverer from Miami, Florida to the Peruvian upwelling was the major activity of the Working Group. Seventeen scientists and technicians participated: 5 from the U.S., 2

Australian, 2 French, 2 from the U.S.S.R., 2 Japanese, 2 Danish, 1 UK and 1 Norwegian. The work of the Group provided a need for the exchange of information on techniques for performing in-water measurements, on instrument design and calibration, and on nomenclature standards. The Group also produced a very large body of valuable *in situ* spectral irradiance data. But perhaps as important as anything was the opportunity provided the participants to collaborate with, and get to know our colleagues from around the world. Many of the contacts made through SCOR WG 15 were lasting and important to participants during the subsequent 20 years.

3. Ocean Color Remote Sensing: PRE CZCS.

With the launch of Sputnik in October of 1957, attention became focused on space and what one could do from this new perspective of the earth. Frankly, not much thought seems to have been given to the possibilities of utilizing ocean color as a means of assessing parameters of oceanographic concern until color photographs of the ocean became available from the manned Gemini and Apollo earth orbital missions. Some of these photographs were quite tantalizing and provoked considerable interest in the possibilities for identifying water masses, assessing wind fields, studying surface expression of internal waves, assisting bathymetric surveys and some even speculated on the possibility of trying to assess chlorophyll concentrations from space.

In August of 1964, NASA convened a "Conference on the Feasibility of Conducting Oceanographic Explorations from Aircraft, Manned Orbital and Lunar Laboratories" at Woods Hole Oceanographic Institution, with Gifford Ewing as Chairman. This conference was important for stimulating thinking and exchanging ideas, but did not present many ideas that led to ocean color remote sensing as we know it today.

At this point in time, most oceanographers were not interested in remote sensing, particularly ocean color remote sensing. They looked upon it as competition for funds, and they were in need of funding for salaries, equipment and shiptime for their existing programs. And, indeed, the potential for ocean color providing benefit to conventional oceanography had not been demonstrated.

Much of the research in the U.S. was supported by the various NASA centers with no central or coordinated oceanography program. In 1966 NASA provided funding to the U.S. Navy Oceanographic Office to initiate the Spacecraft Oceanography (SPOC) Program Office. Its purpose was to foster increased interest in, and to coordinate support for, research in all forms of remote sensing of the oceans. In 1967 J.W. Sherman became head of the SPOC program and was the driving force in supporting much of the innovative research in ocean remote sensing, certainly up to the time actual satellite sensors were approved. One of the reasons for success in ocean remote sensing was his long term support and encouragement of research in academia, industry and in government laboratories. In May of 1972, the SPOC program was moved to the National Oceanic and Atmospheric Administration (NOAA).

There really were not very many members of the ocean color remote sensing community in those early days. Certainly George Clarke and Gifford Ewing were among the early participants. They were originally interested in mounting sensors in aircraft. Using a spectroradiometer (provided by TRW Systems, Inc.) with a 3° field of view, scanning 400 to 700 nm in 1.2 seconds, they flew at 305 meters (1000 feet) over a variety of ocean areas with chlorophyll concentrations from <0.1 to 3.0 mg/m^3. They obtained excellent differentiation of the signals from the various water types (Clarke *et al.*, 1970). They became keenly aware of the degradation of the water-leaving signal as they increased altitude, however.

Other participants in the ocean color arena were Peter White and Dick Ramsey at TRW who not only provided the excellent research airborne spectroradiometric equipment used by Clarke and Ewing and by J.L. Mueller (1976), but also proposed innovative techniques for analysis of spectral signatures such as the use of second derivative processing to extract information from high altitude data (White, 1971). At the Environmental Research Institute of Michigan (ERIM), there were a number of very productive research programs in ocean color remote sensing including those directed by Fabian Poulson who was particularly interested in the bathymetric applications. ERIM also had their own very excellent capability in aircraft remote sensing with a number of aircraft and very capable scanning multispectral radiometers.

Don Ross and Reese Jensen founders of International Imaging Systems (I^{2}S), were interested in the applications of multispectral aerial photography to a variety of ocean color problems. They developed a modification to a 9 inch aerial survey camera that used four matched lens in lieu of the original single lens. Each lens was fitted with a different spectral filter, and film magazines were available for either a single roll of 9 inch film or for 2 rolls of 5 inch aerial film, should different emulsions be required. This camera was used for a number of studies, but the difficulties of photographic radiometry and the advent of high spectral resolution scanning radiometers rendered the photographic techniques obsolete.

One of Ross's projects was an attempt to infer bathymetry of the northern part of the Gulf of California from a color photograph obtained by the Gemini 5 astronauts (Ross, 1969). While it was true that interesting correlations existed between image iso-density contours and reported depths, Austin (1972) performed in situ measurements of the water attenuation properties and demonstrated that no reflected signal from the bottom could be seen from the surface because of thick layers of opaque sediment laden water that underlay the clearer surface waters. That part of the Gulf of California has strong tidal currents due to 8 to 10 meter tidal variations and the highly turbid layers were presumably due to the resuspension of bottom sediments originally brought to the area by the Colorado River.

In the NOAA National Marine Fisheries Service, Bill Stevens, Andy Kemerer, Mike Laurs and others were engaged in the study of applications of remotely sensed ocean color to fisheries management and related problems.

At the S.I.O. Visibility Laboratory, Duntley and his colleagues were very active from the start in ocean color research, and they were able to make other contributions as well because of the active atmospheric optics program at the Laboratory. Numerous aircraft experiments were supported with in-water spectral radiometric measurement campaigns. These included measurements in the Western Atlantic for Ewing and Clarke, and in 1971 in California coastal waters in support of the High Altitude Ocean Color Experiment (HAOCE). In the HAOCE, Warren Hovis, from GSFC, flew a

8

spectroradiometer mounted in a Lear Jet at two altitudes over a surface ship while the Visibility Laboratory documented the in-water upwelling spectral radiance, the chlorophyll pigment concentration, and the appropriate atmospheric optical properties. The effort of preparing the report on that experiment for our Laboratory gave me the opportunity to carry through the calculation of the complete history of the signal from the solar input to the interaction of that radiation with the atmosphere, the water surface, its reflectance by the ocean water, and its transmission back up through the atmosphere. A version of that effort was later presented to the Optical Aspects of Oceanography Symposium in 1972 in Copenhagen (Austin, 1974).

In about 1973, the Coastal Zone Color Scanner became an approved program. About the same time, the Ocean Color Scanner was built by Goddard Space Flight Center (GSFC) for use in the NASA U-2 aircraft. That instrument was to provide additional input and pre-launch experience for the CZCS program. Flying at 19 km, it was above most of the atmosphere, so the received signal could be expected to be similar to that in the proposed CZCS. Unfortunately, the instrument as originally built was sensitive to the polarization of its input radiance and the results were consequently disappointing.

In 1975, the Experiment Team for the CZCS was formed. It consisted of members from academia, government laboratories and had two non-U.S. members. Figure 2 lists the team members.

<table>
<tr><td>Warren Hovis (Sensor Scientist)</td><td>NASA / GSFC</td></tr>
<tr><td>C. S. Yentsch</td><td>Bigelow Lab. for Ocean Sciences</td></tr>
<tr><td>D. K. Clark</td><td>NOAA/NESS</td></tr>
<tr><td>E. T. Baker</td><td>NOAA/PMEL</td></tr>
<tr><td>S. Z. El−Sayed</td><td>Texas A&M University</td></tr>
<tr><td>H. R. Gordon</td><td>University of Miami</td></tr>
<tr><td>R. C. Wrigley</td><td>NASA/Ames</td></tr>
<tr><td>R. W. Austin</td><td>SIO/Vis Lab</td></tr>
<tr><td>F. Anderson</td><td>NRIO, Capetown, South Africa</td></tr>
<tr><td>B. Sturm</td><td>CEC/JRC, Europe</td></tr>
</table>

Figure 2. CZCS Nimbus Experiment Team (NET).

The NET was charged with providing NASA with algorithms for correcting the received image data for atmospheric effects, for devising algorithms for deriving pigment concentrations and diffuse attenuation coefficients from the atmospherically corrected image data, to validate the data so produced and to generally provide advice and support to the project from the scientific and user community. The Team conducted a major oceanographic campaign in the Gulf of Mexico using the NOAA R/V Researcher from AOML in Miami and the Texas A&M R/V Gyre from Galveston Texas. The campaign was conducted 12-30 October 1977, approximately a year before the launch of the Nimbus 7 satellite which was to carry the CZCS. NET team members and supporting staff made extensive optical, biological and physical oceanographic measurements at noon time stations as well as a limited set of measurements along track while underway. These data were used in the development of the in-water algorithms and to gain procedural experience for the teams in anticipation of the post launch validation cruises.

4. Ocean Color Remote Sensing: Post CZCS

In October 1978, the Nimbus 7 spacecraft was successfully placed in a sun synchronous polar orbit. It carried the CZCS along with a number of other sensors. The orbit was tilted 9.28° west of a true south-to-north orbit and crossed the equator slightly before noon (11:50 Local Apparent Time). Figure 3 shows orbit tracks for February 27, 1979. The shaded areas show the swath coverage for the periods when the CZCS was turned on. Because of the power demands of the several sensors carried on Nimbus 7, the CZCS was limited to 2 hours of operation each day.

This time was apportioned to ocean areas of interest to the oceanographic community by the NASA Goddard Space Flight Center (GSFC), with inputs from the CZCS Experiment Team. The "on" periods were usually increments of 2 to 10 minutes each. The received data were reviewed for indications of cloud cover, sun glint contamination and other factors which might impair the data quality. Suitable data was packaged into scenes containing up to 970 lines (2 minutes of along-track data). Each scan line represented a swath 78.68° wide and was sampled to provide 1968 picture elements (pixels). At the satellite altitude of 955 km, the swath was 1659 km wide when the scan was without tilt The scanner could be tilted ± 20° in 2° steps to avoid sun glint; the swath width increased with northward tilt angle and decreased with

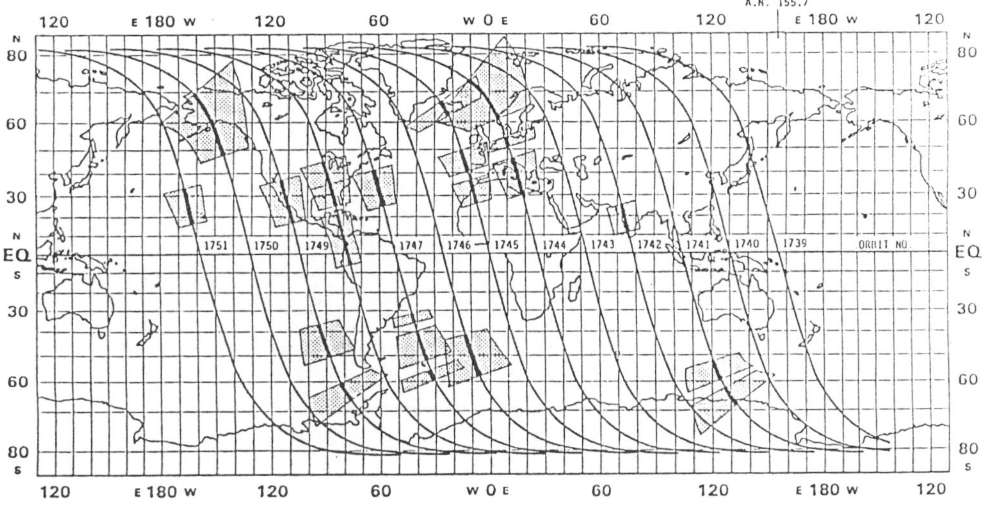

Figure 3 Example of CZCS orbit tracks for 1 day (February 27, 1979).

southward tilt angle. The scanner IFOV (instantaneous file of view; ˜0.05°) yielded pixel sizes which varied from 825 meters at the nadir to 1653 meters at the edge of the swath (with 0° tilt). The orbital velocity of the spacecraft made each successive line advance 793 meters. Thus the successive scan lines slightly overlap, and the 970 lines which comprised a full 2 minute scene represented 769 km along the track. Figure 4 shows schematically the scanning arrangement and the scan pattern.

There were 6 co-registered channels on the scanner. Four were 20 nanometer wide channels in the visible spectrum, one was a near-IR, 100 nanometer wide channel, and one channel in the thermal infrared measured apparent sea surface temperature. Table 1 summarizes the spectral and radiometric characteristics of the CZCS channels.

Most previous satellite remote sensing programs had followed an empirical approach for developing algorithms. Users of LANDSAT data for forestry, agriculture, and urban studies, for example, developed algorithms by noting the type of signals received when observing pine forests, wheat, crops

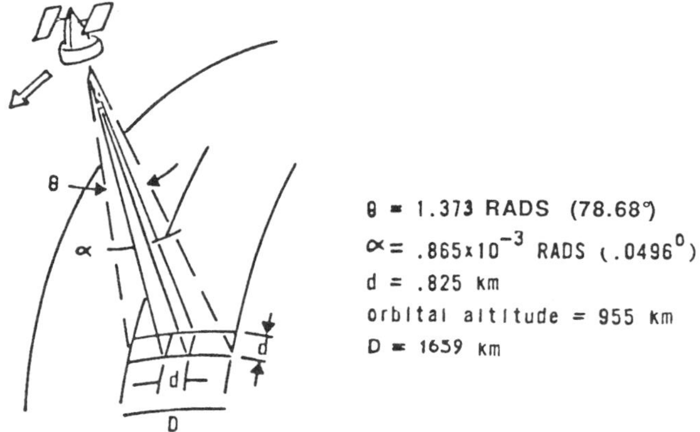

Figure 4a. CZCS Scanning Arrangement.

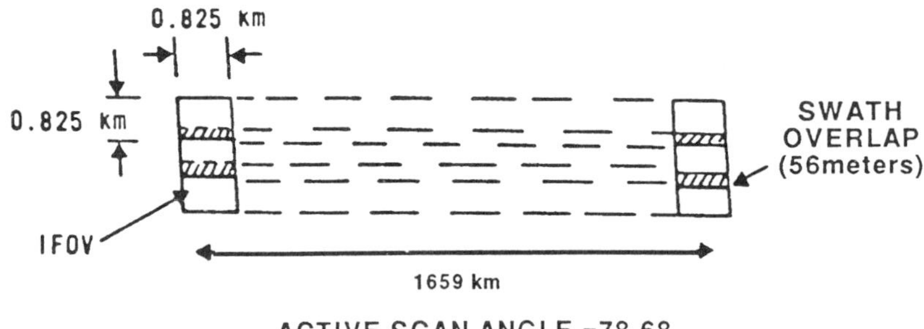

Figure 4b. CZCS Earth Scan Pattern.

TABLE 1 Characteristics of the CZCS

Channel	Wavelengths (nm)	Gain	Saturation Radiance ($\mu W.cm^{-2}.sr^{-1}.nm^{-1}$)	Measured Signal/Noise
1	433-453	4	5.41	158/1
		3	7.64	
		2	9.23	
		1	11.46	
2	510-530	4	3.50	200/1
		3	5.10	
		2	6.20	
		1	7.64	
3	540-560	4	2.86	176/1
		3	4.14	
		2	5.10	
		1	6.21	
4	660-680	4	1.34	118/1
		3	1.91	
		2	2.32	
		1	2.88	
5	700-800		23.9	350/1
6	10,500-12,500			0.22 K*

* Noise equivalent temperature difference at 270 K

suffering from certain diseases, etc., and devised schemes to maximize the differentiation between the various types scenes and "recognize" the particular vegetation or land use. This technique of "training" the system to accomplish the desired task was not used by the CZCS Experiment Team for a variety of reasons.

First, it had been demonstrated that over ocean water, with its inherently low reflectance, 80 to 90% of the signal received by a sensor at satellite altitudes could be expected to be due to the atmosphere and not the water (see, for example, Austin, 1974). It was, therefore, important to devise a method, based on the physics of the radiative processes involved in the transmission of the water-leaving signal through the atmosphere, that could remove the effects of atmospheric transmission and backscatter of solar radiation in the best possible manner.

Second, as opposed to the more stable situation, (both spatially and temporally) encountered in terrestrial scenes, the relatively subtle patterns of water color change on scales of days and kilometers. It is, therefore, necessary to compare the satellite signal with contemporaneously obtained measurements from a surface vessel whose position can be accurately determined in the satellite image (*i.e.*, within a pixel in many situations) in order to validate the overall data retrieval process.

The CZCS Experiment Team conducted a series of six post-launch cruises in the Western Atlantic, Gulf of Mexico, Gulf of California and in California coastal waters. Using data obtained on these and the pre-launch cruises, and data from contemporaneous CZCS images where available, the Team developed in-water algorithms for the determination of chlorophyll-like pigment concentrations (Clark, 1981) and of the diffuse attenuation coefficient of the surface waters (Austin and Petzold, 1981).

These and similar data sets obtained by European and South African investigators, were also used to define and validate the various atmospheric algorithms. In the US, the principle atmospheric algorithm development was due to the work of Gordon (1978), with an interesting iterative technique due to Smith and Wilson (1981) being particularly helpful in turbid water. In Europe, there were various important contributions to the atmospheric correction problem.

In 1979, an international workshop was held at the Joint Research Centre of the CEC in Ispra, Italy (Sorensen, 1979) to assess the various efforts. Additional important work was reported by Quenzel and Kaestner (1980), Tanre, *et al.* (1979), and Viollier *et al.* (1979). In South Africa, Walters (1985) developed atmospheric algorithms which were successfully used in the South African ocean color and upwelling experiment. Reviews of the algorithms and their development are provided by Gordon and Morel (1983) and by Gordon et al (1985).

The atmospheric correction algorithms have undergone significant improvements, as for example, to include actual (vice climatological) atmospheric pressure and ozone fields, and to include the effects of multiple scattering (Gordon *et al.*, 1988). With these improvements in place, NASA has reprocessed the majority of the eight years of CZCS data. As a consequence, there now exists a remarkable data base providing the global distribution of phytoplankton pigment distribution as a function of time and location.

The applications for the CZCS ocean color data are of interest to both the biological and the physical oceanographer. The biologists, many of whom had been reluctant to acknowledge the utility of remote sensing to their research, rapidly recognized that the ability to have a synoptic view of the distribution of near surface pigments provided insights into ocean processes that they could never obtain otherwise. The physical oceanographers also saw that by using plant pigment as a tracer, they could visualize frontal activity, gyres, jets, currents, etc., in a fashion that was heretofore impossible. The naysayers became believers, and passive, if not active supporters.

A new term was introduced into the oceanographic literature by Smith and Baker (1978) *i.e.* the "bio-optical" state of ocean water: the combined effect of absorption and scattering due to suspended and dissolved biogenous material. Biologists became very interested in the optical properties of the ocean, and their measurement, as it became obvious that ocean optics, and ocean color in particular, could provide valuable insights into biological processes.

The recent "discovery", and increased awareness of ocean optics is readily measured by the large increase in the number of papers presented at national society meetings and symposia, and appearing in journals. Certainly, the advent of the CZCS, and the success of its derived data products in depicting ocean phenomenology, has been the greatest single episode in the advancement of the study of ocean color.

References

Austin, R.W. (1972) 'Surface truth measurements of optical properties of the waters in the Northern Gulf of California', 4th Annual Earth Resources Program Review, V. IV NOAA and NRL Programs, Sec. 106, 21pp.

Austin, R.W. (1974) 'Remote sensing of spectral radiance from below the ocean surface', in N. G. Jerlov and E. Steeman Nielsen (eds.), Optical Aspects of Oceanography, Academic Press, New York, pp. 317-344.

Austin, R.W. and Petzold, T.D. (1981) 'The determination of the diffuse attenuation coefficient of sea water using the Coastal Zone Color Scanner', in J.R.F. Gower (ed.), Oceanography from Space, Plenum, New York, pp. 239-256.

Cialdi, Comm. Alessandro (1986) ' Sul moto ondoso del mare e su le correnti de esso specialmente anquelle littorali', (2nd Ed.s Rome 1866), in ONI Transl. A-655, pp 35 USN Hydrographic Office, 1955, pp. 258-288.

Clark, D.K. (1981) 'Phytoplankton algorithms for the Nimbus-7 CZCS', in J.R.F. Gower (ed.), Oceanography from Space, Pleum, New York, pp. 227-238.

Clarke, G.L., Ewing, G.C.,and Lorenzen, C.J. (1970) 'Spectra of backscattered light from the sea obtained from aircraft as a measure of chlorophyll concentration', Science 167, 1119-1121.

Clarke, George, L. (1933) 'Observations on the penetration of daylight into mid-Atlantic and coastal waters', Biol. Bull. 65, 317-337.

Clarke, George L. (1936) 'Light penetration in the western North Atlantic and its application to biological problems', Consiel Perm. Intern. p. l'Explor. de la Mer., Rapp. et Proc. Verb., v 101, pt. 2, no. 3, 14pp, 1936.

Clarke, George L. and Harry, R. James (1939) 'Laboratory analysis of the selective absorption of light by sea water', Journal Opt. Soc. Am., 29, 43-55.

Ewing, G.C. (1965) 'Oceanography from Space', Proceedings of conference on the Feasibility of Conducting Oceanographic Exploration from Aircraft, Manned Orbital and Lunar Observations, W.H.O.I., Ref. no. 65-10, Woods Hole.

Forel, F.A. (1890) 'Une nouvelle forme, de la gamme de couleur pour l'etude de l'eau des lacs', Soc. Vaud. Bull., Lausanne, 25, vi (sic), 1890.

Gordon, H.R. and Morel, A.Y. (1983) 'Remote Assessment of Ocean Color for Interpretation of Satellite Visible Imagery; A Review', Springer-Verlag, New York.

Gordon, H.R., Austin, R.W., Clark, D.K., Hovis, W.A., and Yentsch, C.S. (1985) 'Color measurements', Chapter 8, in B. Saltzman (ed.), Advances in Geophysics, v 27, Satellite Oceanic Remote Sensing, Academic Press, Orlando, pp. 297-333.

Gordon, H.R., Brown, J.W., and Evans, R.H. (1988) 'Exact Rayleigh scattering calculations for use with the Nimbus-7 coastal zone color scanner', Applied Optics 27 (5), 862-871.

Gordon, H.R. (1978) 'Removal of atmospheric effects from satellite imagery of the oceans', Applied Optics 17, 1631-1636.

Jerlov, N.G. (1968) 'Optical Oceanography', Elsevier Oceanography Series 5, Elsevier, Amsterdam, 1st ed., 194pp.

Jerlov, N.G. (1976) 'Marine Optics', Elsevier Oceanography Series 14, Elsevier, Amsterdam, 2nd ed., 231pp.

Kalle, K. (1938) 'Zum Problem der Meereswasserfarbe', Ann. d. Hydrogr. Und Mar. Meteor., Bd 66 5, pp. 1-13.

Preisendorfer, R.W. (1986) 'Secchi disk science: visual optics of natural waters', Limnology and Oceanography 31 (5), 909-926.

Preisendorfer, R.W. (1976) 'Hydrologic Optics', 5 Vols., Washington, W.S. Dept. of Commerce.

Quenzel, H. and Kaestner, M. (1980) 'Optical properties of the atmosphere: calculated variability and appliction to satellite remote sensing of phytoplankton', Applied Optics 19, 1338-1344.

Smith, R.C. and Wilson, W.H. (1981) 'Ship and satellite bio-optical research in the California Bight', in J.R.F. Gower (ed.), Oceanography from Space, Plenum, New York, pp. 281-294.

Sorensen, B.M. (1979) 'Recommendations of the International Workshop on Atmospneric Correction of Satellite Observation of Sea Water Colour',Ispra, Italy, March 1979.

Smith, R.C. and Baker, K.S. (1978) 'The bio-optical state of ocean waters and remote sensing', Limnology and Oceanography 19, 1-12.

Tyler, J.E. and Smith, R.C. (1970) 'Measurements of Spectral Irradiance Underwater', New York, Gordon and Breach, 1979.

Tanre, D. M. Herman, Deschamps, D.Y., and de Leffe, A. (1979) 'Atmospheric modeling for space measurements of ground reflectances, including bidirectional properties', Applied Optics 18, 3587-3594.

Viollier, M., Ballois, M.Y., and Lecompte, Pl. (1979) 'Wavelength dependence of aerosol optical thickness', in Proceedings of the Workshop on the EURASEP Ocean Colour Experiment, Ispra, Italy, October 1979.

White, P.G. (1971) 'High altitude remote sensing of the ocean', in Y. H. Katz (ed.), Remote Sensing of Earth Resources and the Environment, SPIE Seminar Proc. 27, 111-114.

Walters, N.M. (1985) 'Algorithms for the determination of near surface chlorophyll and semi-quantitative total suspended solids in South Africa coastal waters from Nimbus-7 CZCS data', Chapter 12 in South African Ocean Colour and Upwelling Experiment (L.V. Shannon, ed.), Capetown, Sea Fisheries Research Institute, pp. 175-182.

CZCS: ITS ROLE IN THE STUDY OF THE GROWTH OF OCEANIC PHYTOPLANKTON

C. S. YENTSCH
Bigelow Laboratory for Ocean Sciences
McKown Point
West Boothbay Harbor
ME 04575 USA

ABSTRACT. The so called nutrient-light hypothesis, and its basic model assumptions, are used to interpret the seasonal change in CZCS derived chlorophyll observed in regions of the North Atlantic. The combination of P:R and nutrient entrainment in the mixed layer give results which agree with satellite imagery. This agreement supports the idea that ocean color imagery can be used in conjunction with regional databases in a fashion similar to the terrestrial GIS systems. One of the major achievements of CZCS to date has been to reintroduce the major role played by ocean circulation on phytoplankton growth and distribution.

1. Introduction

The use of satellites to study ocean processes and in particular ocean color, has not always been popular. In 1960, the advocates of remote sensing found it an uphill battle to change current thinking. Three things accounted for this turnaround : environmental concerns for planet Earth, the observed close linkage between ocean color and ocean fluid dynamics and the successes of the CZCS mission, namely the algorithms of H.R. Gordon and R.W. Austin. In the latter case, the algorithms provided biological oceanographers the means of testing hypotheses that concern phytoplankton growth. The specific question to be asked is "Can changes observed in ocean color be explained by traditional concepts gathered from sea surface observations?" This paper attempts to demonstrate that satellite remote sensing of ocean color is a dialectical tool, by this I mean that it provides a means of not only observing, but studying ocean processes.

This discussion addresses the processes which are believed to regulate phytoplankton growth in the open North Atlantic Ocean. I begin with a "bare bones" outline of the principal processes believed to regulate growth -- I refer to this as the "nutrient-light hypothesis" (see Yentsch 1990). The concepts embodied in this hypothesis are used to interpret the time/space changes in phytoplankton abundance as observed in North Atlantic images. There is a danger in putting forth concepts that have molded our thinking for over fifty years -- the first is dogmatic belief, the second is not clearly identifying who did what, where and when. Present day wisdom is a corporate body of ideas arising from many observations. To offset the problem of identification of

17

V. Barale and P.M. Schlittenhardt (eds.),
Ocean Colour: Theory and Applications in a Decade of CZCS Experience, 17–32.

sources, let me explain very briefly how this corporate body of wisdom came about.

To my knowledge, Gran and Braarud (1935) were the first to suggest the idea that deep mixing could adversely affect phytoplankton growth. H. U. Sverdrup (1953), and D. H. Cushing (1962) modeled these concepts. G. A. Riley (1963) used these and nutrient enrichment ideas of Harvey at the Plymouth laboratory and others to model phytoplankton growth, and produced a diagnostic model for optimal growth in a mixed layer. These ideas were put into broad oceanographic context by Yentsch (1981), and modelled by Wroblewski, *et al.* (1988), and Yentsch (1990).

The concepts behind the baroclinic effects on primary production stems from H.U. Sverdrup's (1955) knowledge of geostrophic currents and interest in primary production. A.C. Redfield , after attending a lecture by Carl Rossby, realized the importance of the Gulf Stream in the nutrient enrichment of slope waters (Redfield 1936). Following Redfield, I attempted (Yentsch 1974) to model phytoplankton growth in this major frontal region.

2. The Nutrient-Light Hypothesis

The abundance of phytoplankton in a water mass is believed to be due to the primary processes of growth, (*i.e.* photosynthesis and respiration) and external factors which remove phytoplankton such as grazing and sinking. The primary processes are under the control of two substrates, light and nutrients (figure 1). We believe that the change in growth seasonally is closely tied to the change in the depth of the mixed layer which in turn is regulated by climatological processes such as wind stress, air-sea temperature changes and changes in solar radiation.

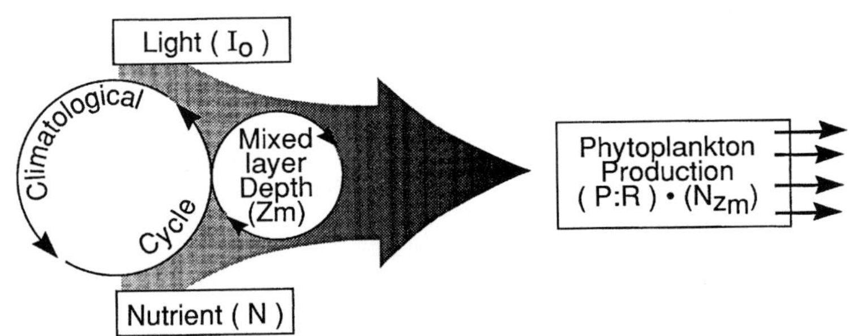

Figure 1. Conceptual model of the interaction between climatological cycles, mixed layer depth and substrates for phytoplankton growth.

In a stratified water mass covered by a mixed layer, the availability of light (I) is antagonistic to the availability of nutrients (N) (figure 2). Light decreases exponentially with depth while nutrients generally increase with depth following the density vs. depth profile. When the surface layers are stirred by wind, the ratio of light to nutrients in the mixed layer is changed. The role of the mixed layer is to couple the two substrates, light and nutrients, thus the extremes of deep or shallow mixing can stress phytoplankton growth. Therefore there must be some depth of mixing where growth is optimal.

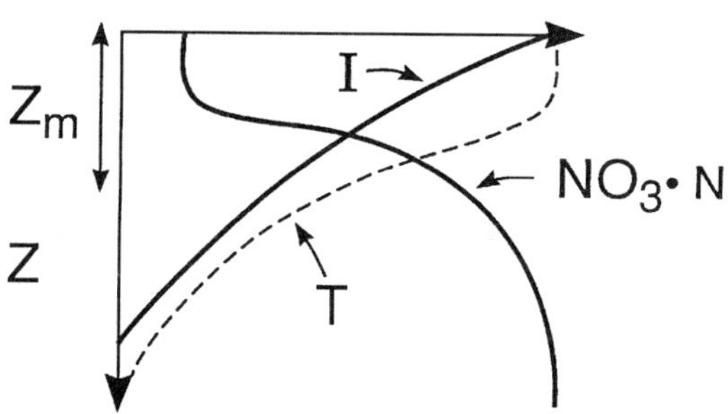

Figure 2. Conceptual model of the vertical distribution of light, nutrients and temperature in a stratified water mass.

The optimal mixing depth can be estimated by knowing the depth of the surface mixed layer, the vertical distribution of nutrients, and the ratio of photosynthesis to respiration ($P{:}R$). Figure 3 conceptually shows the relationship between photosynthesis (P) and respiration (R) for a population distributed uniformly over the euphotic zone. The shape of the photosynthetic curve is due to the manner in which photosynthesis responds to light. The values for R shown in figure 3 represent about 10% of P_{max} -- this is an assumption (Yentsch 1981). As the mixed layer deepens, $i.e.$ extends below the euphotic zone the role of respiration is enlarged. That is, as the mixed layer deepens the proportion of respiration to photosynthesis increases. When the mixed layer is sufficiently deep (5X the euphotic depth), the value for $P{:}R$ will be 1.0. This means there is no net phytoplankton growth -- Sverdrup termed this depth as "critical" to growth.

The optimal mixing depth P_{opt} can be modelled as follows: With the assumption that R is 10% of P_{max} and given the P vs. I curve described by Jassby and Platt (1976) which is,

$$P = P_{max} \, tanh \, a \frac{I}{P_{max}} \tag{1}$$

where, P_{max} is the photosynthetic rate at light saturation and alpha (a) is the slope of the P vs. I (irradiance) curve in the light limited portion of the curve. The integral of P vs. I and R vs. I yields a value of $P{:}R° = 6.0$ which is assumed to be maximal. As the mixed layer (Z_m) deepens $P{:}R$ at any depth can be calculated by,

$$P{:}R = P{:}R^\circ \left(\dfrac{Z_e}{Z_m} \right) \qquad (2)$$

Figure 4 shows two examples how $P{:}R$ of phytoplankton changes in two euphotic zones (Z_e) as the mixed layer shoals and deepens. The initial stages of deepening of the mixed layer are more effective in changing $P{:}R$. The change in $P{:}R$ as it approaches the critical depth, is more gradual.

Remembering the concepts shown in figures 1 and 2; when mixed layers deepen nutrient is stirred into the euphotic zone. From the vertical distribution of the nutrient one can estimate the amount of nitrate swept into the mixed layer by,

$$NZm = \int_{Om}^{Zm} \left(NO_3 - N \right) dz \qquad (3)$$

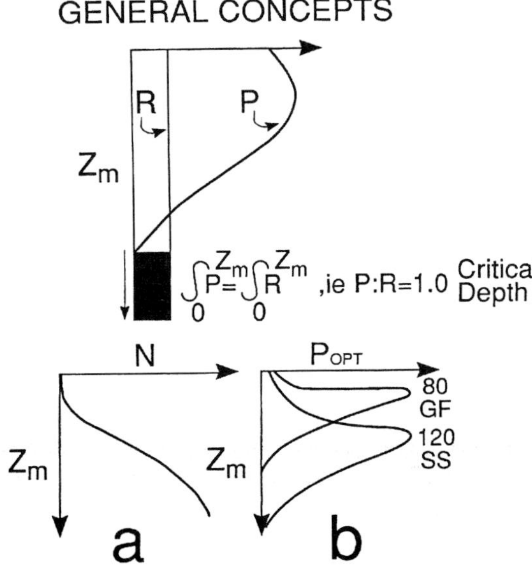

Figure 3. Integral model of photosynthesis and respiration for a uniform population in the euphotic zone. The shaded area under R represents the amount of respiration added to the euphotic zone. The curves shown in the lower part of this figure illustrate the components needed to estimate the optimal mixing depth (P_{opt}). The equations for $P{:}R$ and NZ_m are given in the text.

Summarizing, phytoplankton mixed vertically by changes in the depth of the mixed layer are provided the benefits of nutrient but also the penalty of light limitation. Optimal growth conditions occur for mixing depths where light and nutrients provide maximal growth. The optimal depths can be estimated for different areas of the oceans by combining equations (1) and (2). For example, the products of the two equations for the Gulf of Maine and the Sargasso Sea are shown in figure 4. The former has an optimal depth of 60 meters, the later 120 meters. These results represent differences in the depth structure of the nutrient density field; *i.e.* the nutricline is shallower in the Gulf of Maine than in the Sargasso Sea.

Therefore the differences between the optimal depth for regions of the oceans support the idea that the density structure associated with ocean currents is very important in setting the patterns of phytoplankton abundance in time and space: the presence of ocean currents markedly increases the baroclinicity of water masses, in these regions the isopycnals do not parallel the isobars. The isopycnals steepen in the frontal regions of the ocean currents which provides a gradient in nutrient across the current. This

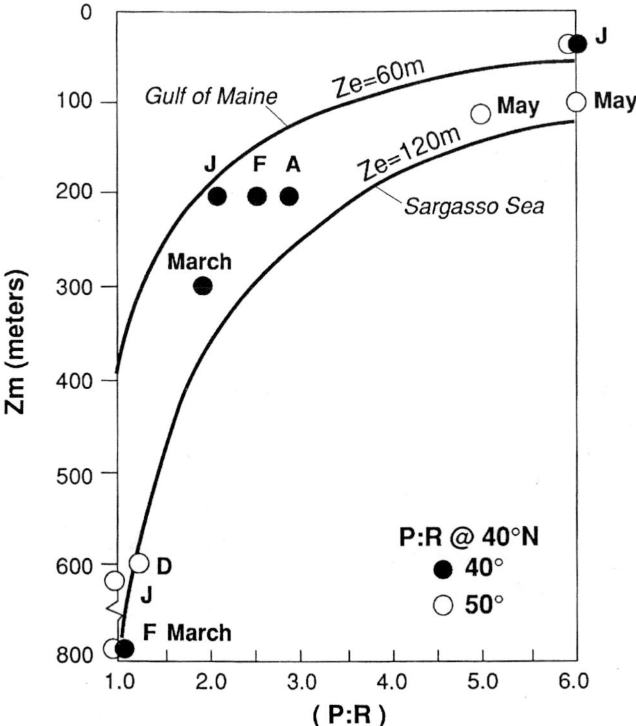

Figure 4. *P*:R change as a function of mixed depth (Z_m) in the Gulf of Maine and Sargasso Sea. The symbols represent monthly *P*:*R* values at two latitudes.

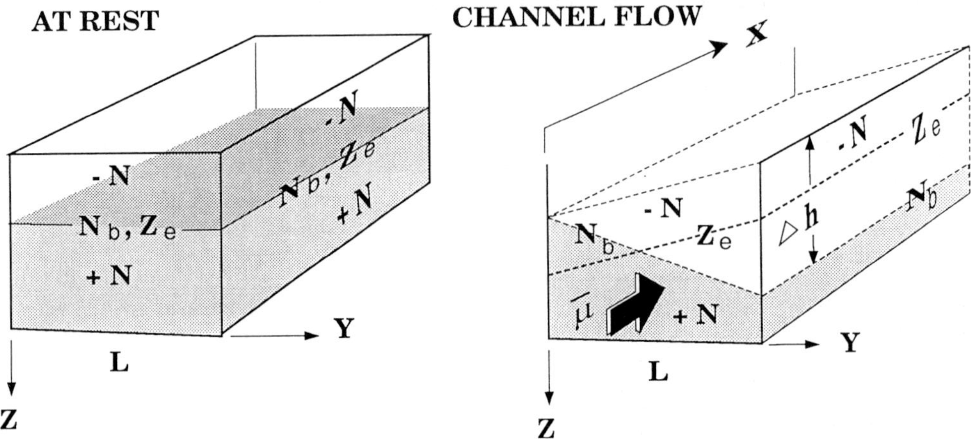

Figure 5. Schematic representation showing the relationship between nutrient (N) the nutrient boundary (N_b) and the euphotic zone (Z_e) in an ocean water mass at rest and flowing.

gradient is coupled to the euphotic zone by the mixed layer. To illustrate, examine the diagram shown in figure 5.

This block represents a water mass first shown at rest, *i.e.* no flow on the x-axis. The mass is stratified having two layers with extremes in nutrient concentration separated by a boundary (Nb). This boundary coincides with the depth of the euphotic zone (Ze). When a flow (current) moves through the block along the x axis, the geostrophic forces (Earth's rotation and pressure gradient) cause the sea surface to slope from right to left. To balance this slope height, dense nutrient-rich water moves closer to the surface on the left side of the water mass. Superimposing a mixed layer over this mass, vertical mixing will supply a higher amount of nutrients to the left side than to the right in the water mass. Therefore, a gradient in the rate of nutrient supply to the euphotic zone will occur from left to right in this water mass.

To sum up, early attempts to accurately portray light and nutrient interactions, *i.e.* the concepts described above, are hampered by incomplete knowledge of the nutrient-density field covering the world's oceans and/or accurate estimates of the depth of mixing by the upper layers. In recent years, important atlases have been assembled from ship observations which have helped us advance the nutrient/light hypothesis. These atlases when used with satellite imagery, provide an oceanographic information system similar to the geographical information systems for terrestrial environments.

In this context, we can attempt to predict how vernal blooming of phytoplankton progresses as the sun moves back and forth seasonally, say for example in the North Atlantic Ocean (Yentsch 1990), which can be coupled with the use of satellites. By studying the patterns of change in ocean color in the North Atlantic, we can test our hypothesis by examining color/chlorophyll changes in relationship to oceanographic parameters such as changes in mixed layer depth.

3. Ocean Information System and North Atlantic Sequences

Since the successes of CZCS is the theme of this workshop, I will concentrate on the North Atlantic series of images provided by Gene Carl Feldman at Goddard (figure 6). I will mostly focus on the months of January through May which covers the onset and appearance of the spring bloom. The CZCS imagery shows that the months of January and February are periods of low abundance of phytoplankton. These are the months where, at temperate and high latitudes, air and water temperatures are lowest, winds are strong, and solar radiation low such that the mixed layer is the deepest. By March, things are changing and a color/chlorophyll front is established along the 35° N meridian. As the season progresses, this front moves slightly northward; and major concentrations of chlorophyll appear at higher latitudes.

The depth of the mixed layer deepens from the equator northward during the winter to spring sequence (figure 7). Between latitude 0°N to 20°N it hovers around 50 to 75 meters. From 40°N to 60°N it is greater than 200 meters. The maximum depth is almost 1000 meters at latitude 60°N. The monthly sequence of the temperate bloom at latitudes 40°N to 50°N is now examined in the context of $P{:}R$ (see equation 2), and changes in the depth of the mixed layer. The predicted net growth ($P{:}R > 1.0$) for most of the winter months is low (see figure 4) between $P{:}R$ 1.8 to 2.7 at 40°N. At latitude 50°N the mixed layer is deeper than at 40°N and in the winter months, the $P{:}R$

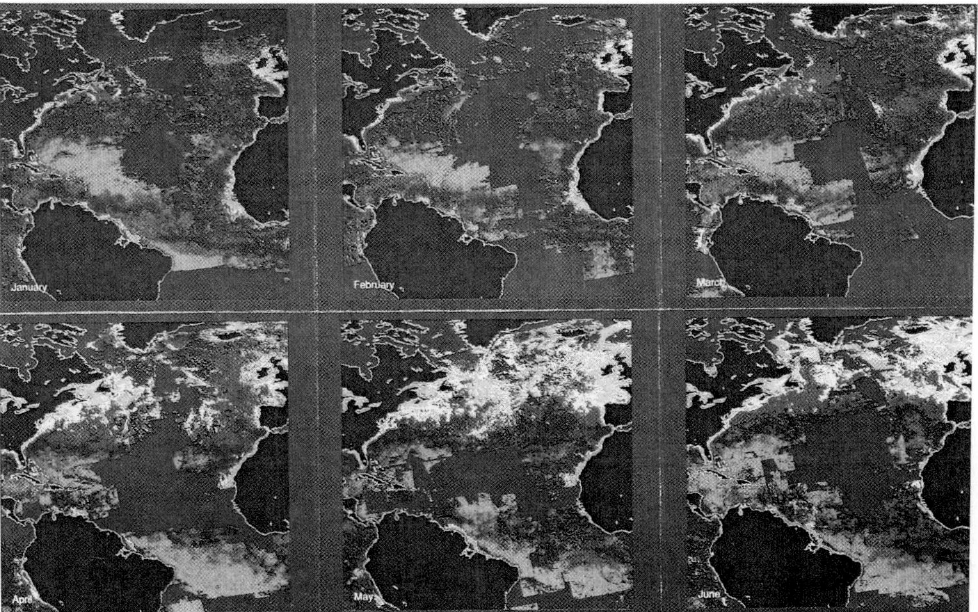

Figure 6. Primary production in the North Atlantic during the spring months (Source: G. Feldman, Goddard Space Flight Center). Yellow indicates high concentrations of chlorophyll, blue is intermediate and purple-low chlorophyll. Colorphotograph on p. 349

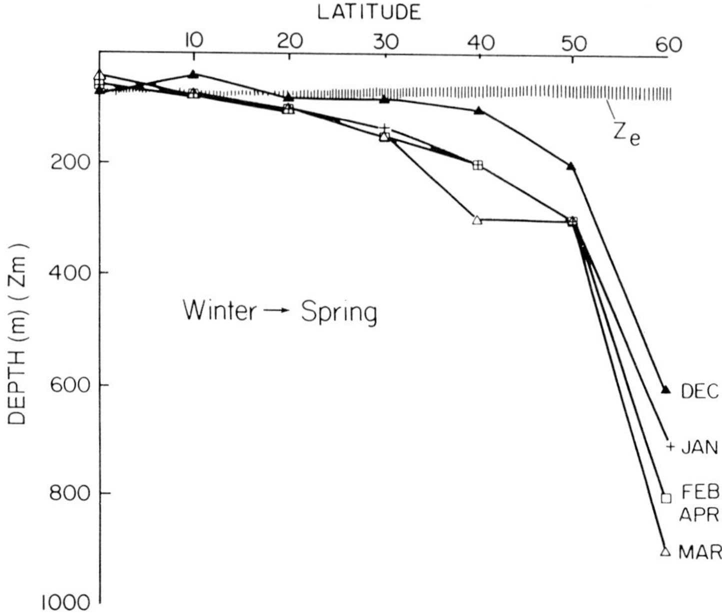

Figure 7. Depth of the mixed layer, winter to spring in the North Atlantic, 40°W longitude. Data taken from Robinson *et al.* (1979).

values all range between 1.0 to 1.2. At both latitudes, by early spring, $P{:}R$ has increased to greater than 2.0. By May and June, $P{:}R$ values are at the theoretical maximum for both latitudes (figure 4).

This analysis demonstrates that by combination of oceanographic information with the $P{:}R$ model, the sequence of blooming as viewed by the satellite is predicted. Therefore, satellite observations corroborate the empirical model. In summary, the transition seasonally from months where phytoplankton abundance is low to those months where it is high, is due to the shoaling of the mixed layer. Perhaps the most amazing thing is the rapidity of the change from light limited growth ($P{:}R < 6.0$) to conditions of blooming. Between April and May, there is a twofold increase in P:R and chlorophyll concentration at both latitudes.

On a basin scale, the effect of the blooming is shown by satellite to reduce the size of the oligotrophic regions in the North Atlantic. This can be seen by following the seasonal migration of the major chlorophyll/color front (figure 6). During the winter months there is little evidence of this front, most of the region can be characterized as oligotrophic. By March, a chlorophyll front appears slightly above latitude 30°N. By April and May, the front has intensified, *i.e.* chlorophyll has increased, and resides between latitudes 40-50°N.

The $P{:}R$ mixed layer analysis shows that over the entire basin during February, between latitudes 40-50°N, the $P{:}R$ values reside between 1.0 to 2.0

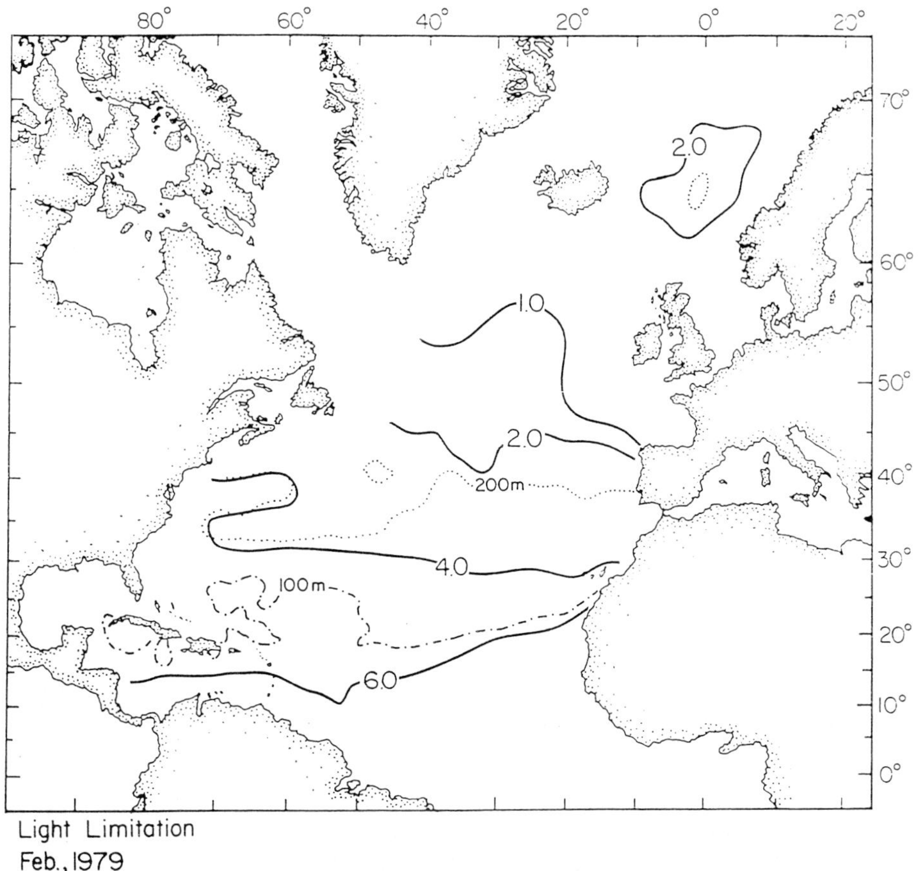

Light Limitation
Feb., 1979

Figure 8. $P{:}R$ values in the North Atlantic during February. $P{:}R$ is calculated by Eq. 2 using data on Z_e from Yentsch (1990) and Z_m from Robinson *et al.* (1979).

(figure 8). The only area where $P{:}R$ values are optimal are near the equator centered around 10°N. This demonstrates that growth is light limited throughout most of the basin during winter months. By the month of May, the values for $P{:}R = 6.0$, which is the optimal ratio, has moved northward residing between latitude 40-50°N (figure 9). At higher latitudes, the $P{:}R$ analysis shows that these regions are still light limited. Utilizing oceanographic data I will attempt to predict the onset of the bloom at 40-50°N as observed by CZCS. This will require knowing the depths of the mixed layer, nitrate-nitrogen concentrations for the water column, and $P{:}R$ ratios. In figure 10 these parameters are compared with chlorophyll pigment values recorded by CZCS.

The changes in the depth of the mixed layer are a mirror image of the amount of nitrate-nitrogen (figure 10). This indicates that the resident concentration of nitrate in the mixed layer is dependent on the depth of the mixed layer. *P:R* values are low until April (figure 10), after which they increase to 6.0 by June. Chlorophyll concentrations closely follow the *P:R* values. At both latitudes the bloom appears to commence shortly after the relaxation of deep mixing in March. It should be noted that the values of chlorophyll observed at 50°N are considerably higher than those at 40°N. The difference is probably due to the fact that the nutricline is higher at 50°N than at 40oN therefore, the mixed layer at 50°N contains more nitrate than that at 40°N (Yentsch, 1990). Figure 11 shows the apparent influence of ocean currents on the nitrate structure of the North Atlantic Basin. The two "ridges" of high nitrate concentration are associated with the baroclinic features of the Gulf Stream and the Northern Equatorial Current systems. In the case of the Gulf Stream, the direction of flow is toward the reader whereas the reverse is true for the Equatorial Current system.

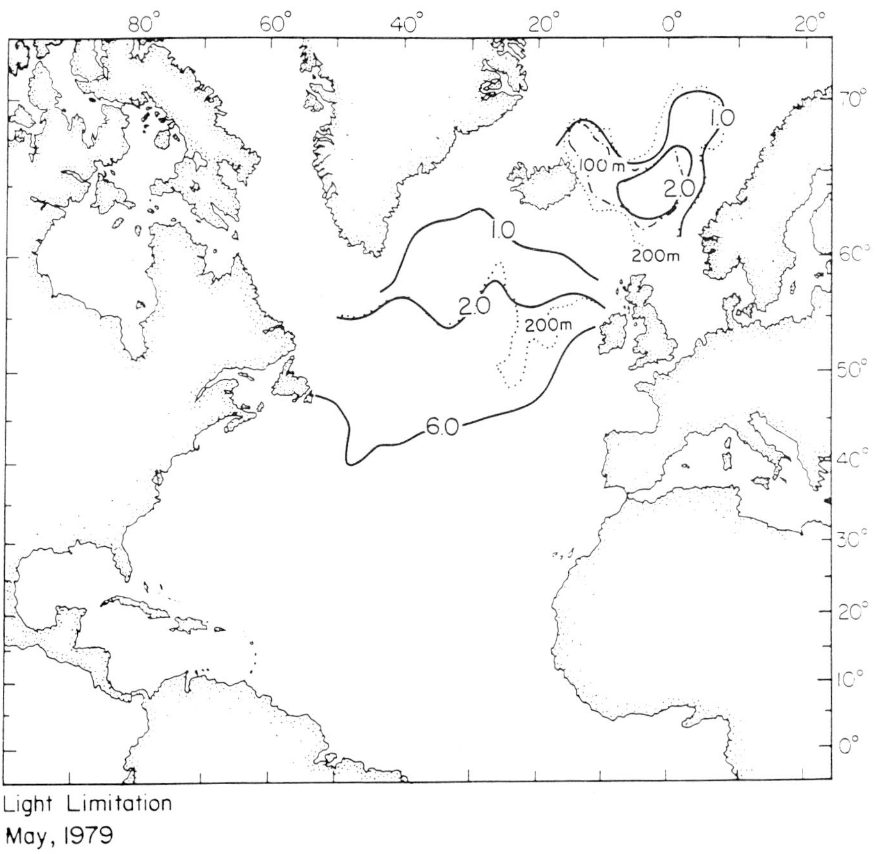

Figure 9. Same as Figure 8 for the month of May.

On a basin scale (figure 12), the 200 m nitrate concentrations produce a pattern which resembles a large eddy. The central region of the eddy is the oligotrophic nutrient-poor waters of the Sargasso Sea. This region is bounded by nutrient ridges where the nitrate concentrations are 10 micromoles or higher. The bloom at high latitudes shown in the imagery appears to coincide with the 10 micromolar isopleth.

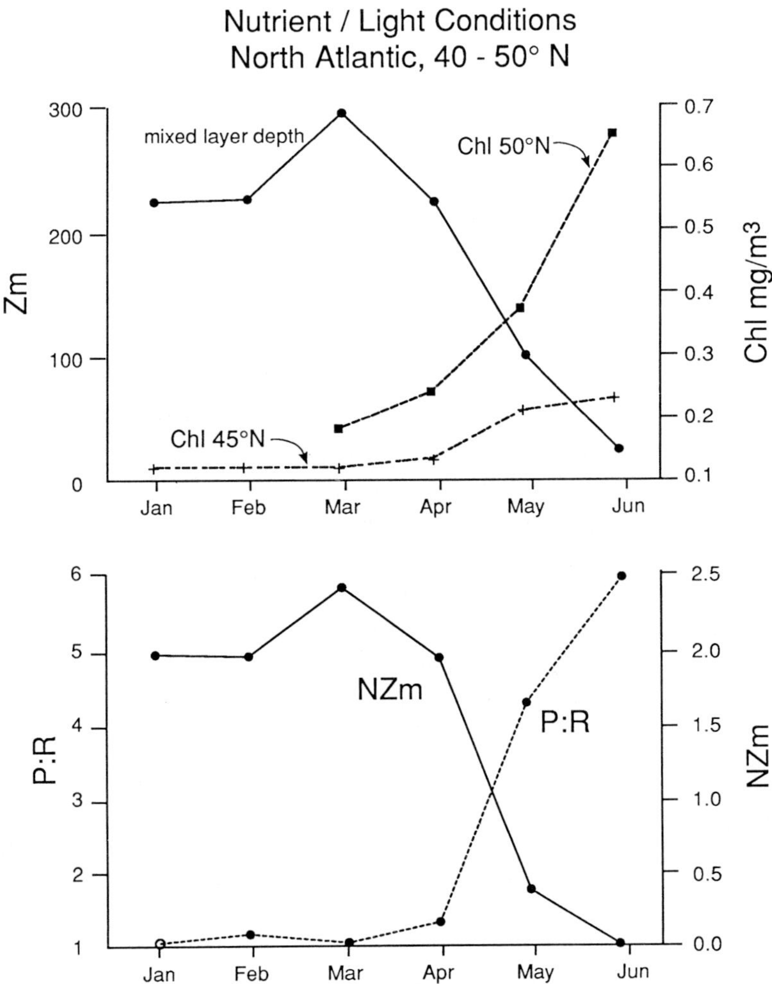

Figure 10. Anatomy of the spring bloom at 40°N. NZ_m is calculated from Eq. 3, $P{:}R$ from Eq. 2 in the text. Zm is taken from data in Robinson *et al.* (1979). Chlorophyll values for 40°N and 50°N are derived from CZCS images.

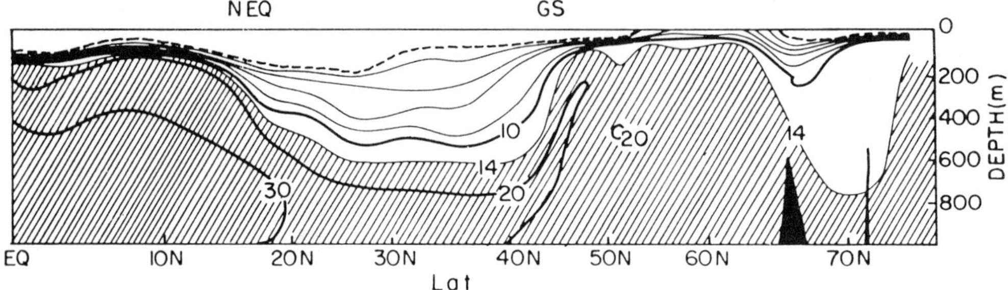

Figure 11. Vertical distribution of nitrate-nitrogen (µm kg-1) along 40°W longitude. Data from Geosecs 1972-1973.

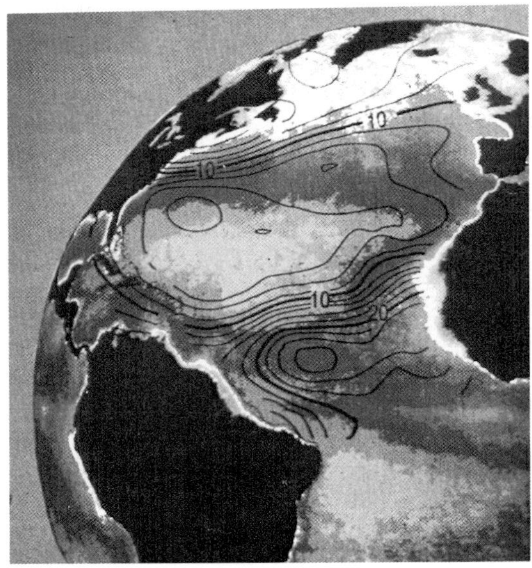

Figure 12. The pattern of nitrate-nitrogen (µm kg-1) at 200 meters in the North Atlantic, from Wroblewski *et al.* (1988) superimposed on ocean color composite of the North Atlantic (Goddard/Feldman series). Yellow indicates a high concentration of chlorophyll, blue is intermediate, and purple is low chorophyll. Colorphotograph on p. 349

4. The Role of Ocean Currents in the Regulation of Phytoplankton Production

Our concepts postulate that phytoplankton growth in the ocean is regulated by the admixture of two layers - the upper barotropic layer known as the mixed layer overlying a baroclinic layer which is the main body of the ocean containing the nutrient sources. Because of the movement of ocean currents, the degree of baroclinicity varies throughout the North Atlantic Ocean as well as other regions. The pattern of this baroclinicity in the North Atlantic is due to the eastward flow of the Gulf Stream and the westward flow of the equatorial current giving rise to the toroidal pattern apparent in figure 12.

If one examines the levels of primary production sliced through a section of the mid Atlantic toroid; the extremes in baroclinicity along this section are reflected by two major peaks in primary production (figure 13).

The major peak is concentrated between 40°-50° N which coincides with a portion of the spring bloom, the other is at latitude 10°N (Yentsch 1990). Superimposed in figure 13 is the depth of $\partial_t = 27.0$ isopleth. At this density, the maximum concentration of nitrate-nitrogen is observed (GEOSECS Atlas data source). The correspondence between the peaks of productivity and the level of the maximum concentration of nitrate-nitrogen suggests that the high latitudinal levels of productivity are due to the proximity of the maximum source of nitrate-nitrogen to the surface layer.

Sverdrup predicted that geostrophic ocean currents are a first order cause of the distribution of global production. Sverdrup constructed a semi- intuitive map of primary production based largely on his knowledge of geostrophic currents, this map closely resembles the satellite data composite of the Earth's biosphere shown in figure 14.

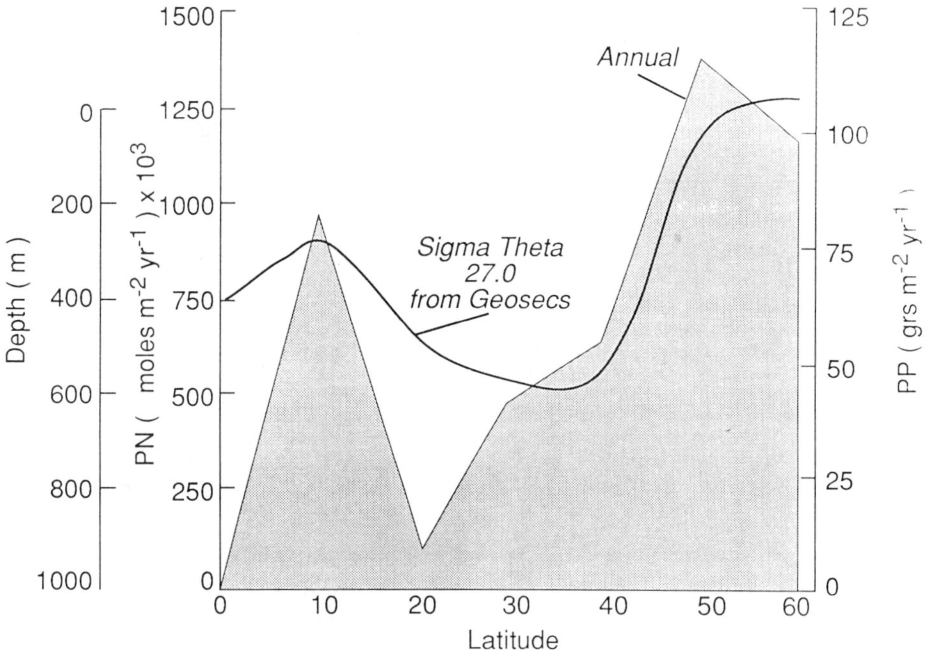

Figure 13. Annual latitudinal primary production of particulate nitrogen (PN) and carbon (PC) at Long. 40°W (Yentsch 1990).

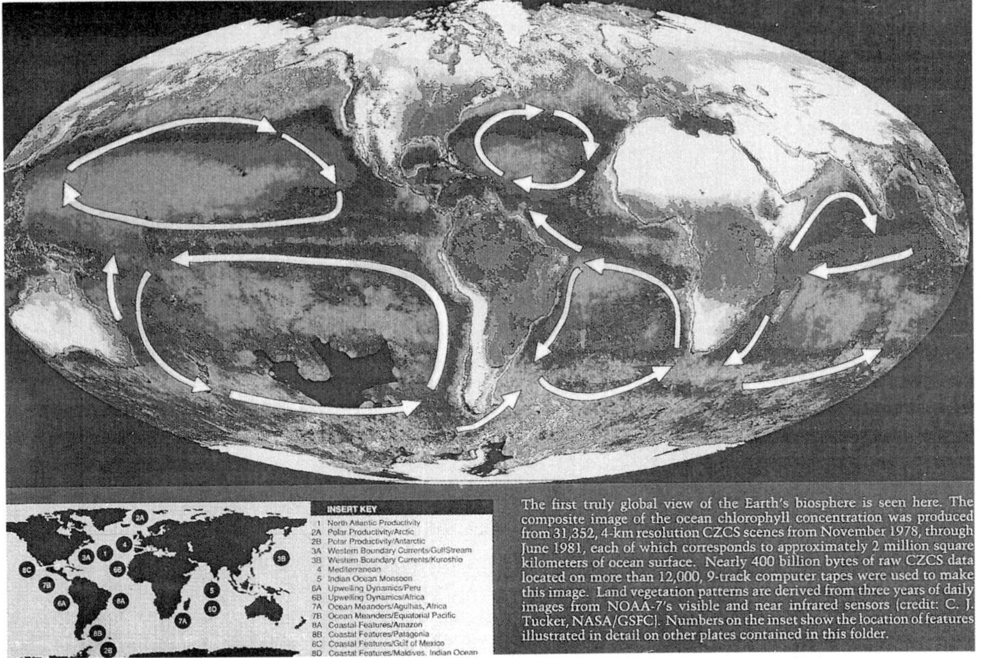

Figure 14. The global biosphere (G. Feldman, private communication). Superimposed are the pathways of major ocean currents. Colorphotograph on p. 350

5. CZCS Impact on Biological Oceanography

The Earth's biosphere composite of figure 14 provides the first step in the construction of a global estimate of primary productivity with specific reference to how climatological factors might influence the pattern on earth. In this context the importance of the CZCS mission is the rediscovery of the importance of ocean circulation on the distribution of primary production in the oceans.

In the introduction I stated there's a danger in relying too much on dogma. Yet I believe it is a tribute to the pioneers, who proposed these concepts, that CZCS observations tend to support their ideas. You could ask "if so what's left?" My answer is "plenty." All of the mechanisms of physical/chemical interaction portrayed are at best semiquantitative. For example the mechanism of how nutrient is supplied along or across isopynals is still vague. With regard to phytoplankton physiology the respiration rate of open ocean species is not well known. Our ability to predict will be retarded until this is changed.

The major broad benefit of CZCS is that it provided the biological oceanographer a view of the domain of phytoplankton, but it also provided a side benefit which is very important. CZCS observations brought the research interests of ocean biologists, chemists and physicists closely together under a unified theme of understanding the oceanographic processes which regulate primary production.

One other point - the major difference between the scientific approach by the earth science biologist and the cellular biologist is that for the most part, the former is confined to observation and modeling, whereas, the latter has the additional option of experimentation. The earth scientist envies the cellular biologist in having the ability to place the object under the microscope to probe and poke. But perhaps the satellite observations are changing this advantage. It has been said that by way of human activities we are performing an experiment on the planet. If true then the role of the satellite is to study the earth's response to human activities.

Reference

Bain Bridge, A. E. (1981) GEOSECS Atlantic Expedition, Hydrographic DATA 1972-1973, U.S. Goverment Printing Office #038-000-00491-3.

Cushing, D.H. (1962) 'An alternative method for estimating the critical depth', Journal du Conseil Perm. International pour l'Exploration de la Mer 27, 131-140.

Gran, H.H. and Braruud, T. (1935) 'A quantitative study of the phytoplankton in the Bay of Fundy and the Gulf of Maine', J. Biol. Bd. Can. 1, 279-467.

Jassby, A.D. and Platt,T. (1976) 'Mathematical formulation of the relationship between photosynthesis and light for phytoplankton', Limnology and Oceanography 21, 540-547.

Redfield, A.C. (1936) 'An ecological aspect of the Gulf Stream', Nature 138, 1013.

Riley, G.A. (1963) 'Theory of food-chain relations in the ocean', in M.N. Hill (ed.), The Sea 2, John Wiley, New York, pp. 438-463.

Robinson, M.K., Bauer, R.A. and Schroeder, E.H. (1979) 'Atlas of North Atlantic - Indian Ocean monthly mean temperatures and mean salinites of the surface layer', Naval Oceanographic Office N00- RP18.

Sverdrup, H.U. (1953) 'On the conditions for the vernal blooming of phytoplankton', Journal du Conseil Perm. International pour l'Exploration de la Mer 18, 287-295.

Sverdrup, H.U. (1955) 'The place of physical oceanography in oceanographic research', Journal of Marine Research 14, 287-294.

Wroblewski, J.S., Sarmiento, J.L. and Flierl, G.R. (1988) 'An ocean basin scale model of plankton dynamics in the North Atlantic. I. Solutions for the climatological oceanographic conditions in May', Global Biogeochem. Cycles 21, 199-218.

Yentsch, C.S. (1974b) 'The influence of geostrophy on primary production', Tethys 6, 111-118.

Yentsch, C.S. (1981) 'Vertical mixing, a constraint to primary production: an extension of the concept of an optimal mixing zone', in J.C.J. Nihoul (ed.), Ecohydrodynamics, Elsevier Oceanography Series 32, pp. 67-78.

Yentsch, C.S. (1990) 'Estimates of 'new production' in the Mid-North Atlantic', Journal of Plankton Research 12 (4), 717-734.

RADIATIVE TRANSFER IN THE ATMOSPHERE
FOR CORRECTION OF OCEAN COLOR REMOTE SENSORS

H. R. GORDON
Department of Physics
University of Miami
Box 248046
Coral Gables, FL 33124
USA

ABSTRACT. In this paper I have tried to provide a more or less self-contained discussion on the optical properties of the atmosphere and radiative transfer theory to provide the reader with an understanding of atmospheric correction of satellite ocean color remote sensing data. The absorption properties of the optically important gases (H_2O, O_2, and O_3) in the atmosphere have been presented in the form of spectral transmittance curves. The scattering properties of the aerosol have been described with examples taken from Mie scattering theory applied to aerosol models. The development of the CZCS algorithm has been described in detail starting from the single scattering solution of radiative transfer theory. A critical evaluation of the model is then carried out and efforts to circumvent the difficulties specific to the CZCS band set are presented. Processes ignored in the original algorithm but included in later versions, *e.g.*, multiple scattering, polarization, and variations in the O_3 concentration and the surface atmospheric pressure, are briefly examined. Finally, the question of atmospheric correction of future, more sensitive, ocean color sensors, such as SeaWiFS, is considered. An improved correction algorithm is proposed and the remaining problems, along with suggested approaches for solving them, are described.

1. Introduction

Following the work of Clarke, Ewing, and Lorenzen (1970) showing that the chlorophyll concentration in the surface waters of the ocean could be deduced from aircraft measurements of the spectrum of upwelling light from the sea - the "ocean color" - NASA launched the Coastal Zone Color Scanner (CZCS) on Nimbus-7 in late 1978 (Hovis *et al.*, 1980; Gordon *et al.*, 1980). The CZCS was a proof-of-concept mission with the goal of measuring ocean color from space. It was a scanning radiometer that had four bands in the visible at 443, 520, 550, and 670 nm with bandwidths of 20 nm, one band in the near infrared (NIR) at 750 nm with a band width of 100 nm, and a thermal infrared band (10.5 to 12.5 µm) to measure sea surface temperature. The four visible bands possessed high radiometric sensitivity (well over an order of magnitude higher than other sensors designed for earth resources, *e.g.*, the Multi Spectral Scanner (MSS) on Landsat) and were specifically designed for ocean color. Further technical details concerning CZCS are given in the Appendix. The CZCS experience demonstrated the feasibility of the measurement of

33

V. Barale and P.M. Schlittenhardt (eds.),
Ocean Colour: Theory and Applications in a Decade of CZCS Experience, 33–77.

plankton pigments, and possibly even productivity (Platt and Sathyendranath, 1988; Morel and André, 1991), on a *global* scale. This feasibility rests on two observations: (1) there exists a more or less universal relationship between color of the ocean and plankton pigment concentration for most open ocean waters; and (2) it is possible to develop algorithms to remove the interfering effects of scattering in the atmosphere. In this paper we will review the atmospheric effects associated with CZCS.

The paper is structured in the following way. First, the basic concepts of radiometry that are required to understand radiative transfer are presented. Second, the relevant optical properties of the atmosphere are reviewed in detail. Next, the radiative transfer equation is introduced, a method of solution is described, and a first order solution is derived for a scattering atmosphere.

This solution is then applied to the development of the CZCS atmospheric correction algorithm. The shortcomings of the algorithm are then discussed along with the modifications that improve its performance with CZCS. Finally, I provide some indication of the modifications to the correction algorithm that will be required for the new, more sensitive, ocean color instruments, *e.g.*, Sea-viewing Wide Field-of-view Sensor (SeaWiFS) to be launched in late 1993.

2. Atmospheric Optical Properties

2.1 RADIANCE

Light of wavelength λ (SI unit: nm) can be considered to be composed of a stream of photons with each photon possessing an energy hc/λ, where h is Planck's constant and c is the speed of light. A basic concept of radiometry is that of the *spectral radiant power* $P(\lambda)$. Let light pass through a filter transmitting a spectral bandwidth $\Delta\lambda$ centered on λ and fall on a radiation detector. If the detector records N photons per second, the spectral radiant power is defined by

$$P(\lambda) = \frac{hc}{\lambda} \frac{N}{\Delta\lambda} \tag{1}$$

The SI unit for $P(\lambda)$ is Watts/nm. In remote sensing it is important to record the direction in which the light is propagating as well as the associated power. This is accomplished with a quantity called the radiance. Consider a detector of spectral radiant power having an physical area A. Place a spectral filter, which passes a range of wavelengths $\Delta\lambda$ centered on λ, over the detector and equip the detector with an optical system which restricts its field of view to a small solid angle $\Delta\Omega$ (SI unit: Ster). Such an arrangement is called a radiometer. If the detector records a power $P(\lambda, \xi)$ when the radiometer aimed in a direction to receive photons traveling in a direction indicated by the unitvector ξ, it records, at its position, a *radiance* $L(\lambda, \xi)$ defined by

$$L(\lambda, \xi) = \frac{P(\lambda, \xi)}{A \, \Delta\Omega \, \Delta\lambda} \tag{2}$$

The SI unit for radiance is Watts/m^{2}nm Ster. In practice, $\Delta\Omega$ and $\Delta\lambda$ need to be sufficiently small so that a further reduction in their size does not change the radiance. All satellite and airborne ocean color remote sensing instruments measure spectral radiance.

2.2 FUNDAMENTAL QUANTITIES

The fundamental optical properties of a medium are defined and measured by probing samples of the medium with a well defined beam of light. Consider a small volume Δv of length Δl illuminated by a parallel beam of light traveling in a direction specified by the unit vector ξ. Let $P_0(\lambda)$ be the radiant power entering the volume. As the photons pass through the volume some are removed from the beam by *absorption* within Δv. Others are removed from the parallel beam by a change in their direction *scattering* within Δv, and they will exit Δv traveling in directions other than ξ, e.g., ξ'. If $\Delta P(\lambda)$ is the spectral radiant power removed from the parallel beam by virtue of scattering *and* absorption, then the attenuation or *extinction coefficient* $c(\lambda)$ is defined by

$$c(\lambda) = \frac{1}{\Delta l}\left[\frac{\Delta P(\lambda)}{P_0(\lambda)}\right] \tag{3}$$

This is the fraction of the power removed from the beam per unit length. The SI unit for $c(\lambda)$ is m^{-1}. If $\Delta^2 P(\lambda,\xi')$ is the spectral radiant power scattered into a small solid angle $\Delta\Omega(\xi')$ containing the direction ξ', the *volume scattering function* $\beta(\lambda, \xi\rightarrow\xi')$ is defined according to

$$\beta\left(\lambda,\xi\rightarrow\xi'\right) = \frac{1}{\Delta l \Delta\Omega(\xi')}\left[\frac{\Delta^2 P(\lambda,\xi')}{P_0(\lambda)}\right] \tag{4}$$

The volume scattering function is the fractional power scattered from ξ into the direction ξ' per unit length per unit solid angle around ξ'. [The "2" on $\Delta^2 P(\lambda,\xi')$ indicates that it is of second order in smallness, *i.e.*, small because Δl is small and also small because $\Delta\Omega(\xi')$ is small.] β is the differential scattering cross section per unit volume. The SI unit for β is m^{-1}Ster^{-1}. In Equations (3) and (4) Δl and $\Delta\Omega(\xi)$ must be sufficiently small that photons have a negligible probability of scattering more than once in Δv. For particles in random orientation, β depends on direction only through the angle a between ξ and ξ' given by $a = \cos^{-1}(\xi\bullet\xi')$. If we sum the contributions from each $\Delta\Omega(\xi')$ over the entire sphere surrounding Δv, *i.e.*, sum the light scattered into all directions, the result is called the *scattering coefficient $b(\lambda)$*:

$$b(\lambda) = \int_{4\pi} \beta\left(\lambda,\xi\rightarrow\xi'\right) d\Omega(\xi') \tag{5}$$

The 4π on the integral in Equation (5) means that it is to be taken over 4π Ster. The scattering processes we deal with here are *elastic*, *i.e.*, there is no wavelength (energy) change upon scattering. Finally, since light that is removed from the beam, but not scattered, must have been absorbed, we can define *absorption coefficient* $a(\lambda)$ through

$$a\left(\lambda\right) = c\left(\lambda\right) - b\left(\lambda\right) \tag{6}$$

The quantities $a(\lambda)$, $b(\lambda)$, $c(\lambda)$, and $\beta(\lambda,\ \xi \rightarrow \xi')$ are referred to as the inherent optical properties (IOP's) of the atmosphere (Preisendorfer, 1961; Preisendorfer, 1976). From the definitions of the IOP's (which require the absence of multiple interactions within Δv) they must be additive over the constituents of the medium. It is useful to introduce two auxiliary IOP's: the single scattering albedo

$$\omega_0\left(\lambda\right) = \frac{b\left(\lambda\right)}{c\left(\lambda\right)}$$

which is the probability that when a photon interacts with the atmosphere it will be scattered; and the scattering phase function

$$P\left(\lambda, \xi' \rightarrow \xi\right) = \frac{\beta\left(\lambda, \xi' \rightarrow \xi\right)}{b\left(\lambda\right)}$$

i.e., the VSF normalized to the total scattering coefficient.

Absorption and scattering in the atmosphere is usually very small and measurements of the optical properties are often carried out utilizing very long paths, *i.e.*, kilometers. A measurement of particular importance is the transmittance. Let a radiometer view an extended source, *e.g.*, the sun. Then the *transmittance* T of the path between the source and the observer is defined to be the measured radiance (L_m) divided by the radiance of the source (L_s):

$$T = \frac{L_m}{L_s}$$

We shall see later that if there is no absorption or scattering over the path the transmittance is unity. If there are N scattering and absorbing species in the path, then $T = \Pi_{i=1}^{N} T_i$, where T_i is the transmittance of the ith constituent alone. Much information concerning the optical properties of the atmosphere is obtained by measuring its transmittance using the sun as a source.

2.3 OPTICAL PROPERTIES OF THE ATMOSPHERE

The atmospheric constituents that produce significant absorption in the visible portion of the spectrum are O_2, O_3, and H_2O. An example of the transmission of the atmosphere from 400 to 1000 nm looking toward the zenith, derived using LOWTRAN 7 for a particular model of the atmosphere (the 1976 U.S. Standard Atmosphere (NASA, 1976)), is provided in figure 1. LOWTRAN 7 is a computer program developed at the U.S. Air Force Geophysics Laboratory. It combines the results of laboratory measurements

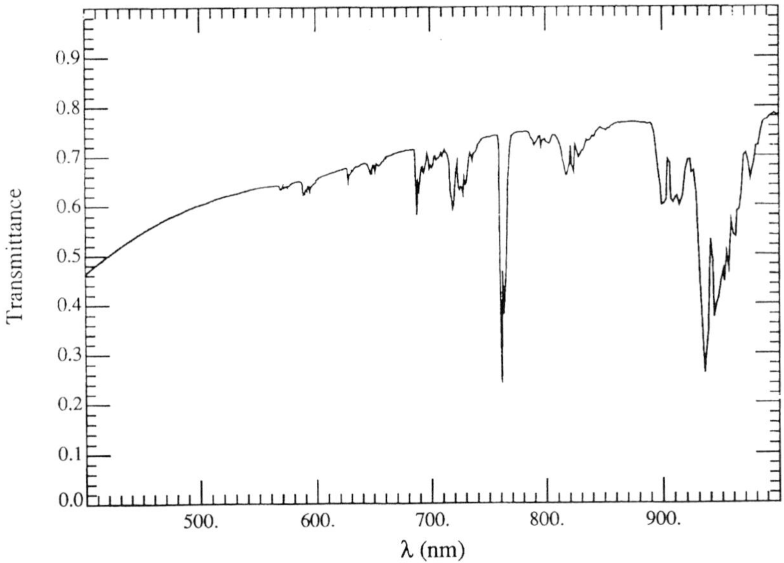

Figure 1. Vertical path transmittance of the atmosphere including the contributions of air, water, vapor and ozone.

on gases of interest in the atmosphere with climatological mean vertical distributions of the gases and other constituents to allow one to compute the transmittance of the atmosphere for any path (Kneizys *et al.*, 1988). Figure 1 includes the effects of all of the gases above as well as scattering by molecules and the aerosols (small particles suspended in the air). The absorption features near 686 nm and 759 nm are due to O_2. The rest of the distinct features are due to H_2O, with the exception of a weak absorption by O_3 extending through most of the visible spectrum. The individual transmittances of H_2O and O_3 in the visible, derived from LOWTRAN 7, is shown in figures 2 and 3, respectively. Clearly, it is desirable to place the spectral bands on ocean color remote sensing instruments away from the absorption bands of atmospheric gases, particularly gases with highly variable concentrations, *e.g.*, H_2O and O_3. However, this is not always possible (figure 3).

The transmittance in figure 1 is also affected by scattering by the atmospheric constituents. The scattering of light by the molecules themselves is referred to as *Rayleigh* scattering. The volume scattering function for Rayleigh scattering is

$$\beta_r(a) = b_r \frac{3}{8\pi} \left[\frac{1+\delta}{2+\delta} \right] \times \left[1 + \frac{1-\delta}{1+\delta} cos^2 a \right]$$

where b_r is the total scattering coefficient (proportional to the density of the

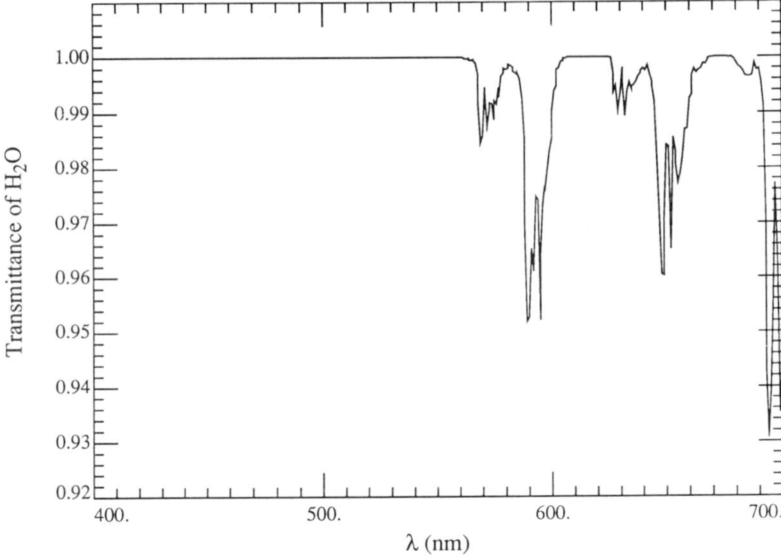

Figure 2. Contribution to the vertical path transmittance of the atmosphere by water vapor absorption alone.

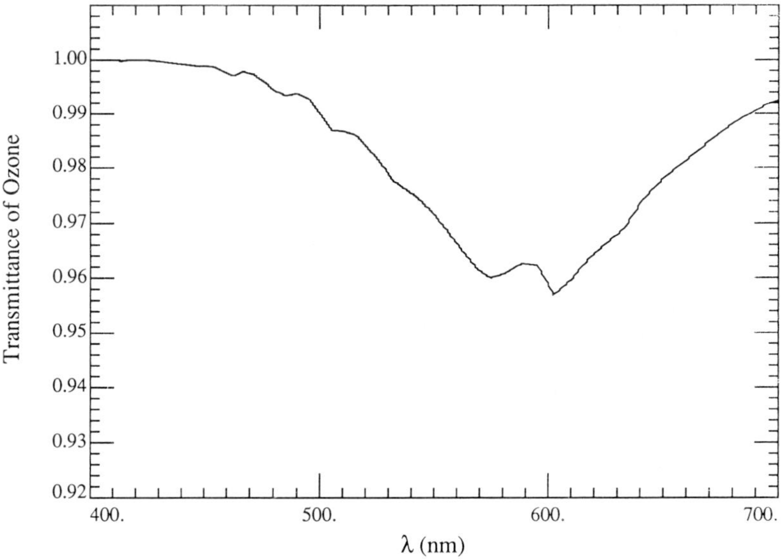

Figure 3. Contribution to the vertical path transmittance of the atmosphere by ozone absorption alone.

air) and δ is the depolarization ratio, the ratio of the radiance scattered at $a = 90°$ with polarization parallel to that with polarization perpendicular to the scattering plane, when the incident beam is unpolarized. Its value is taken by Hansen and Travis (1974) to be $\delta = 0.031$. From the form of β_r we see that Rayleigh scattering is symmetric with respect to a scattering angle of $90°$, $i.e.$, $\beta(a) = \beta(\pi-a)$. The $optical$ $thickness$ for Rayleigh scattering in the atmosphere is defined by

$$\tau_r \equiv \int_0^\infty b_r(h)\, dh \tag{7}$$

where h is altitude. Hansen and Travis (1974) give

$$\tau_{r_0} = 0.008569\, \lambda^{-4} \left(1 + 0.0113\, \lambda^{-2} + 0.00013\, \lambda^{-4} \right) \tag{8}$$

where τ_{r_0} is the optical thickness at the standard atmospheric pressure P_0 of 1013.25 mb and λ is the wavelength in µm. Note that τ_{r0} varies nearly as λ^{-4} with wavelength. Since b_r is $proportional$ to the air density, τ_r will be $proportional$ to the surface pressure, $i.e.$, at any surface pressure P,

$$\tau_r = \frac{P}{P_0}\tau_{r_0} \tag{9}$$

figure 4 shows the effect of Rayleigh scattering on the transmittance in the visible part of the spectrum.

The transmittance is also influenced by scattering by aerosols - solid and/or liquid particles suspended in the air. In fact, in the "windows" between the absorption features in figure 1, the attenuation is due to a combination of molecular scattering and aerosol attenuation. Since molecular scattering only weakly influences the transmittance in the red (figure 4), most of the attenuation in the windows there is due to aerosols. For aerosol particles, the value of ω_0 is typically near unity so their principal effect in the atmosphere is to scatter light. We will need to understand the scattering properties of the aerosol and their variability.

Aerosols are typically modeled as a collection of homogeneous spheres with a range of sizes. If we let D represent the diameter of the particles, and $dn(D)$ the number of particles per unit volume with diameters between D and $D + dD$, then the $size$ $distribution$ is defined to be $dn(D)/dD$. If N is the total number of particles per unit volume, $i.e.$,

$$N = \int_0^\infty \frac{dn(D)}{dD}\, dD$$

then the $normalized$ size distribution or $size$ $frequency$ $distribution$ is defined to be $dn(D)/NdD$. The electromagnetic parameter that governs the optical properties of the particles is the complex index of refraction, $m = n_r - in_i$, where n_r is the real part of the index and governs refraction, and n_i is the imaginary part of the index and is proportional to the absorption coefficient

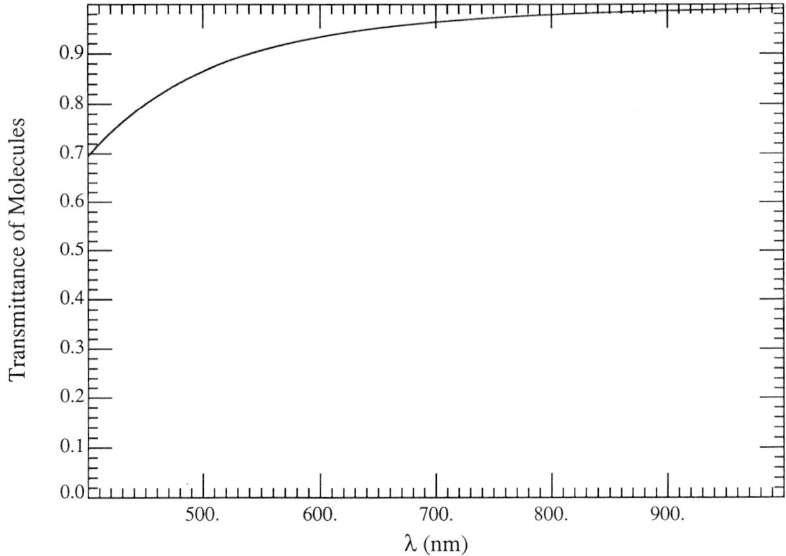

Figure 4. Contribution to the vertical path transmittance of the atmosphere by molecular scattering alone.

a_m of the material of which the particle is composed ($n_i = a_m\lambda/4\pi$). Given $dn(D)/NdD$, the fact that the particles are assumed to be *homogeneous* and *spherical*, and their refractive index m, it is possible to compute their scattering and absorbing properties from Mie theory (Mie, 1908; Bohren and Huffman, 1983; Van de Hulst, 1957). To demonstrate some of the properties of the aerosol scattering and absorption, I have computed the scattering phase functions for particles distributed according to

$$\frac{dn(D)}{dD} = K, \qquad D_0 < D < D_1$$

$$= K\left(\frac{D_1}{D}\right)^{\nu+1}, \quad D_1 < D < D_2$$

$$= 0, \qquad D > D_2$$

The parameters D_0, D_1, D_2, ν, and m are provided in Table 1. (Note that N is *proportional to K*, so K is not specified.) These models define what I refer to as an *aerosol type, i.e.,$dn(D)/NdD$ and m.*

The models with $m=1.50$ define a typical continental aerosol (Deirmendjian, 1969), while those with $m=1.333$ refer to a continental aerosol size distribution composed of water particles. To model a marine

TABLE 1. Parameters of the aerosol models.

Model	D_0 (μm)	D_1 (μm)	D_2 (μm)	ν	m
Haze C	0.06	0.20	20.0	2.5-4	1.50
Haze C	0.06	0.20	20.0	2.5-4	1.33
HMF7	0.20	0.40	17.5	2.95	$1.45 - 0.020i$
HMF9	0.20	0.60	17.5	2.95	$1.37 - 0.004i$

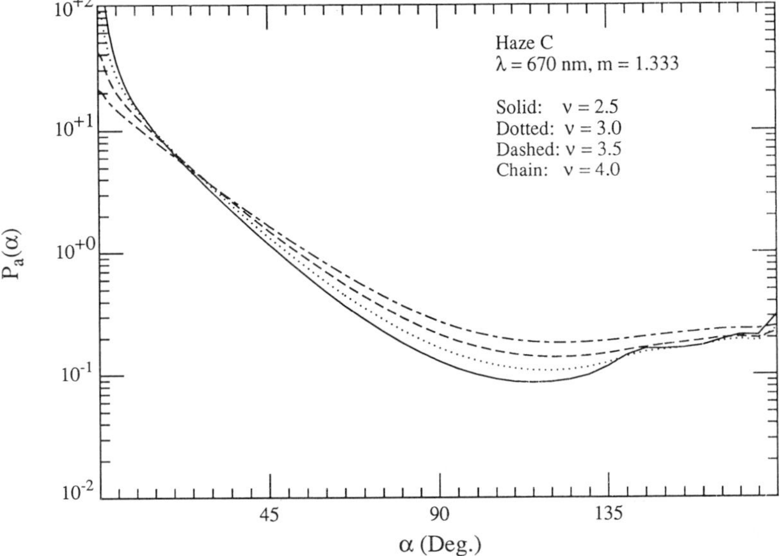

Figure 5. Aerosol phase functions for the Haze C model with $m = 1.333$.

aerosol (very small and very large particles missing) I have chosen size distributions and refractive indices similar to Quenzel and Kastner (1980). HMF7 models an aerosol for a relative humidity of 70%, while HMF9 models an aerosol for a relative humidity of 90%. Note the increased water content (m closer to that of water) and the particle swelling which takes place as the relative humidity increases from 70 to 90%.

Samples of the computed scattering phase functions at 670 nm are shown in figures 5 and 6 for the continental aerosol models. We note that these phase functions are very strongly peaked in the forward direction ($a = 0°$). As the particle size distributions become more concentrated toward smaller particle sizes (larger v), the phase function shows less pronounced forward scattering and more enhanced scattering at angles greater than about 30°. However, for

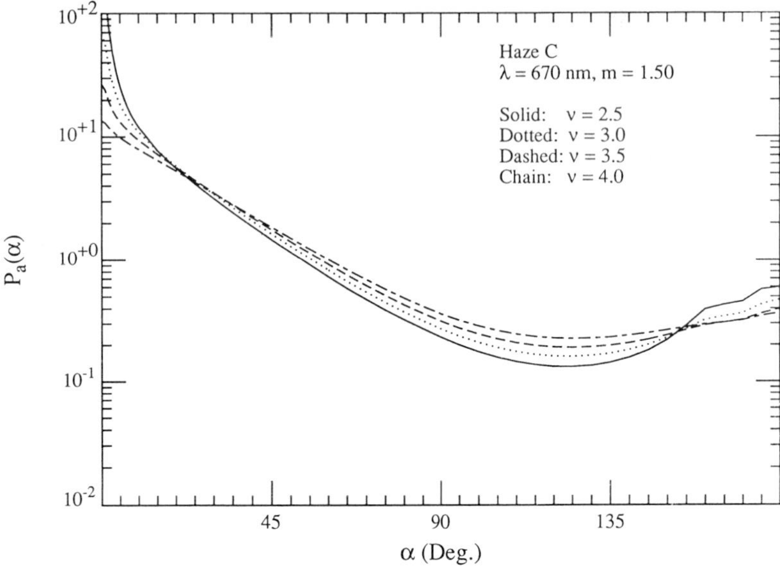

Figure 6. Aerosol phase functions for the Haze C model with $m = 1.50$.

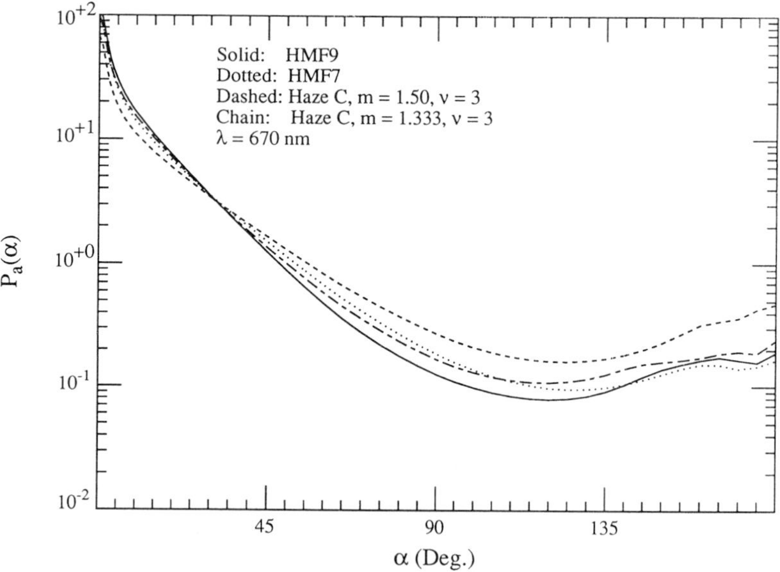

Figure 7. Aerosol phase functions for the marine aerosol models.

$m=1.50$ the back scattering, *i.e.*, $a \geq 150°$ is more pronounced for the distributions favoring the larger particle sizes. Figure 7 compares all of the models, with $v \approx 3$, at 670 nm, providing examples of how the phase function changes with m for models with nearly the same size distribution. (Note, however, that unlike the marine aerosol models, the Haze *C* distributions have particles in the size range $0.06 \leq D \leq 0.2$ μm.) It suggests that even when the size distribution is known, a realistic range of refractive indices can result in the variation of $P_a(a)$ of over a factor of two for $a \geq 45°$. Figures 8 and 9 show examples of the variation of the aerosol phase function with wavelength. Clearly, the aerosol phase function depends only *weakly* on wavelength throughout the visible. One might expect that the phase function should increase with wavelength for angles greater than about 30°, since the particle's diameter becomes a smaller fraction of the wavelength. The Haze *C* model follows this (figure 8); however, in the case of the marine aerosol this tendency is reversed for angles greater than about 150° (figure 9).

Figures 10, 11, and 12 provide the computations of the variation of the scattering coefficient b_a with λ, *i.e.*, the quantity $b_a(\lambda)/b_a(670)$ for $\lambda = 443$, 520, 550, and 670 nm. If the aerosol *type* is independent of altitude, and the aerosol is *nonabsorbing*, this is

$$\frac{b_a(\lambda)}{b_a(670)} = \frac{\tau_a(\lambda)}{\tau_a(670)}$$

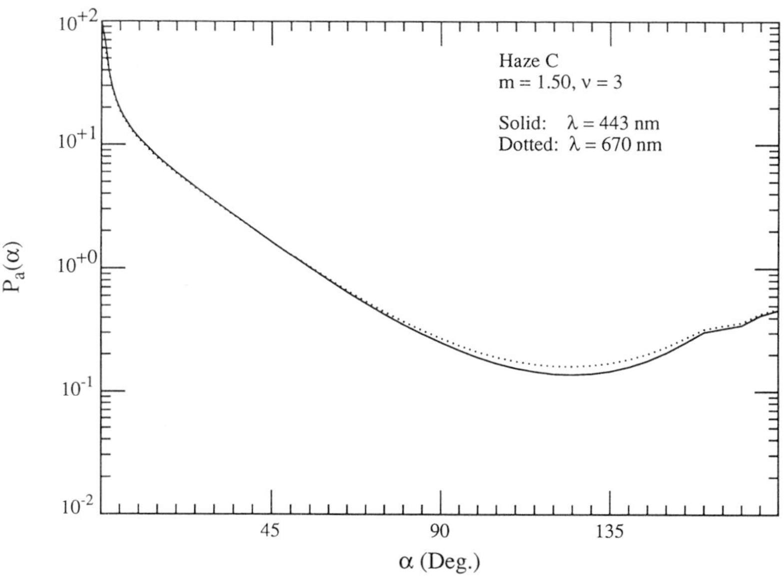

Figure 8. Aerosol phase functions for the Haze C model with $m = 1.50$ and $v = 3$.

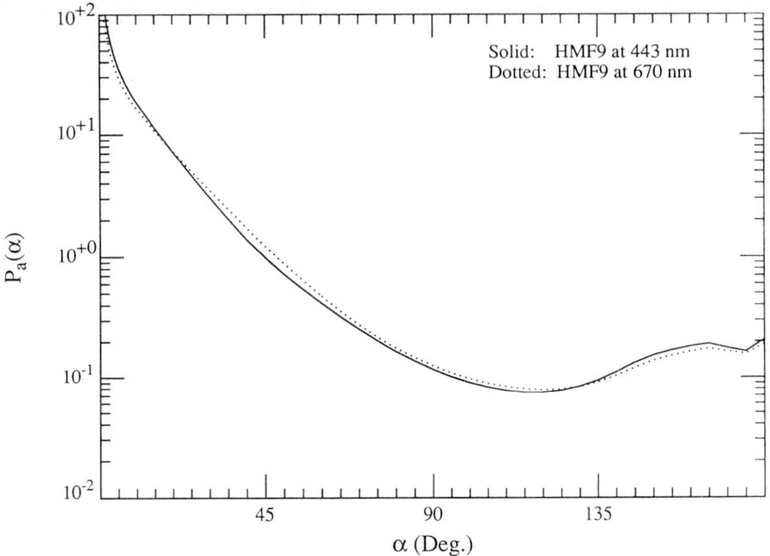

Figure 9. Aerosol phase functions for HMF9 marine model.

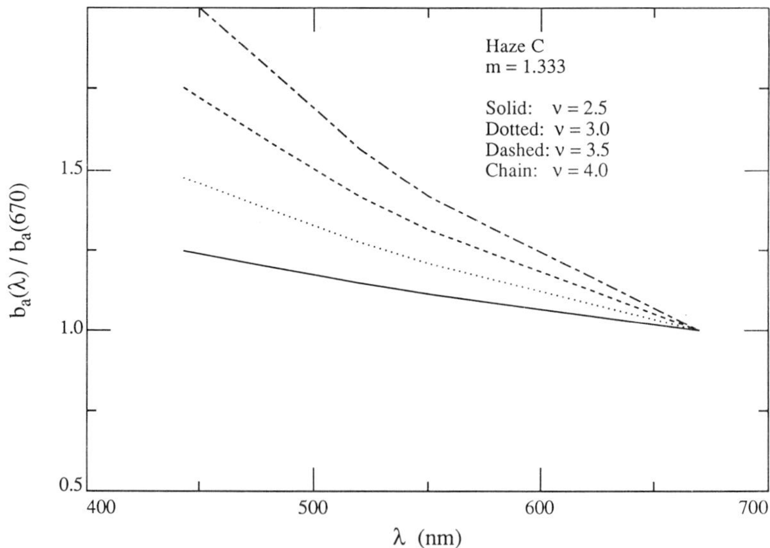

Figure 10. Spectral variation of b_a for the Haze C models with $m = 1.333$.

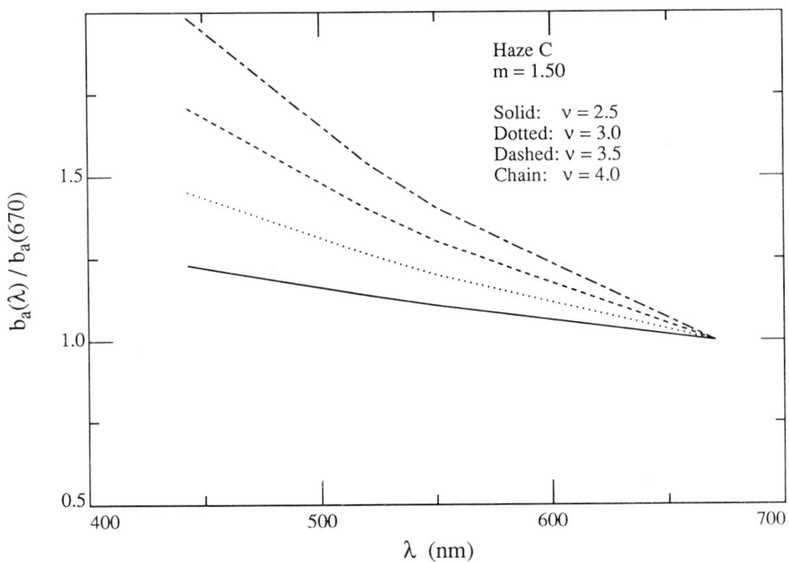

Figure 11. Spectral variation of b_a for the Haze C models with $m = 1.50$.

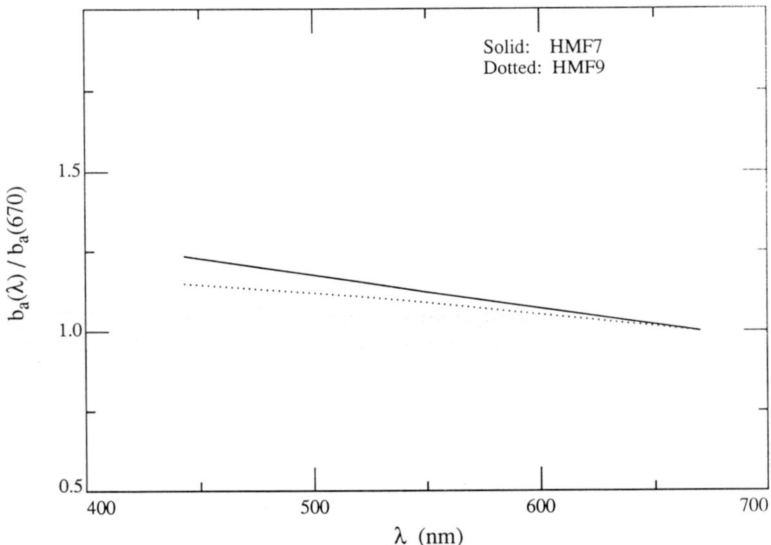

Figure 12. Spectral variation of the aerosol scattering coefficient for the marine aerosol models.

where $\tau_a(\lambda)$ is the optical thickness of the aerosol component. These figures suggest that

$$\frac{\tau_a(\lambda)}{\tau_a(670)} \approx \left(\frac{670}{\lambda}\right)^n \tag{10}$$

where for both the Haze C models $n \approx \nu\text{-}2$. Furthermore, for this range of refractive indices, the spectral variation of b_a is more strongly influenced by the size distribution than by the actual value of m (compare figures 10 and 11).

It was mentioned earlier that ω_0 for the aerosols is near reunity. In fact, if $n_i = 0$, $i.e.$, the refractive index is real, there is no absorption and $\omega_0 = 1$, exactly, otherwise, $\omega_0 < 1$. Thus, the Haze C aerosol models here have $\omega_0 = 1$. For the marine aerosol models, the Mie computations provide $0.832 \leq \omega_0(\lambda) \leq 0.843$ for HMF7 and $0.939 \leq \omega_0(\lambda) \leq 0.950$ for HMF9, with $443 \leq \lambda \leq 670$. Note the model with the larger $|n_i|$ yields smaller ω_0. These computations suggest that as long as n_i is constant, $\omega_0(\lambda)$ will vary very little with λ over the visible spectrum.

The above models are introduced to provide examples of the scattering properties of the aerosols and their variation. More sophisticated models of the aerosol now exist (see for example the aerosol models in LOWTRAN 7); however, the simple models I have discussed provide sufficient background to understand the influence of aerosols on ocean color remote sensing.

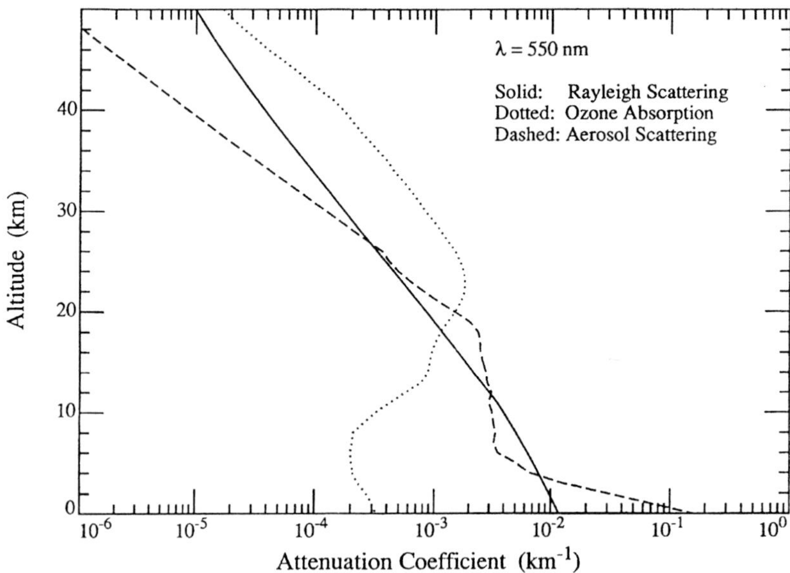

Figure 13. Vertical distribution of the attenuation coefficient of air, aerosols and O_3

Finally, we consider the vertical distribution of the optically important constituents. To compare one component with another it is more instructive to use profiles of attenuation rather than concentration. This is effected in figure 13, which provides the attenuation coefficient of air, aerosols, and O_3.

These species are distributed with altitude according to mean profiles determined by Elterman (1968). Water vapor is not considered since ocean color sensors avoid the water vapor bands. It is seen that the aerosols provide the dominant attenuation near the surface, Rayleigh scattering above the immediate surface layer, and O_3 above about 20 km. It is important to note that the aerosol profile can differ significantly from that in the figure. For example, a major volcanic eruption, such as El Chichón in 1982, can eject enough aerosol into the stratosphere that the aerosol, rather than O_3, determines the optical properties there.

It is rare, however, that the aerosol concentration is not largest near the surface.

3. Radiative Transfer

Consider a cartesian coordinate system with its origin at the "top" of the atmosphere. The x and y axes parallel to a plane tangent to the earth's surface below the origin and the z axis is directed toward the earth. A photon's direction is specified by the polar angle (θ) and the azimuth angle $(\varnothing)$ of a spherical coordinate system built on the cartesian system, $i.e.$, if a photon is traveling in the direction specified by the unit vector ξ, the components of ξ are $(x,y,z) = (sin\theta cos\varnothing, sin\theta sin\varnothing, cos\theta)$.
Thus, photons traveling toward the earth have $0 \leq \theta < 90°$, while photons traveling toward space have $90 < \theta \leq 180°$.
The atmosphere is assumed to be plane parallel and horizontally homogeneous, so L is taken to be a function of altitude, direction, and wavelength, $i.e.$, $L = L(z,\theta,\varnothing,\lambda)$.

3.1 THE RADIATIVE TRANSFER EQUATION

The propagation of radiance is governed by the radiative transfer equation (RTE). In an atmosphere in which the IOP's depend only on altitude, the RTE is

$$cos\theta \frac{dL\left(z,\lambda,\theta,\varnothing\right)}{dz} = -c\left(z,\lambda\right) L\left(z,\lambda,\theta,\varnothing\right) +$$

$$+ \int_{4\pi} \beta\left(z,\lambda,\theta',\varnothing' \rightarrow \theta,\varnothing\right) L\left(z,\lambda,\theta',\varnothing'\right) d\Omega' \tag{11}$$

where $d\Omega' = sin\theta' d\theta' d\varnothing'$, and the 4π on the integral means that the integration is to be carried out over all θ' and $\varnothing'$.
The first term on the right-hand-side represents the loss of radiance in the direction $(\theta,\varnothing)$ by scattering and absorption.

The second term is the gain in radiance due to scattering of radiance from all other directions $(\theta',\varnothing')$ into the direction $(\theta,\varnothing)$. In this formulation the polarization of the light is ignored. In what follows, we shall omit the explicit dependence of the various quantities on λ except where necessary to avoid confusion. In terms of these auxiliary IOP's, ω_0 and P, the RTE becomes

$$cos\theta\frac{dL(z,\theta,\varnothing)}{c(z)\,dz} = -L(z,\theta,\varnothing) +$$

$$+\omega_0(z)\int_{4\pi} P(z,\theta',\varnothing'\to\theta,\varnothing)L(z,\theta',\varnothing')\,d\Omega' \qquad (12)$$

Case (1957) has shown that, given the radiance incident on the upper and lower boundaries of the atmosphere, the solutions to the RTE are unique if $\omega_0 < 1$, *i.e.*, there is *some* absorption.
Note that in Equation (12), $c(z)$ occurs only in the combination $c(z),dz$, so a dimensionless depth τ - the *optical depth* - such that $d\tau = c(z)dz$, is introduced. In a homogeneous atmosphere, $\tau = cz$, and the RTE becomes

$$cos\theta\frac{dL(\tau,\theta,\varnothing)}{d\tau} = -L(\tau,\theta,\varnothing) +$$

$$+\omega_0\int_{4\pi} P(\theta',\varnothing'\to\theta,\varnothing)L(\tau,\theta',\varnothing')\,d\Omega' \qquad (13)$$

Analytical solutions to the RTE are possible only in the simplest case,*e.g.*, $\omega_0 = 0$, so one must be satisfied with numerical solutions.

3.2 THE SUCCESSIVE ORDER OF SCATTERING SOLUTION

The successive order of scattering technique is the most straightforward technique for solving the RTE. The basic idea is to successively compute the radiance that is scattered once, twice, etc., and then to sum these contributions to obtain the total radiance (Hansen and Travis, 1974; Van de Hulst, 1980).
The development is simplified if we consider a homogeneous atmosphere with $\omega_0 < 1$.
We then assume that the radiance can be expanded in a power series in ω_0 (Gordon, 1992), *i.e.*,

$$L(\tau,\theta,\varnothing) = L^{(0)}(\tau,\theta,\varnothing) + \omega_0 L^{(1)}(\tau,\theta,\varnothing) +$$

$$+\omega_0^2 L^{(2)}(\tau,\theta,\varnothing) + ...,$$

put this into the RTE, and group like powers of ω_0. The RTE is then satisfied if the individual $L^{(n)}$'s satisfy

$$\cos\theta \frac{dL^{(0)}}{d\tau} = -L^{(0)} \, ,$$

$$\cos\theta \frac{dL^{(1)}}{d\tau} = -L^{(1)} + \int PL^{(0)\prime} \, d\Omega' ,$$

$$\cos\theta \frac{dL^{(2)}}{d\tau} = -L^{(2)} + \int PL^{(1)\prime} \, d\Omega' ,$$

$$\vdots$$

$$\cos\theta \frac{dL^{(n)}}{d\tau} = -L^{(n)} + \int PL^{(n-1)\prime} \, d\Omega' ,$$

$$\vdots$$

(14)

These represent a simplification in that the single integral-differential equation has been transformed into a set of ordinary differential equations (the integrals can now be evaluated in principle since each integrand is furnished by solving the preceding equation). If the atmosphere is illuminated from above by a radiance $L_{inc}(0,\theta,\varnothing)$, we will choose the simplest way of satisfying the boundary condition at $z=0$:

$$L^{(0)}(0,\theta,\varnothing) = L_{inc}(0,\theta,\varnothing)$$

$$L^{(n)}(0,\theta,\varnothing) = 0 \quad for \quad n > 0 \quad and \quad \theta < 90°.$$

In the case of interest, the atmosphere is bounded below by a Fresnel-reflecting sea surface. If the surface is *flat*, the lower boundary condition (at $z = z_1$) is

$$L^{(n)}(z_1,\theta_r,\varnothing_r) = \rho(\theta_i) L^{(n)}(z_1,\theta_i,\varnothing_i) \quad for\ all\ n, \tag{15}$$

where $(\theta_i,\varnothing_i)$ is the direction of the incident photon, $(\theta_r,\varnothing_r)$ is the direction of the reflected photon ($\theta_r = \pi - \theta_i$, $\varnothing_r = \varnothing_i$), and $\rho(\gamma)$ is the Fresnel reflectance of the flat ocean surface for an incident angle of γ with respect to the normal.

50

It is given by

$$\rho(\gamma) = \frac{1}{2} \left[\frac{tan^2(\gamma-\gamma')}{tan^2(\gamma+\gamma')} + \frac{sin^2(\gamma-\gamma')}{sin^2(\gamma+\gamma')} \right]$$

where γ and γ' are related by Snell's law: $m_w sin\gamma' = sin\gamma$, with m_w representing the real part of the index of refraction of water.

3.3 THE SINGLE SCATTERING APPROXIMATION

In the *single scattering* approximation the series is terminated at $n=1$, *i.e.*, $L = L^{(0)} + \omega_0 L^{(1)}$. We will now develop this solution for a homogeneous atmosphere with no upward radiance incident on the lower boundary, which we take to be located at $z=z_1$ or $\tau = \tau_1$. The radiance incident on the top of the atmosphere is that due to the solar beam, *i.e.*,

$$L_{inc}(0,\theta,\varnothing) = F_0 \delta\left(cos - cos\theta_0\right)\delta\left(\varnothing - \varnothing_0\right) \tag{16}$$

where θ_0 and $\varnothing_0$ are the solar zenith and azimuth angles, respectively, and λ is the Dirac delta function. F_0 is the solar irradiance (the power per unit area per unit $\Delta\lambda$) on a plane normal to the sun's rays at the top of the atmosphere. It depends on the position of the earth in its orbit and is given by

$$F_0(\lambda) = \left\langle F_0(\lambda) \right\rangle \left(1 + 0.016cos\left[\frac{2\pi(D-3)}{365} \right] \right)^2$$

where D is the day of the year (D=1 on 1 Jan. and D=365 on 31 Dec.). $<F_0>$ as a function of wavelength, measured by Neckel and Labs (1984), is presented in figure 14. The equation for $L^{(0)}$ can be solved yielding

$$L^{(0)}(\tau,\theta,\varnothing) = F_0 \delta\left(cos - cos\theta_0\right)\delta\left(\varnothing - \varnothing_0\right) exp\left(\frac{-\tau}{cos\theta_0} \right) \tag{17}$$

Note that the transmittance defined earlier is given by $T(\theta_0,\varnothing_0) = (-\tau / cos\theta_0)$, and measurement of T yields τ, the optical thickness of the atmosphere. The equation for $L^{(1)}$ then becomes

$$cos\theta \frac{dL^{(1)}}{d\tau} = -L^{(1)} + \int PL^{(0)'} d\Omega'$$

$$= -L^{(1)} + P\left(\theta_0, \varnothing_0 \to \theta, \varnothing\right) F_0 exp\left(\frac{-\tau}{cos\theta_0} \right)$$

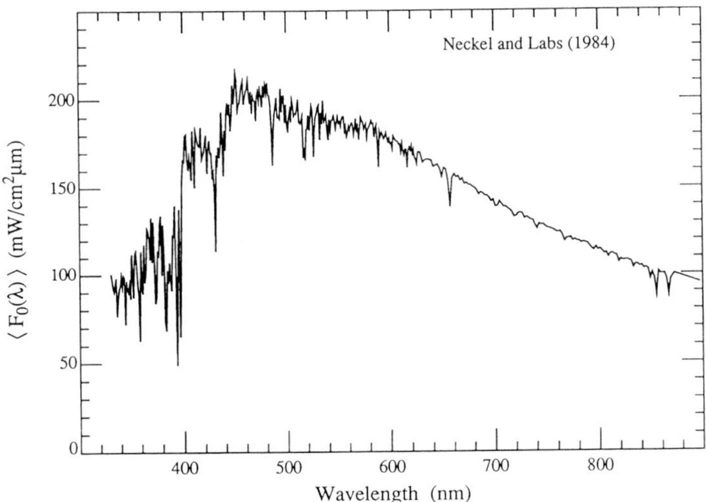

Figure 14. Mean extraterrestrial solar irradiance.

The solution to this equation for the radiance in the atmosphere is

$$L_u^{(1)}(\tau,\theta,\varnothing) = \frac{F_0 \cos\theta_0 P(a)}{\cos\theta_0 - \cos\theta} exp\left(\frac{-\tau}{\cos\theta}\right) \times$$

$$\times \left(exp\left[\tau\left(\frac{1}{\cos\theta} - \frac{1}{\cos\theta_0}\right)\right] - exp\left[\tau\left(\frac{1}{\cos\theta} - \frac{1}{\cos\theta_0}\right)\right]\right) \qquad (18)$$

and

$$L_d^{(1)}(\tau,\theta,\varnothing) = \frac{F_0 \cos\theta_0 P(a)}{\cos\theta_0 - \cos\theta}\left(exp\left[\frac{-\tau}{\cos\theta_0}\right] - exp\left[\frac{-\tau}{\cos\theta}\right]\right) \qquad (19)$$

where

$$\cos a = \cos\theta \cos\theta_0 + \sin\theta \sin\theta_0 \cos\varnothing$$

and $\varnothing_0$ is taken to be zero, *i.e.*, the sun's rays are parallel to the *x-z* plane. The subscripts u and d in Equations (18) and (19) mean "up" ($\theta > 90°$) and "down" ($\theta < 90°$), respectively. To this order, $L = L^{(0)} + \omega_0 L^{(1)}$, so if $\theta \neq \theta_0$, $L_u(\tau,\theta,\varnothing)$ and $L_d(\tau,\theta,\varnothing)$ are given by Equations (18) and (19) with P replaced by $\omega_0 P$, respectively. If the atmosphere is "optically thin," *i.e.*, $\tau_1 \ll 1$, the exponentials can be expanded in a power series in τ_1 to obtain,

$$L_u(0,\theta,\varnothing) = -\frac{F_0\,\omega_0\,P(a)\tau_1}{\cos\theta}, \qquad \theta > 90°$$

$$L_d(\tau_1,\theta,\varnothing) = +\frac{F_0\,\omega_0\,P(a)\tau_1}{\cos\theta}, \qquad \theta < 90° \tag{20}$$

Thus, the radiance exiting a thin layer of atmosphere is directly proportional to $\omega_0 P(a)\tau_1$.

Now we consider the case of a thin atmosphere bounded below by a specularly reflecting, flat ocean surface. (We ignore for the present the contribution to the radiances from photons which are backscattered *out* of the ocean.) We wish to compute the radiance leaving the atmosphere $L_u(0,\theta,\varnothing)$. Equation (20) provides the contribution to $L_u(0,\theta,\varnothing)$ from single scattering of the direct solar beam in the medium; however, there are two other single scattering contributions to $L_u(0,\theta,\varnothing)$. The solar beam can first scatter in the atmosphere generating $L_d(\tau_1,\theta,\varnothing)$ at the lower boundary which is then Fresnel-reflected by the sea surface back into the atmosphere and propagates to the top. Alternatively, the solar beam can propagate to the surface without scattering, Fresnel-reflect from the sea surface back into the atmosphere, scatter in the atmosphere toward the direction $(\theta,\varnothing)$ and propagate to the top. In the thin layer approximation (first order in τ_1) these two processes yield a contribution to $L_u(0,\theta,\varnothing)$ of

$$-\frac{F_0\,\omega_0\,P(a_-)\tau_1}{\cos\theta}\left[\rho(\theta) + \rho(\theta_0)\right]$$

where

$$\cos a_- = -\cos\theta\cos\theta_0 + \sin\theta\sin\theta_0\cos\varnothing$$

Thus, for a thin homogeneous atmosphere of optical thickness τ_1 with a Fresnel reflecting lower boundary, the radiance leaving the top of the atmosphere is, to first order in τ_1,

$$L_u(0,\theta,\varnothing) = -\frac{F_0\,\omega_0\,\tau_1}{\cos\theta}\left(P(a_+) + \left[\rho(\theta) + \rho(\theta_0)\right]P(a_-)\right) \tag{21}$$

where

$$\cos a_\pm = \pm\cos\theta\cos\theta_0 + \sin\theta\sin\theta_0\cos\varnothing$$

Note that $\omega_0 P\tau_1$ is just $\beta\,z_1$, so if there are several scattering constituents in the atmosphere, the contributions from each are linearly additive in the single scattering approximation (note that each component will have a different τ_1).

4. First Order Radiance Model for Atmospheric Correction

We can use the approximation of an optically thin atmosphere to model the ocean color remote sensing situation depicted schematically in figure 15. To effect this we must calculate the radiance reflected from the atmosphere-ocean system. Consider first just the atmosphere and the flat air-sea interface. In this case, we can use Equation (21) to estimate the upward radiance at the top of the atmosphere resulting from scattering by molecules and aerosols. This is

$$L_r + L_a$$

with

$$L_x(0,\theta,\varnothing) = -\frac{F_0 \omega_0 \tau_1}{\cos\theta}\left(P_x\left(a_+\right) + \left[\rho(\theta) + \rho\left(\theta_0\right)\right]P_x\left(a_-\right)\right) \quad (22)$$

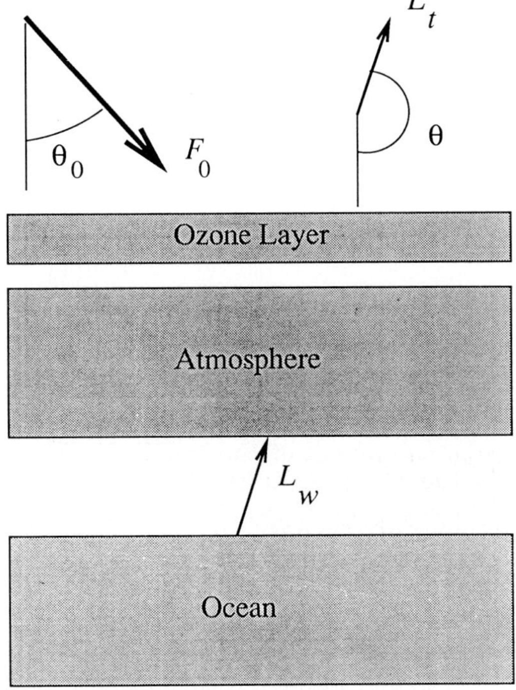

Figure 15. Schematic of the sensing geometry.

54

where x refers to the particular component, *i.e.*, $x = r$ or a for the "Rayleigh" and aerosol components, respectively. However, this radiance is not correct because the influence of O_3 has been ignored. In figure 13 we see that the major contribution from O_3 comes at altitudes above 20 to 25 km, where the Rayleigh and aerosol scattering coefficients are small. Thus, a simple way to include O_3 in this first order model is to confine it in a layer above the thin atmosphere. The scattered radiance will then have to make two trips through the O_3 layer to reach the receiver as shown in figure 15. This will attenuate the radiance by a factor

$$exp\left[-\tau_{Oz}\left(\frac{1}{cos\theta_0} - \frac{1}{cos\theta}\right)\right]; \qquad 0 > 90°$$

where τ_{Oz} is the optical thickness of O_3 in the atmosphere. Thus, the radiances given above can be modified to include O_3 by letting

$$F_0 \to F_0' = F_0\, exp\left[-\tau_{Oz}\left(\frac{1}{cos\theta_0} - \frac{1}{cos\theta}\right)\right]; \qquad 0 > 90° \qquad (23)$$

Another component of the radiance at the top of the atmosphere is due to the reflection of the direct solar beam from the sea surface, and its subsequent transmission to the top of the atmosphere. It is given by

$$L_g(0,\theta,\varnothing) = F_0\, \rho\left(\theta_0\right)\delta\left(\theta + \theta_0 - \pi\right)\delta(\varnothing)\, exp\left[-\tau\left(\frac{1}{cos\theta_0} - \frac{1}{cos\theta}\right)\right]$$

where $\tau = \tau_r + \tau_a + \tau_{Oz}$ is the total optical thickness of the atmosphere. The subscript "g" is used for this component since, when the sea surface is ruffled by the wind (and no longer flat), this term will produce what is commonly called sun glint.

Finally, there is a component of the radiance due to light that has been backscattered *out* of the water. At the sea surface this component is called the *water-leaving radiance*, L_w. In propagation to the top of the atmosphere, it is attenuated by a factor t yielding tL_w at the sensor. The attenuation factor t depends on the angular distribution of L_w. If L_w were large in a single direction (like L_{inc} at the top of the atmosphere) then t would be the "direct" transmittance (similar to $T(\theta_0,\varnothing_0)$). If L_w were totally diffuse, *i.e.*, if L_w were independent of θ and $\varnothing$, then t would be the transmittance function for irradiance (Chandrasekhar, 1950; Tanre *et al.*, 1979; Deschamps *et al.*, 1981), *i.e.*, the *diffuse* transmittance. Irradiance is defined to be the spectral radiant power per unit area per $\Delta\lambda$ falling on a horizontal surface. Since L_w is much closer to being totally diffuse than beam-like, we use the diffuse transmittance for t. It is given by (Gordon *et al.*, 1983)

$$t = exp\left[\frac{\left(\dfrac{\tau_r}{2} + \tau_{Oz}\right)}{cos\,\theta}\right] t_a, \qquad 0 > 90°$$

where

$$t_a = exp\left[\frac{\left(1 - \omega_a F_a\right)\tau_a}{cos\,\theta}\right] \tag{24}$$

and F_a is the probability that a photon scattered by the aerosol will be scattered through an angle less than 90°. The total radiance at the top of the atmosphere, L_t, is then

$$L_t = L_r + L_a + L_g + tL_w \tag{25}$$

The basic problem of atmospheric correction is to extract L_w from L_t. To provide an appreciation for the importance (and difficulty) of atmospheric correction, we present in figure 16 simulated spectra of the total radiance L_t at the top of the atmosphere and the desired water-leaving radiances L_w for low and high pigment concentrations in Case 1 waters (Gordon *et al.*, 1988).

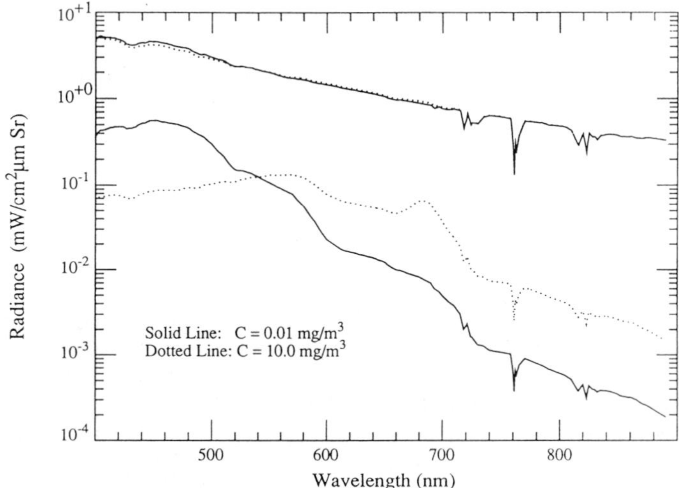

Figure 16. Simulated spectra of the upward radiance at the sea surface (lower curves) and the top of the atmosphere (upper curves) for low (solid) and high (dotted) pigment concentrations.

Note that Case 1 waters (Morel and Prieur, 1977; Gordon and Morel, 1983) are defined to be waters for which the optical properties are controlled principally by the water itself and/or by phytoplankton and their immediate degradation products. The optical properties of the biogeneous component can be parameterized by the pigment concentration (Gordon and Morel, 1983; Morel, 1988). The pigment concentration is defined to be the sum of the concentrations (in mg/m^3) of chlorophyll a and phaeophytin a.

We note that the variations in L_w are nearly masked by the atmospheric scattering in L_t. Also, we see that for low pigment concentrations L_w ranges from about 15% of L_t in the blue to 1% in the red, while for the high pigment concentration the range is from about 2% in the blue to as much as 5% in the red. In the near infrared (NIR) L_w is always less than 1% of L_t. Algorithms for extracting pigment concentration from ocean color measurements usually use the ratio of L_w's in the blue and green regions of the spectrum (Morel and Prieur, 1977; Gordon and Morel, 1983, Gordon and Clark, 1980; Clark, 1981), e.g., in the case of CZCS (spectral bands at 443, 520, 550, and 670 nm, referred to as λ_1, λ_2, λ_3, and λ_4, respectively) the pigment concentration is estimated from the ratios

$$r_{1,3} = \frac{L_w\left(\lambda_1\right)}{L_w\left(\lambda_3\right)} \quad and/or \quad r_{2,3} = \frac{L_w\left(\lambda_2\right)}{L_w\left(\lambda_3\right)} \tag{26}$$

Since the red and NIR portions are least affected by L_w it is natural to use this spectral region to assess the effects of the atmosphere and sea surface.

To estimate L_w in Equation (26) we need estimates of L_r, L_g and L_a. Knowing the optical thicknesses τ_r and τ_{Oz}, we can compute L_r using Equation (22). We note that τ_r depends on wavelength as given in Equation (8) and on the surface atmospheric pressure as in Equation (9). Gordon, Brown, and Evans (1988) show that the neglect of the surface pressure variations will result in at most an error of $\pm 1.5\%$ and usually much less, and at this point we will ignore the surface pressure variation. Also, although the O_3 concentration varies significantly in space and time, at this point we will simply use a climatological mean to determine τ_{Oz}.

L_g in Equation (25) is large only near the specular image of the sun, $i.e.$, for a flat sea surface L_g is traveling in the direction $(\pi\text{-}\theta_0, \varnothing_0)$. For a rough surface L_g is also large for directions close to this. In the case of ocean color sensors, this term is minimized by providing the instrument with the capability of tilting the scan plane away from the above direction. Because of this we ignore L_g. Note that, for a flat ocean, L_g is orders of magnitude larger than the other radiances in Equation (25). For a rough ocean it is smaller, but at its maximum could still be one or two orders of magnitude larger than the other terms in the equation. Thus, the region of large L_g must *be avoided*: it cannot be estimated with sufficient accuracy to utilize imagery acquired in the sun glint region.

The diffuse transmittance t can be estimated by using the approximation $t_a \approx 1$. This is possible because the aerosol is strongly forward scattering, so F_a is near unity, and even for mildly absorbing aerosols ω_a is usually greater than about 0.85. Thus, the product $\omega_a F_a > \sim 0.75$ in Equation (24) and $t_a \approx 1$

as long as τ_a is not too large. (Recall that the assumption that the atmosphere is thin still prevails, and this requires $\tau_a \ll 1$.)

The remaining quantity, L_a, cannot be computed because τ_a is a strong function of space and time. Also, even given τ_a, e.g., from surface measurements of $T(\theta_0,\varnothing_0)$, computation of L_a requires the aerosol phase function which is very difficult to obtain. Thus, this term must be estimated in some way from measurements made at the sensor. A scheme for carrying this out was first proposed by Gordon (1978) in the late 1970's. The basic idea is to use the fact that the water-leaving radiance in the red and/or NIR portion of the spectrum is very small compared to the other terms in Equation (25). Let $\lambda_i < \lambda_j$ be two spectral bands in this region of the spectrum. Then Equation (25) can be used to estimate L_a, i.e., $L_a = L_t - L_r$, since $L_w = 0$ in this region and the sensor tilt renders $L_g = 0$ in situations where it might be a problem.

Given $L_a(\lambda_i)$ and $L_a(\lambda_j)$ we can form

$$\frac{L_a\left(\lambda_i\right)}{L_a\left(\lambda_j\right)} = \varepsilon\left(\lambda_i,\lambda_j\right)\frac{F_0'\left(\lambda_i\right)}{F_0'\left(\lambda_j\right)} \tag{27}$$

where

$$\varepsilon\left(\lambda_i,\lambda_j\right) = \frac{\omega_a\left(\lambda_i\right)\tau_a\left(\lambda_i\right)}{\omega_a\left(\lambda_j\right)\tau_a\left(\lambda_j\right)}$$

$$\times \frac{\left(P_a\left(a_+,\lambda_i\right)+\left[\rho(\theta)+\rho\left(\theta_0\right)\right]P_a\left(a_-,\lambda_i\right)\right)}{\left(P_a\left(a_+,\lambda_j\right)+\left[\rho(\theta)+\rho\left(\theta_0\right)\right]P_a\left(a_-,\lambda_j\right)\right)} \tag{28}$$

For a given aerosol *type*, the aerosol optical thickness is proportional to the concentration, so the concentration cancels out of Equation (28) and $\varepsilon\,(\lambda_i,\lambda_j)$ is independent of the aerosol concentration. Also, for the aerosol models described earlier, ω_a is very nearly independent of λ, so the $\omega_a(\lambda)$ terms should cancel as well. If the aerosol phase function were independent of λ, then the P_a terms would also cancel, yielding

$$\varepsilon\left(\lambda_i,\lambda_j\right) = \frac{\tau_a\left(\lambda_i\right)}{\tau_a\left(\lambda_j\right)} \tag{29}$$

For a given *aerosol type* $\tau_a(\lambda_i)/\tau_a(\lambda_j)$ is a constant, so $\varepsilon\,(\lambda_i,\lambda_j)$ would be *constant* everywhere in an image in which the aerosol type does not change from one position to another. However, figures 8 and 9 show that P_a is weakly dependent on λ, so Equation (29) is only an approximation, and we can expect some variations in ε over an image even when the aerosol type does not change. Equations (29) and (10) suggest that it is reasonable to assume that ε varies with wavelength according to λ^{-n}, *i.e.*,

$$\varepsilon\left(\lambda_i,\lambda_j\right)=\left(\frac{\lambda_j}{\lambda_i}\right)^n \tag{30}$$

where n is *approximately* constant. Using the value of n estimated in this manner, we can extrapolate $\varepsilon\ (\lambda_i,\lambda_j)$ from the red/NIR into the green and blue regions of the spectrum. This then enables the estimation of $\varepsilon(\lambda,\lambda_j)$ and from this $L_a(\lambda)$:

$$L_a\left(\lambda\right)=\varepsilon\left(\lambda,\lambda_j\right)\frac{F_0'\left(\lambda\right)}{F_0'\left(\lambda_j\right)}L_a\left(\lambda_j\right) \tag{31}$$

The procedure described above yields $L_w(\lambda)$ given two other wavelengths for which L_w can either be assumed to be zero or has a known value. When we apply it to the CZCS our goal is to determine L_w at 443, 520, and 550 nm in order to estimate the pigment concentration (Gordon and Morel, 1983). We note from figure 16 that for the lower pigment concentration the approximation $L_w(670)\approx0$ is reasonable in that $L_w(670) \leq 1\%$ of $L_t(670)$. However, L_w at the other bands is a significant portion of L_t and is unknown. Thus, there is not enough information to estimate the values of $\varepsilon(\lambda,670)$ required to execute the procedure. One method for obtaining some information about ε was developed by Gordon and Clark (1981), who showed that when $C\leq0.25$ mg/m^3, the *normalized water-leaving radiance*, $[L_w]_N$, defined through

$$L_w\left(\lambda\right)=\left[L_w\left(\lambda\right)\right]_N cos\,\theta_0\,exp\left[\frac{-\left(\frac{\tau_r}{2}+\tau_{Oz}\right)}{cos\,\theta_0}\right] \tag{32}$$

was independent of C and has the values 0.498, 0.30, and less than 0.015 mW/cm^2 μm Ster for 520, 550, and 670 nm, respectively. In contrast, at 440 nm even for $C\leq0.25$ mg/m^3, $[L_w]_N$ depends very strongly on the actual value of C. The normalized water-leaving radiance is approximately the radiance that would exit the ocean if the sun were at the zenith and the atmosphere were removed. This clear water radiance concept was utilized by Gordon *et al.* (1983) to process CZCS imagery in the following manner. First, a region of an image for which $C < 0.25$ mg/m^3 was located. Next, the procedure described above was used to determine $\varepsilon(520,670)$, $\varepsilon(550,670)$, and $\varepsilon(670,670)$, and $\varepsilon(443,670)$ for this region by extrapolation using Equation (30). Finally, these values determined for the ε parameters were used throughout the entire image. There are, however, several serious difficulties with this procedure. First, there may be no "clear water" in the image of interest. Second, the aerosol type may vary over the image in which case the ε's are expected to depend strongly on position (Bricaud and Morel, 1987; André and Morel, 1991). Third, even if the aerosol type remains constant, our models (figures 8 and 9) show that the aerosol phase function depends weakly on wavelength which implies that the ε's will depend on position in the image even if all of the assumptions inherent in the single scattering approximation are valid (figure 17) (Gordon, 1984). Forth, the single scattering approximation is not sufficiently accurate. The necessity of clear water in a scene will be circumvented in future sensors by virtue of additional spectral bands with $\lambda > 700$ nm (figure 16); however, for CZCS it must be faced head on. The most promising approach is that described for Case 1 waters by Morel and co-workers (Bricaud and Morel, 1987; André and Morel, 1991) based on earlier ideas of Smith and Wilson (1981) (See also, Gordon *et al.* (1988)). In these waters $[L_w]_N$ for a given wavelength is modeled as a function of C. Thus, L_w at each wavelength can be written in terms of C, *i.e.*, ignoring L_g in Equation (25),

$$L_t(\lambda) = L_r(\lambda) + L_a(\lambda) + tL_w(\lambda, C)$$

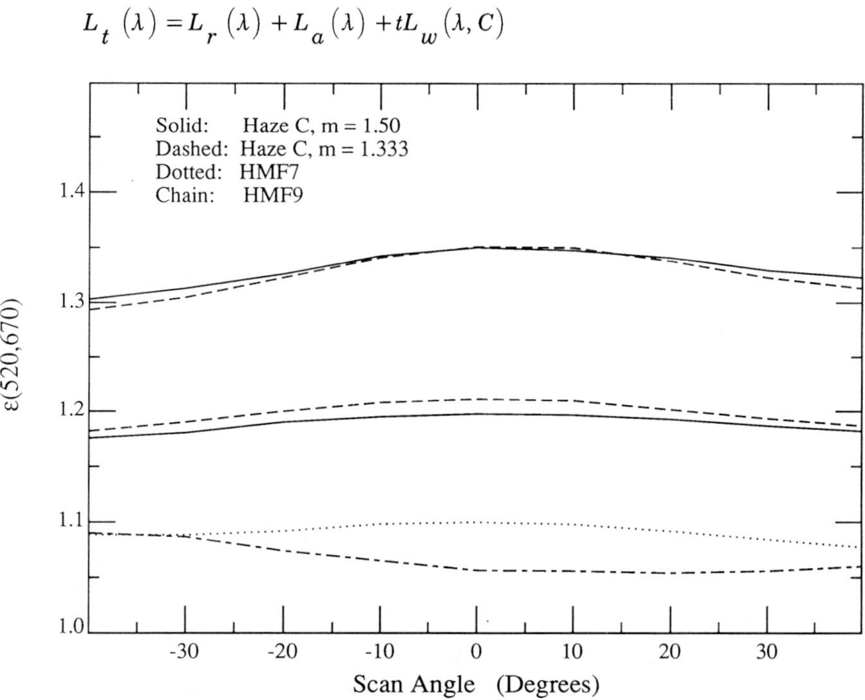

Figure 17. Computed variation of $\varepsilon(520,670)$ across a CZCS scan line for the geometry of Orbit 130 over the Gulf of Mexico (November 2, 1978) using Eq. (28). The upper (lower) solid and dashed curves are for $\nu = 4\,(3)$.

Then using Equations (30) and (31), we have

$$tL_w(\lambda, C) = L_t(\lambda) - L_r(\lambda) - \left(\frac{\lambda}{\lambda_4}\right)^n \frac{F_0'(\lambda)}{F_0'(\lambda_4)}\left[L_t(\lambda_4) - L_r(\lambda_4) - tL_w(\lambda_4, C)\right] \quad (33)$$

where $\lambda = \lambda_1$, λ_2, or λ_3. Given $L_w(\lambda, C)$, this equation can be solved for three values of n corresponding to λ_1, λ_2, and λ_3. These are averaged to obtain a single value. An iterative procedure is used to find $L_w(\lambda, C)$ and n as follows:

(1) start by setting $L_w(\lambda_4, C)$ and n to zero and compute $L_w(\lambda, C) \equiv L_w^{(0)}(\lambda, C)$ using Equation (33);

(2) use the ratio $r_{1,3}$ or $r_{2,3}$ to estimate C;

(3) use the $[L_w]_N$ model for Case 1 waters to estimate $L_w(\lambda, C) \equiv L_w^{(1)}(\lambda, C)$;

(4) use $L_w^{(1)}(\lambda, C)$; in Equation (33) to estimate a mean $n \equiv n^{(1)}$;

(5) use $n^{(1)}$ in Equation (33) to estimate $L_w(\lambda, C) \equiv L_w^{(2)}(\lambda, C)$; and

(6) repeat steps (2) through (5) until $n^{(k)}$ and $L_w^{(k)}(\lambda, C)$ converge.

In this manner we arrive at values of C and $L_w(\lambda, C)$ which are consistent with the model for Case 1 waters. Schematically, this procedure is represented by

$$\cdots L_w^{(k)}(\lambda, C) \xrightarrow{Eq.\ (33)} n^{(k)} \xrightarrow{Eq.\ (33)} L_w^{(k+1)}(\lambda, C) \xrightarrow{r_{1.3}} C$$

$$C \xrightarrow{Model} L_w^{(k+2)}(\lambda, C) \xrightarrow{Eq.\ (33)} + n^{(k+2)} \cdots,$$

where "Model" refers to the radiance model for Case 1 waters. The virtues of this procedure are (1) that the maximum value of C can be quite large, *i.e.*, 1.5-2.0 mg/m^3, so the clear water requirement can be relaxed, and (2) that the procedure can be applied to images on a pixel-by-pixel basis so the variability of ε is irrelevant. However, it must be recalled that the procedure is based on Equation (30) which is an approximation. Direct comparison between ship-measured and CZCS-derived pigment concentrations can determine the efficacy of this assumption, but only in a very indirect manner. Also, the procedure requires a radiance model of Case 1 waters which is probably *not* representative of *all* Case 1 situations. Case 2 waters, *i.e.*, waters for which other constituents such as dissolved organic material from river runoff or resuspended sediments can influence the optical properties of the medium, must of course be excluded from such a procedure. With these caveats, the only remaining criticism of the atmospheric correction is the forth - that the single scattering approximation is not accurate enough. However, we have made other assumptions that also require examination, *e.g.*, that the sea surface is flat, that a mean O$_3$ concentration is sufficient, etc. We examine these in the next section.

5. Second Order Processes

Examination of the influence of the higher order processes on the basic correction algorithm (Equations (22) - (31)) have been under way since the mid 1980's. Topics that have been addressed include, the influence of multiple scattering on the basic algorithm (Deschamps *et al.*, 1983; Gordon and Castaño, 1989), the error incurred by ignoring multiple scattering and polarization in the computation of L_r (Gordon, 1988), the impact of high-altitude volcanically-generated aerosols, *e.g.*, from El Chichón, on the algorithm (Gordon and Castaño, 1988), the error incurred by ignoring the spatial-temporal variation in the ozone concentration and the surface atmospheric pressure (André and Morel, 1989), and the impact of the assumption that the sea surface is flat (Gordon and Wang, 1991a ; Gordon and Wang, 1991b). We now discuss these individually.

5.1 MULTIPLE SCATTERING

The influence of multiple scattering on the basic CZCS correction algorithm was studied by Deschamps *et al.* (1983) and Gordon and Castaño (1989). In both studies a realistic model of the atmosphere was employed and L_t was derived from radiative transfer equation including all orders of multiple scattering. Deschamps *et al.* (1983) investigated the validity of the equation $L_t = L_r + L_a$ for a model with the sea surface absent. This is analogous to examining the validity of Equation (25). They computed L_r and L_a separately, *i.e.*, L_r for the case of no aerosols and L_a for the case with no air molecules, and exactly. They concluded that this approximation is in error by an amount that is only slightly above the limit of detectability with CZCS, but will have to be dealt with in future, more sensitive, instruments. Gordon and Castaño included the effects of a flat Fresnel-reflecting sea surface in their computations of L_t. They derived values for L_r which included multiple scattering, and then applied the CZCS algorithm, Equations (25), (27), (30), and (31), to a situation in which L_w was known at $\lambda_2, \lambda_3,$ and $\lambda_4,$ and derived $L_w(\lambda_1)$. Computations were carried out for several CZCS orbital geometries. As in the Deschamps *et al.* study, errors in the derived $L_w(\lambda_1)$ were detected, and explained by an interaction between Rayleigh and aerosol scattering, *i.e.*, by photons scattering from *both* molecules and aerosols. It was found that the simple procedure of reducing the value of $\varepsilon(\lambda_1,\lambda_4)$, found from extrapolation using Equation (30), by 5%, usually reduced the error in $L_w(\lambda_1)$ to 1-2 CZCS digital counts. This procedure was also tested by Gordon and Castaño (1988) for situations with a high-altitude aerosol, *e.g.*, produced by the volcano El Chichón in 1982. They showed that the presence of this aerosol should not degrade the atmospheric correction. The message from these studies, however, is clear: the standard CZCS correction algorithm will not be sufficiently accurate to utilize fully the more sensitive instruments proposed for future missions.

5.2 MULTIPLE SCATTERING AND POLARIZATION EFFECTS ON L_r

Thus far the polarization properties of the light have been ignored; however, a correct treatment of radiative transfer requires that polarization be considered. Kattawar, Plass, and Hitzfelder (1976) have shown that ignoring the polarization properties of the light can result in significant errors in the radiance reflected from a Rayleigh scattering atmosphere. Thus, even when the multiple scattering effects discussed in the previous section are included in the algorithm, if L_r is incorrectly computed by virtue of ignoring polarization, the resulting L_w may still contain significant errors.

When the polarization state of the light is included, the radiance L in the radiative transfer equation is replaced by a vector I, and the transport equation becomes (Chandrasekhar, 1950)

$$\cos\theta \; \frac{d I(\tau;\theta,\varnothing)}{d\tau} = -I(\tau;\theta,\varnothing) +$$

$$+ \omega_0 \int_0^\pi d\theta' \int_0^{2\pi} d\varnothing' \, \sin\theta' \, Z(\tau;\theta',\varnothing'\to\theta,\varnothing) \, I(\tau;\theta',\varnothing') \quad (34)$$

The phase function, $p(\tau;\theta',\varnothing'\to\theta,\varnothing)$, in the scalar case (Equation (13)) is replaced by a 4×4 phase matrix $Z(\tau;\theta',\varnothing'\to\theta,\varnothing)$ in the vector theory. Hansen and Travis (1974) show in detail how Z can be derived using Mie theory given the properties of the scattering particles or molecules. The scalar phase function is actually included in Z, i.e.,

$$p(\tau;\theta',\varnothing'\to\theta,\varnothing) = Z_{11}(\tau;\theta',\varnothing'\to\theta,\varnothing)$$

The Stokes vector (Van de Hulst, 1957; Van de Hulst, 1980; Chandrasekhar, 1950; Shurcliff, 1962) I can be written

$$I = \begin{pmatrix} I \\ Q \\ U \\ V \end{pmatrix}$$

where I is the radiance measured by an instrument that is *insensitive* to the polarization state of the light (denoted by L in the *scalar* theory). The polarization characteristics of the light are determined by the other components of I, for example, the degree of polarization of the radiation is

$$P = \frac{\sqrt{Q^2 + U^2 + V^2}}{I}$$

where $0 \le P \le 1$. $P=0$ corresponds to completely unpolarized light and $P=1$ completely polarized light. Light with an intermediate value of P is partially polarized. When $V=0$ and $P\neq0$ the light is said to be linearly polarized, otherwise it is elliptically polarized.

A complete multiple scattering solution of Equation (34) can be obtained by the method of successive orders of scattering in a manner similar to Equation (13). Gordon, Brown, and Evans (1988) have carried out such a solution for unpolarized sunlight falling on a Rayleigh scattering atmosphere bounded below by a flat Fresnel-reflecting ocean. Their computations showed that the single scattering solution for L_r, i.e., Equation (22) with $x=r$, was typically about 3-5% too low over the center 80% of the CZCS scan for geometries with

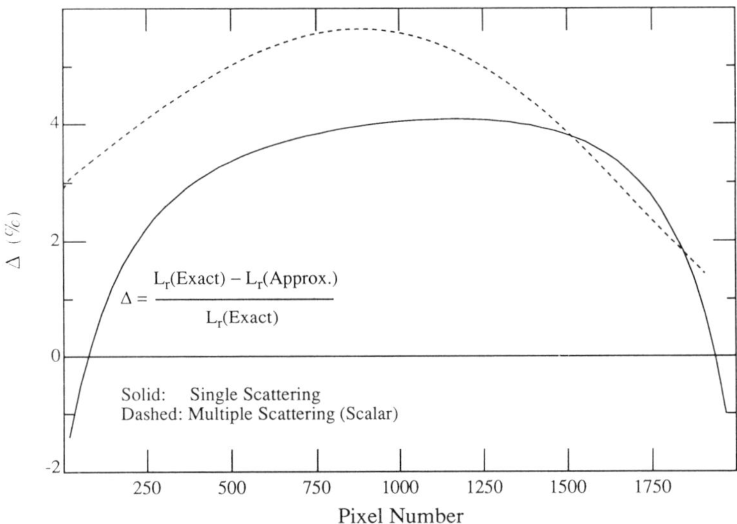

Figure 18. Percent difference between the exact Rayleigh contribution, L_r (Exact), and that computed using approximate methods L_r (Approx.), for CZCS Band 1.

$\theta_0 \leq 50°$. An example of this is given in figure 18 which provides the % difference between the exact Rayleigh contribution, L_r (Exact), for CZCS Band 1, computed by considering all orders of multiple scattering including polarization, *i.e.*, solving Equation (34), and that computed using approximate methods, L_r (Approx.). The geometry is that of Orbit 3226 over the Middle Atlantic Bight (Gordon *et al.*, 1983; Gordon *et al.*, 1988). For the solid line L_r (Approx.) is computed using single scattering, *i.e.*, Equation (22), while for the dashed line it is computed by accounting for all orders of multiple scattering but *neglecting* polarization, *i.e.*, solving Equation (13). The small variability of the error in the single-scattering L_r with scan and sun angle explains its success: the error was effectively removed by adjusting the sensor calibration near the start of the mission (Gordon *et al.*, 1983; Gordon *et al.*, 1988). For large values of θ_0, the single-scattering L_r becomes larger and the error can reach as much as 15% for $\theta_0 \approx 70°$. However, when the solution to Equation (34) is used in the computation of L_w, atmospheric corrections have been demonstrated for $\theta_0 \leq 70°$. So far I have compared only single scattering L_r with the exact value computed including all multiple scattering and polarization effects. It is natural to ask if it is possible to ignore polarization and find L_r by solving the scalar transport equation, Equation (13). Gordon, Brown, and Evans (1988) also investigated this question and found that in some cases, calculating L_r by including multiple scattering but ignoring polarization actually produced *larger* errors than the single scattering approximation (figure 18). Tables of L_r including multiple scattering and polarization are available and are being used at many locations. Eckstein and Simpson (1991) have suggested that use of these tables can result in very large errors in atmospheric correction due to large differences ($\geq 20\%$) they claim to observe between the single scattering and the multiple scattering values of L_r. However, this *incorrect* conclusion is the result of their *misinterpretation* of the geometry employed for the multiple scattering results in Gordon, Brown, and Evans (1988). In most cases, unless $\theta_0 \geq 65°$, the differences between the exact and single scattering values of L_r are similar to that shown in figure 18.

The accuracy in the computation of L_r for CZCS is now limited by the accuracy with which the O_3 concentration and surface atmospheric pressure are known.

5.3 OZONE AND SURFACE PRESSURE VARIATIONS

The surface atmospheric pressure varies as weather systems move across the globe. Similar (noncorrelated) variations are observed in the total Ozone concentration as stratospheric "weather" systems move over the globe. In fact, the Ozone concentration over a given location can vary by more than 100 DU over a period of a few days (Bhartia *et al.*, 1984). This is an enormous variation considering the mean O_3 concentration is only about 350 DU. The surface pressure variation causes variations in τ_r (Equation (9)) which in turn cause variations in L_r.

From Equations (9) and (22) we see that if $P_0 \to P_0 + \Delta P$, then $L_r \to L_r + (\Delta L_r)_P$, where

$$\frac{\left(\Delta L_r\right)^P}{L_r} = \frac{\Delta P}{P_0}$$

For a variation of 15 mb around $P_0 = 1013$ mb, this gives $(\Delta L_r)_P/L_r = \pm 1.5\%$. The O_3 variation influences L_r through its influence on F_0' (Equations (22) and (23)). It is easy to see that if $\tau_{Oz} \to \tau_{Oz} + \Delta \tau_{Oz}$, then $L_r \to L_r + (\Delta L_r)_O$, where

$$\frac{\left(\Delta L_r\right)^O}{L_r} = -\Delta \tau_{oz}\left(\frac{1}{\cos \theta_0} - \frac{1}{\cos \theta}\right)$$

or

$$\frac{\left|\left(\Delta L_r\right)^O\right|}{L_r} = \left|\Delta \tau_{oz}\right|\left(\frac{1}{\cos \theta_0} - \frac{1}{\cos \theta}\right) > 2\left|\Delta \tau_{oz}\right|$$

Now, $\tau_{Oz} = a_{Oz}{}^* C_{Oz}$, where $a_{Oz}{}^*$ is the specific O_3 absorption coefficient and C_{Oz} is the O_3 in DU. At 550 nm, figure 3 suggests that $\tau_{Oz} \approx 0.03$ for $C_{Oz} = 350$ DU. Thus, for a ± 50 DU variation around 350 DU we can expect $|(\Delta L_r)_O|/L_r > 0.008$ or $\sim 1\%$. Although these variations in L_r are small, they can make a significant impact on the retrieved pigment concentration (André and Morel, 1989) because they vary strongly with wavelength. They lead to errors in L_w which also vary strongly with wavelength. For example, in the ratio r_{13}, O_3 concentration variations principally influence $L_w(\lambda_3)$, while variations in P principally influence $L_w(\lambda_1)$. The magnitude of the errors in the pigment concentration resulting from P and O_3 variations has been thoroughly discussed by André and Morel (1989). The only way to correctly address these problems is to have fields of P and O_3 available for inclusion in the data processing. In the case of CZCS the O_3 concentration is available from other sensors on NIMBUS-7 and they have been included in the processing stream; however, in the case of variations in P there are no data fields simultaneous with CZCS and thus far these variations have been ignored.

In the case of future instruments, *e.g.*, Sea-viewing Wide Field-of-view Sensor (SeaWiFS), Ocean Colour and Temperature Scanner (OCTS), and Moderate-Resolution Imaging Spectrometer (MODIS), pressure fields derived from numerical weather models will be used to incorporate estimates of P.

5.4 ROUGH SURFACE EFFECTS

In all of the computations described so far, it has been assumed that the sea surface is flat. Neglecting direct sun glitter L_g, in the application of the algorithm, this assumption is utilized *only* in the computation of L_r. The basic effect of the surface roughness is to change the boundary condition at the lower boundary of the atmosphere (the sea surface). In contrast to the flat ocean case (Equation (15)), reflection from the rough surface is described by a bi-directional reflectance distribution function (BRDF), and is given by

$$L\left(z_1, \theta_r, \varnothing_r\right) = \int_{2\pi} r\left(\theta_i, \varnothing_i \to \theta_r, \varnothing_r\right) L\left(z_1, \theta_i, \varnothing_i\right) d\Omega_i \quad (35)$$

where the 2π on the integral indicates that the integration is to be taken over 2π Ster of Ω_i. The quantity $r(\theta_i, \varnothing_i \to \theta_r, \varnothing_r)$ specifies the BRDF of the surface. Gordon and Wang (1991a) carried out multiple scattering computations, including polarization, for a Rayleigh scattering atmosphere assuming that the wind-ruffled sea surface consists of a collection of individual facets obeying the slope statistics derived by Cox and Munk (1954).

For simplicity, the surface slope distribution was assumed to be independent of the wind direction. They found that, in the case of CZCS, the effect of surface roughness was usually below the detectable range for wind speeds up to 16.9 m/s when the sun angle was 40°; however, for $\theta_0 = 60°$ the difference between the flat and rough surface, although usually small, could become as large as three CZCS digital counts for large scan angles at 670 nm. They showed that the maximum error in assuming the ocean was flat in the computation of L_r was about the same order of magnitude as that which would arise from a ±15 mb variation in P or a ±50 DU variation in the O_3 concentration; however, the overall effect is smaller than that of variations in O_3 and P because the error is more spectrally neutral.

By simulating the entire correction process (Gordon and Wang, 1991b), they concluded that little would be gained by including surface roughness effects in routine CZCS processing, but demonstrated that such information may be required for more sensitive instruments.

6. Correction of Future Sensors

The next generation ocean color sensors, such as SeaWiFS (NASA, 1987) and MODIS (NASA, 1986; NASA, 1988) will have a radiometric sensitivity that is superior to CZCS. Several effects thus far ignored in the CZCS processing algorithms, but which must be included in order that the improved radiometric sensitivities can be fully utilized are listed below.

- The interaction between Rayleigh and aerosol scattering.
- The curvature of the earth.
- The large ε extrapolation required.
- The presence of whitecaps on the sea surface.
- The residual polarization sensitivity of the sensor.

These will now be discussed individually.

6.1 INTERACTION BETWEEN RAYLEIGH AND AEROSOL SCATTERING

The basic equation of the CZCS atmospheric correction algorithm is Equation (25), which was derived from the single scattering approximation. Although L_r is computed using the multiple scattering method (including polarization) described earlier, an error still remains. Basically, even if L_r and L_a both include all orders of multiple scattering, i.e., L_r is computed using a multiple scattering code with $\tau_a=0$ and L_a computed with $\tau_r=0$, Equation (25) still does not allow for the possibility that photons can scatter from *both* aerosols and air molecules. This is called the Rayleigh-aerosol interaction, and was first described quantitatively by Deschampes *et al.* (1983). To increase the accuracy of the CZCS algorithm to deal with the more sensitive instruments, it is necessary to modify Equation (25) to explicitly include the interaction, *i.e.*,

$$L_t(\lambda) = L_r(\lambda) + L_a(\lambda) + L_{ra}(\lambda) + L_g(\lambda) + tL_w(\lambda) \qquad (36)$$

and to provide a way of computing the interaction, L_{ra}.

Using the ideas of Gordon and Castaño (1989), Wang (1991) has developed a technique for including the Rayleigh-aerosol interaction into the formalism without directly calculating L_{ra}. Briefly, he found through a large number of radiative transfer simulations that a linear relationship exists between L_a+L_{ra} and the single-scattered aerosol radiance, *i.e.*,

$$L_a(\lambda) + L_{ra}(\lambda) = I(\lambda) + S(\lambda) L_{as}(\lambda) \qquad (37)$$

where L_{as} is given by Equation (22) with $x=a$. These simulations included several aerosol models (phase functions), several wind speeds (0 to 16.9 m/s), and aerosol optical thicknesses over the range $\tau_a = 0$ to 0.6. The results showed that, in geometries similar to those employed in ocean color sensing, the values of the "intercept" I and the "slope" S depend strongly on the geometry but very weakly on the aerosol model and the wind speed. Thus, I and S determined from these model computations should be applicable for use in atmospheric correction. The plan is to use Equation (37) to estimate L_{as} in spectral regions in which $L_w \approx 0$, *e.g.*, 750 and 865 nm for SeaWiFS and most other proposed ocean color sensors, and then to use L_{as} in place of L_a in Equations (27) through (31). This procedure would provide L_{as} in the blue and green regions of the spectrum from L_{as} in the red and NIR. Equation (37) would then yield $L_a(\lambda) + L_{ra}(\lambda)$ at the short wavelengths, from which Equation (36) can be solved for L_w in the absence of L_g. Since I and S depend weakly on the surface roughness, knowledge of the wind speed improves the accuracy of the algorithm.

This scheme has been tested for an ocean color instrument, with the CZCS band set, observing clear water, *i.e.*, it is assumed that $L_w(\lambda)$ is given for λ_2, λ_3, and λ_4 and that we want to retrieve $L_w(\lambda_1)$. An example of the resulting error for a relatively turbid atmosphere is provided in figure 19 in which $\Delta\rho_w = \pi\Delta L_w/F_0cos\theta_0$ is the error in the retrieved water-leaving *reflectance*. The aerosol is assumed to be nonabsorbing, to have a scattering coefficient that is independent of λ, and to have a scattering phase function that is approximately the average of HMF7 and HMF9. The wind speed W has been assumed to be unknown, so $W=0$ has been used in the computation of L_r. For reference, 1 DC at Gain 1 (see Appendix) for CZCS would correspond to a reflectance of 0.00076 at $\theta_0=0$ and 0.00153 at $\theta_0=60°$. Thus, in this example, the error in $L_w(\lambda_1)$ would usually be less than 1 CZCS DC even though the surface roughness is ignored. Figure 20 shows the improvement that results when the correct wind speed is used in the algorithm. Wang's analysis suggests that this procedure can provide an atmospheric correction that is nearly an order of magnitude more accurate than the standard CZCS algorithm.

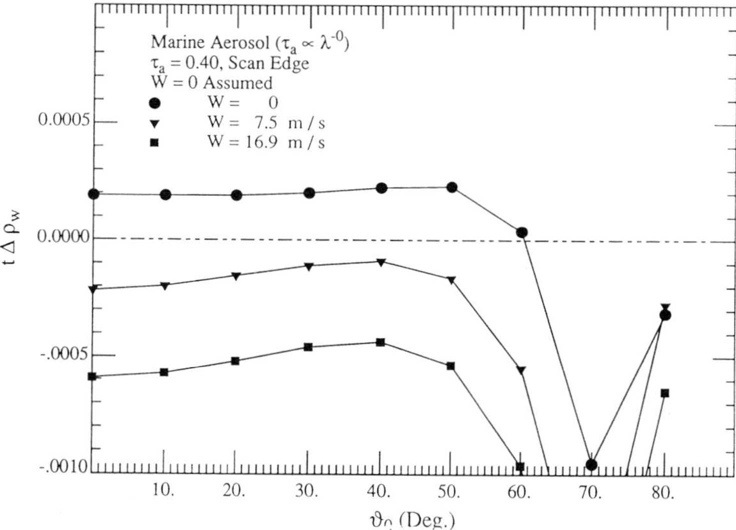

Figure 19. Simulated error in the new atmospehric correction algorithm at the CZCS scan edge as a function of θ_0. the wind speed has been assumed to be unknown and $W = 0$ used in the computation of L_r.

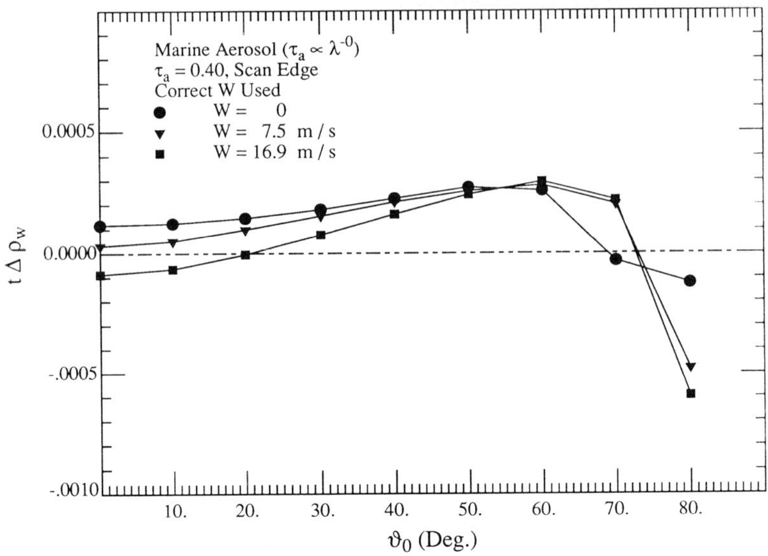

Figure 20. Simulated error in the new atmosphere correction algorithms at the CZCS scan edge as a function of θ_0. The wind speed has been assumed to be known and the correct value is used in the computation of L_r, I, and S.

6.2 THE LARGE ε EXTRAPOLATION

Following the scheme proposed above for the new ocean color sensors, the quantities determined from the NIR bands are $\varepsilon(750,865)$ and $\varepsilon(865,865)$ in Equation (28). These must be extrapolated into the visible to obtain $\varepsilon(\lambda,865)$, where λ can be as small as 410 nm. Thus, it will be necessary to understand the manner in which ε values determined in the NIR relate to those in the visible. This question is being studied now, within the limits imposed by the nonspherical nature of the aerosol, using Mie theory models of aerosol scattering.

6.3 THE EFFECT OF EARTH CURVATURE

All atmospheric corrections algorithms developed thus far ignore the curvature of the earth, *i.e.*, the plane parallel atmosphere (PPA) approximation has always been used in the radiative transfer simulations for ocean remote sensing.

However, at the level of accuracy required to use the full sensitivity of new instruments, it may be necessary to take the curvature of the earth into account (Gordon *et al.*, 1988), particularly at large sun angles. For example, figure 21 shows the error in L_r caused by the assumption of a plane parallel atmosphere at the scan center and the scan edge of a sensor like SeaWiFS (Gordon and Ding, unpublished). For $\theta_0 < 60°$ the error is <1% and can be ignored for CZCS and possibly SeaWiFS; however, for larger θ_0, *e.g.*, high latitudes, the error becomes excessive. Computations by Adams and Kattawar (1987) suggest that nearly all of the difference between the PPA approximation and the true spherical shell atmosphere (SSA) is in the first scattering, *i.e.*, the fraction of the radiance due to multiple scattering is approximately the same for the PPA and SSA. An effort is now under way at the University of Miami to use this to provide a first order correction for the effects of earth curvature.

6.4 WHITECAPS

Gordon and Jacobs (1977) have presented computations suggesting that sea foam (whitecaps) could significantly increase the flux leaving the top of the atmosphere over the oceans. This added radiance, must be considered in the radiance budget at the sensor. The effect of whitecaps can be estimated (Gordon, 1987) by combining Koepke's (1984) determinations of the surface reflectance *increase* due to whitecaps as a function of the wind speed, with extrapolations into the blue green region of the spectrum of laboratory measurements of the foam reflectance spectrum in the red and NIR made by Whitlock *et al.* (1982). The result is provided in figure 22. The figure gives the contribution of whitecaps to the radiance at the top of the atmosphere for the four CZCS bands. The radiance is in CZCS Gain 1 digital counts (See Appendix). It is seen that even wind speeds below *mean* values, which only rarely reach 7-8 m/s (Malkus, 1962; Von Arx, 1962; Hsiung, 1986), whitecaps increase the radiance in the CZCS red band by a measurable amount. This increase is ignored in CZCS processing. With higher radiometric sensitivity and with spectral bands in the NIR, whitecaps will influence L_t for new sensors at even lower wind speeds.

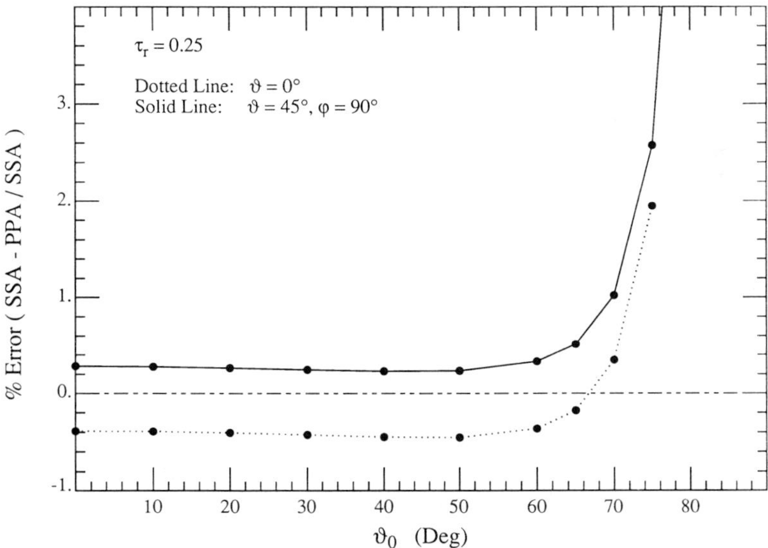

Figure 21. Error in L_r computed under the plane parallel assumption for the atmosphere.

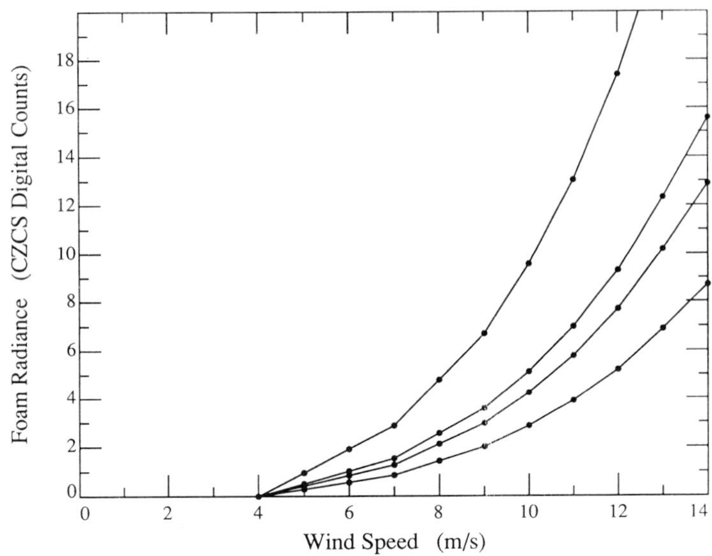

Figure 22. Increase in L_t in CZCS digital counts resulting from whitecaps on the sea surface. Curves from bottom to top correspond to CZCS Bands 1-4, respectively.

The increased radiance will be interpreted as aerosol by the atmospheric correction algorithm yielding incorrect water-leaving radiances. If ignored, the expected improvement over CZCS due to the presence of the NIR "atmospheric correction" bands will be degraded. It is possible to develop a whitecap removal algorithm based on our present knowledge, *i.e.*, given the wind speed, the whitecap reflectance could be estimated in a manner similar to that used to prepare figure 22 and removed from L_t *before* atmospheric correction. However, the present analysis ignores the sea surface temperature, which Bortkovskii (1987) reports can significantly influence the foam coverage of the sea surface.

Thus, it is believed that for an accurate assessment and removal of whitecap effects, more experimental measurements are needed. The issues that must be addressed to deal with whitecaps are as follows: first, the relationship between whitecap coverage, wind speed, and sea surface temperature is not well understood and needs to be; next, the spectral reflectance of individual whitecaps has never been measured in the field (nor in the laboratory in the blue and green regions of the spectrum) and should be; and finally, the increase in surface reflectance due to whitecaps has not been measured directly, it has been deduced from the individual whitecap reflectance and the fraction of the surface covered, *e.g.*, figure 22. For the purposes of remote sensing, field measurements of the actual increase in the average reflectance, over areas with sizes of the order of a km, as a function of variables that can be measured from space or deduced from numerical models, *e.g.*, wind speed and sea surface temperature, would provide the most direct way of developing the necessary removal algorithm.

The typical specification - that the polarization sensitivity of ocean color sensors should be less than 2% - is based on the requirement that the unknown polarization of L_t induced by the aerosols can be corrected using aerosol models (Gordon, 1988). This implies the processing will still be required to remove the the residual polarization effects. We have developed a formalism (Gordon, 1988) which provides the framework for such processing.

7. Appendix: CZCS Radiometry

The CZCS data was digitized to 8-bits on board the satellite and relayed to the ground in digital form. The sensor sensitivity was adjustable to four levels (Gains). The approximate radiance increment corresponding to a change in the output of one digital count (DC) is presented in Table A1 (Ball Aerospace Division, 1979a; Ball Aerospace Division, 1979b). Under ideal operation the sensitivity would vary along the orbit to account for the decrease of incident solar irradiance with latitude away from the solar equator. In practice, most data were acquired at Gains 1 and 2. At typical signal levels the signal-to-noise ratio is above 150 for all bands with the exception of Band 4, for which the measured value was 118 (Hovis *et al.*, 1980). The data provided in the table was applicable at launch (October 1978). The sensitivity of the CZCS degraded in orbit from the values presented above. This degradation has been described by Gordon *et al.* (1983), Mueller (1985), Hovis *et al.* (1985), and Gordon and Evans (1990). It amounts to as much as a 40% drop in sensitivity during the eight years of operation. The author believes that this variation in sensitivity with time on both short and long time scales accounts for much of the difficulties and inconsistencies encountered with CZCS data. For SeaWiFS, the long-term and short-term variability of the sensitivity will be accurately monitored by viewing the moon and an internal reflectance standard, respectively.

TABLE A1. Radiance (DC) in mW/cm$^2\mu$m Ster corresponding to one digital count for the four CZCS sensitivities.

Band	λ	DC			
	(nm)	Gain 1	Gain 2	Gain 3	Gain 4
1	433–453	0.045	0.036	0.030	0.021
2	510–530	0.031	0.025	0.020	0.015
3	540–560	0.025	0.020	0.016	0.012
4	660–680	0.011	0.0090	0.0074	0.0053

The CZCS was designed to be insensitive to the polarization state of the incident radiance, *i.e.*, if the incident radiance was 100% linearly polarized and the direction of polarization was rotated through 180°, the output was supposed to vary less than about 2%.

Acknowledgement

The author is grateful to M. Wang for carrying out the LOWTRAN-7 simulations, and to the National Aeronautics and Space Administration for providing support under Grant NAGW-273.

References

Adams, C. N., and Kattawar, G. W. (1978) 'Radiative Transfer in Spherical Shell Atmospheres I. Rayleigh Scattering', Icarus 35, 139-151.

André, J.-M. and Morel, A. (1989) 'Simulated Effects of Barometric Pressure and Ozone Content Upon the Estimate of Marine Phytoplankton From Space', Journal of Geophysical Research 94C, 1029-1037.

André, J.-M. and Morel, A. (1991) 'Atmospheric Corrections and Interpretation of Marine Radiances in CZCS Imagery', Revisited, Oceanologica Acta 14, 3-22.

Ball Aerospace Division, Boulder CO (1979a) 'Development of the Coastal Zone Color Scanner for Nimbus-7: Volume 1 - Mission Objectives and Instrument Description', Final Report F78 - 11, Rev. A, NASA Contract NAS5-20900.

Ball Aerospace Division, Boulder CO (1979b) 'Development of the Coastal Zone Color Scanner for Nimbus-7: Volume 2 - Test and Performance Data', Final Report F78-11, Rev. A, NASA Contract NAS5-20900.

Bhartia, P. K., Klenk, K. F., Wong, C. K., Gordon, D., and Fleig, A. J. (1984) 'Comparison of the NIMBUS 7 SBUV/TOMS Total Ozone Data Sets With Dobson and M83 Results', Journal of Geophysical Research 89D, 5239-5247.

Bohren, C. F., and Huffman, D. R. (1983) 'Absorption and Scattering of Light by Small Particles', Wiley, New York, pp. 530.

Bortkovskii, R. S. (1987) 'Air-Sea Exchange of Heat and Moisture During Storms', Reidel, Dordrecht, pp. 194.

Bricaud, A., and Morel, A. (1987) 'Atmospheric Corrections and Interpretation of Marine Radiances in CZCS Imagery: Use of a Reflectance Model', Oceanologica Acta 7, 33-50.

Case, K. M. (1957) 'Transfer Problems and the Reciprocity Principle', Reviews of Modern Physics 29, 651-663.

Chandrasekhar, S. (1950) 'Radiative Transfer', Oxford University Press, Oxford, pp. 393

Clarke, G. L., Ewing, G. C., and Lorenzen, C. J. (1970) 'Spectra of

Backscattered Light from the Sea Obtained From Aircraft as a Measurement of Chlorophyll Concentration', Science 167, 1119-1121.

Clark, D. K. (1981) 'Phytoplankton Algorithms for the Nimbus-7 CZCS', in J. R. F., Gower Oceanography from Space, Plenum Press, New York, NY, pp. 227-238.

Cox, C.and Munk, W. (1954) 'Measurements of the Roughness of the Sea Surface from Photographs of the Sun's Glitter', Journal of the Optical Society of America 44, 838-850.

Deirmendjian, D. (1969) 'Electromagnetic Scattering on Spherical Polydispersions', Elsevier, New York, NY, pp. 290.

Deschamps, P. Y., Herman, M., Lenoble, J., Tanre, D., and Viollier, M. (1981) 'Atmospheric Effects in Remote Sensing of Ground and Ocean Reflectances', in A. Deepak (ed.), Remote Sensing of Oceans and Atmospheres, Academic Press, New York, NY, 115-147.

Deschamps, P. Y., Herman, M., and Tanre, D. (1983) 'Modeling of the atmospheric effects and its application to the remote sensing of ocean color', Applied Optics 22, 3751-3758.

Eckstein, B. A., and Simpson, J. J. (1991) 'Aerosol and Rayleigh Radiance Contributions to Coastal Zone Color Scanner Images', International Journal of Remote Sensing 12, 135-168.

Elterman, L. (1968) 'UV, Visible, and IR Attenuation for Altitudes to 50 km', Report AFCRL-68-0153, AFCRL, Bedford, MA.

Gordon, H. R. (1978) 'Removal of Atmospheric Effects from Satellite Imagery of the Oceans', Applied Optics 17, 1631-1636.

Gordon, H. R. (1984) 'Some Studies of Atmospheric Optical Variability in Relation to CZCS Atmospheric Correction', Final Report Contract No. NA-79-SAC-00714, NOAA National Environmental Satellite and Data Information Service.

Gordon, H. R. (1987) 'Calibration Requirements and Methodology for Remote Sensors Viewing the Oceans in the Visible', Remote Sensing of Environment 22, 103-126.

Gordon, H. R. (1988) 'Ocean Color Remote Sensing Systems: Radiometric Requirements', Society of Photo-Optical Instrumentation Engineers, Recent Advances in Sensors, Radiometry, and Data Processing for Remote Sensing 924, 151-167.

Gordon, H. R. (1992) 'Modeling and Simulating Radiative Transfer in the Ocean', in Ocean Optics, edited by R. W. Spinrad, K. L. Carder, and M. J. Perry, Oxford, Submitted.

Gordon, H. R., Brown, J. W., and Evans, R. H. (1988) 'Exact Rayleigh Scattering Calculations for use with the Nimbus-7 Coastal Zone Color Scanner', Applied Optics 27, 862-871.

Gordon, H. R., Brown, J. W., Brown, O. B., Evans, R. H., and Clark, D. K., (1983) 'Nimbus 7 Coastal Zone Color Scanner: reduction of its radiometric sensitivity with time', Applied Optics 22, 3929-3931.

Gordon, H. R., Brown, O. B., Evans, R. H., Brown, J. W., Smith, R. C., Baker K. S., and Clark, D. K. (1988) 'A Semi-Analytic Radiance Model of Ocean Color', Journal of Geophysical Research 93D, 10909-10924.

Gordon, H. R., and Castaño, D. J. (1988) 'The Coastal Zone Color Scanner Atmospheric Correction Algorithm: Influence of El Chichón', Applied Optics 27, 3319-3321.

Gordon, H. R., and Castaño, D. J. (1989) 'Aerosol Analysis with the Coastal Zone Color Scanner: A Simple Method for Including Multiple Scattering Effects', Applied Optics 28, 1320-1326.

Gordon, H. R., and Clark, D. K. (1980) 'Atmospheric effects in the remote sensing of phytoplankton pigments', Boundary-Layer Meteorology 18, 299-313.

Gordon, H. R.,and Clark, D. K. (1981) 'Clear water radiances for atmospheric correction of coastal zone color scanner imagery', Applied Optics 20, 4175-4180.

Gordon, H. R., Clark, D. K., Brown, J. W., Brown, O. B., Evans, R. H., and Broenkow, W. W. (1983) 'Phytoplankton pigment concentrations in the Middle Atlantic Bight: comparison between ship determinations and Coastal Zone Color Scanner estimates', Applied Optics 22, 20-36.

Gordon, H. R., Clark, D. K., Mueller, J. L., and Hovis, W. A. (1980) 'Phytoplankton pigments derived from the Nimbus-7 CZCS: initial comparisons with surface measurements', Science 210, 63-66.

Gordon, H. R. and Evans, R. H. (1990) 'The CZCS: Algorithms and Characterization', OCEANS FROM SPACE, Scuola Grande San Giovanni Evangelista, Venice, Italy, May 22-26, 1990

Gordon, H. R., and Jacobs, M. M. (1977) 'The Albedo of the Ocean-Atmosphere System: Influence of Sea Foam', Applied Optics 16, 2257-2260.

Gordon, H. R., and Morel, A. Y. (1983) 'Remote Assessment of Ocean Color for Interpretation of Satellite Visible Imagery: A Review', Springer-Verlag, New York, pp. 114.

Gordon, H. R., and Wang, M. (1991a) 'Surface Roughness Considerations for Atmospheric Correction of Ocean Color Sensors. 1: The Rayleigh Scattering Component', Applied Optics, 31, 4247-4260.

Gordon, H. R., and Wang, M. (1991b) 'Surface Roughness Considerations for Atmospheric Correction of Ocean Color Sensors. 2: Error in the Retrieved Water-leaving Radiance', Applied Optics 31, 4261-4268.

Hsiung, J. (1986) 'Mean Surface Energy Fluxes Over the Global Ocean', Journal of Geophysical Research 91C, 10,585-10,606.

Hansen, J. E., and Travis, L. D. (1974) 'Light Scattering in Planetary Atmospheres', Space Science Reviews 16, 527-610.

Hovis, W. A., Clark, D. K., Anderson, F., Austin, R. W., Wilson, W. H., Baker, E. T., Ball, D., Gordon, H. R., Mueller, J. L., Sayed, S. Y. E., Strum, B., Wrigley, R. C., and Yentsch, C. S. (1980) 'Nimbus 7 coastal zone color scanner: system description and initial imagery', Science 210, 60-63.

Hovis, W. A., Knoll, J. S., and Smith, G. R. (1985) 'Aircraft Measurements for Calibration of an Orbiting Spacecraft Sensor', Applied Optics 24, 407-410.

Kattawar, G. W., Plass, G. N., and Hitzfelder, S. J. (1976) 'Multiple scattered radiation emerging from Rayleigh and continental haze layers. 1: Radiance, polarization, and neutral points', Applied Optics 15,632-647.

Kneizys, F. X., Shettle, E. P., Abreu, L. W., Chetwynd, J. H., Anderson, G. P., Gallery, W. O., Selby, J. E. A., and Clough, S. A. (1988) 'Users Guide to LOWTRAN 7', Air Force Geophysics Laboratory, AFGL-TR-88-0177.

Koepke, P. (1984) 'Effective Reflectance of Oceanic Whitecaps', Applied Optics 23, 1816-1824.

Malkus, J. S. (1962) 'Large Scale Interactions', The Sea, M. N. Hill,Wiley-Interscience, New York, NY, 88-294.

Mie, G. (1908) 'Beiträge zur Optik trüber Medien, speziell kollidalen Metall-lösungen', Annals of Physics 25, 377-445.

Morel, A. (1988) 'Optical Modeling of the Upper Ocean in Relation to Its Biogenous Matter Content (Case I Waters)', Journal of Geophysical Research 93C, 10,749-10,768.

Morel, A. and André, J. -M. (1991) 'Pigment Distribution and Primary Production in the Western Mediterranean as Derived and Modeled From Coastal Zone Color Scanner Observations', Journal of Geophysical Research 96C, 12,685-12,698.

Morel, A., and Prieur, L. (1977) 'Analysis of Variations in Ocean Color', Limnology and Oceanography 22, 709-722.

Mueller, J. L. (1985) 'Nimbus-7 CZCS: confirmation of its radiometric sensitivity decay rate through 1982', Applied Optics 24, 1043-1047.

NASA (1976) 'U.S. Standard Atmosphere Supplements', U.S. Government Printing Office, Washington, D.C.

NASA, MODIS (Moderate-Resolution Imaging Spectrometer), 'Earth Observing System Volume IIb', 1986.

NASA and the Earth Observations Satellite Company, 'System Concept for Wide-Field-of-View Observations of Ocean Phenomena from Space', 1987.

NASA, Earth Observing System (Eos) Background Information Package, Announcement of Opportunity No. OSSA-1-88, (Announcement of Opportunity No. OSSA-1-88, January 1988).

Neckel, H., and Labs, D. (1984) 'The Solar Radiation Between 3300 and 12500 A', Solar Physics 90, 205-258.

Platt, T. and Sathyendranath, S. (1988) 'Oceanic Primary Production: Estimation by Remote Sensing at Local and Regional Scales', Science 241, 1613-1620.

Preisendorfer, R. W. (1961) 'Application of Radiative Transfer Theory to Light Measurements in the Sea', Union Geodesique et Geophysique Internationale 10, 11-30.

Preisendorfer, R. W. (1976) 'Hydrologic Optics V. 1: Introduction', (NTIS PB-259 793/8ST), National Technical Information Service, Springfield, VA.

Quenzel, H., and Kastner, M. (1980) 'Optical properties of the atmosphere: calculated variability and application to satellite remote sensing of phytoplankton', Applied Optics 19, 1338-1344.

Shurcliff, W. A. (1962) 'Polarized Light', Havard University Press,Cambridge, MA, pp. 207.

Smith, R. C., and Wilson, W. H. (1981) ' Ship and satellite bio-optical research in the California Bight', in J. F. R. Gower (ed.), Oceanography from Space,Plenum, New York, NY, 281-294.

Tanre, D., Herman, M., Deschamps, P. Y., and de Leffe, A. (1979) 'Atmospheric modeling for space measurements of ground reflectances, including bidirectional properties', Applied Optics 18, 3587-3594.

Van de Hulst, H. C. (1957) 'Light Scattering by Small Particles', Wiley, New York, pp. 470.

Van de Hulst, H. C. (1980) 'Multiple Light Scattering', Academic Press, New York, pp 739.

Von Arx, W. S. (1962) 'An Introduction to Physical Oceanography', Addison-Wesley, Reading, MA, pp. 422.

Wang, M. (1991) 'Atmospheric Correction of the Second Generation Ocean Color Sensors', Ph.D. Dissertation, University of Miami, Coral Gables FL, pp. 135.

Whitlock, C. H., Bartlett, D. S., and Gurganus, E. A. (1982) 'Sea Foam Reflectance and Influence on Optimum Wavelength for Remote Sensing of Ocean Aerosols', Geophysical Research Letters 7, 719-722.

UNDERWATER LIGHT FIELD AND PRIMARY PRODUCTION: APPLICATION TO REMOTE SENSING

S. SATHYENDRANATH
Department of Oceanography
Dalhousie University
Halifax, Nova Scotia
B3H 4J1 Canada

T. PLATT
Biological Oceanography Division
Bedford Institute of Oceanography
P.O. Box 1006
Dartmouth, Nova Scotia
B2Y 4A2 Canada

ABSTRACT.   Satellite-based ocean colour sensors can be used to estimate the global distribution of primary production. A general approach is presented, which consists of the development of a local algorithm to compute the photosynthetic rate by phytoplankton, and the application of this local result to estimate the integrated primary production at larger scales. Based on the well known relationship between light and photosynthesis, the local model uses spectral computation of the irradiance field at the ocean surface, and *in situ* observations of photosynthetic parameters. From satellite data and parametrisation of a generalised profile, the local structure of the vertical pigment profile is retrieved allowing then the computation of the submarine light field as a function of depth, wavelength and zenith angle. Finally, perspectives and applications of the model are discussed in parallel with some recent advances in scientific disciplines related to primary production.

1. Introduction

At present, and in the foreseeable future, the photosynthetic biomass of phytoplankton can be measured much more readily than can its rate of primary production. However, the latter is of as much, or more, ecological interest as the former, and it is often required to make an estimate of the rate of primary production given information on the biomass.

In this chapter we discuss how such estimates may be made. The principles are the same whether the given biomass information is based on refined measurement techniques for photosynthetic pigments or on more approximate methods such as *in vivo* fluorescence or remote sensing of ocean colour. Whichever procedure is followed for the estimation of primary production, we shall find that certain steps are unavoidable. It is necessary to specify the light field in some way. The vertical profile of photosynthetic biomass must be specified. Some model for the relationship between

79

V. Barale and P.M. Schlittenhardt (eds.),
Ocean Colour: Theory and Applications in a Decade of CZCS Experience, 79–93.

photosynthesis and light must be adopted, with appropriate parameters. The resolution, accuracy and precision with which these items are specified will determine the quality of the estimates of primary production. For many important questions in modern biological oceanography there is an additional challenge: how to make estimates of primary production that are representative of regional or ocean-basin scales. This problem is tackled in the second part of the chapter.

2. Basic Principles

The rate of primary production that can be achieved in any set of circumstances depends on the flux of photons available and the ability of the phytoplankton to intercept them. The local chlorophyll (*sensu lato*) concentration determines the capacity of the phytoplankton to intercept photons (the absorption cross section) and for this reason we adopt it as our index of phytoplankton biomass B. The problem addressed in this chapter is to estimate the primary production P given information on B.

The biomass will vary from place to place, and in talking about the primary production it will be useful to have an index of P from which these variations have been removed. We therefore define the normalised production P^B as

$$P^B \equiv P\big/B \tag{1}$$

In any application, the absolute primary production can always be recovered by inversion of (1).
Explicitly,

$$P = B\ P^B \tag{2}$$

The normalised production rate P^B is determined by the flux of photons available. Clearly, in the ocean, the photon flux is a function of depth z.
In general also, the pigment biomass depends on depth and (2) may be rewritten as

$$P\left(z\right) = B\left(z\right)\ P^B\left(z\right) \tag{3}$$

The explicit dependence of $P^B(z)$ on photon flux $I(z)$ may be stated formally as

$$P^B\left(z\right) = p\left(I\left(z\right)\right) \tag{4}$$

where the function p remains to be specified.

Empirically, we know the general shape that $p(I(z))$ must take. It is a function that can be described with no more than two parameters, in the absence of photoinhibition (Platt *et al.*, 1977; Platt *et al.*, 1988). Conventionally, these are chosen to be the slope $\alpha\alpha^B$ of $p(I)$ near the origin (called the initial slope), and the height $P_m{}^B$ of the plateau (called the assimilation number). Once an explicit functional form has been adopted for $p(I)$, the magnitudes of α^B and $P_m{}^B$ are all that is required to determine $P^B(z)$ completely in any given case.

It is often required to determine the water column production $\int_Z P$, computed by integration of (3)

$$\int_Z P = \int_Z P(z)\, dz = \int_0^\infty B(z)\, P^B(z)\, dz \tag{5}$$

where z is measured positive downwards.

The value of $P(z)$ given for example by (3) is an instantaneous rate. Of course, $I(z)$ in (4) is a time-dependent variable, such that (3) may be further refined as

$$P(z,t) = B(z,t)\, P^B(z,t) \tag{6}$$

It is frequently required to integrate the instantaneous rates through time to compute primary production for a finite time interval, for example $\int_{Z,T} P$, the daily rate

$$\int_{ZT} P = \int\int B(z,t)\, P^B(z,t)\, dz\, dt \tag{7}$$

The rest of this chapter deals with the implementation of the basic formalism presented in (1) through (7), in which we follow the protocol proposed by Platt and Sathyendranath (1988). It will be useful to consult Table 1 in which the typical units of the relevant variables and parameters are given.

3. The Irradiance Field

Let $I(0^+)$ be the downwelling irradiance field just above the surface of the ocean. In principle, $I(0^+)$ can be computed for any given time and position on the globe (see Bird, 1984; Sathyendranath and Platt, 1988; Platt and Sathyendranath, 1988). One calculates first for the solar irradiance field outside the atmosphere and modifies the result to allow for atmospheric transmission losses.

For a complete description it is essential to take into account the spectral and angular dependence of irradiance:

$$I\left(0^+\right) = I\left(0^+, \lambda, \theta, t\right) \tag{8}$$

where λ is the wavelength, θ the sun zenith angle and t is local time. For the spectral distribution of $I(0^+)$, the wavelength range of interest is from 400 to 700 nm, the so-called photosynthetically-active radiation (PAR). The main difficulty in computing $I(0^+)$ is in allowing for the influence of aerosols and clouds in modifying atmospheric transmission. We can allow for the effect of clouds on the magnitude of $I(0^+)$, but we do not yet have a good parametrisation for their effect on spectral distribution.

We may distinguish two components of $I(0^+)$, the direct sunlight component $I_d(0^+)$ and diffuse component $I_s(0^+)$, also called the skylight:

$$I\left(0^+\right) = I_d\left(0^+\right) + I_s\left(0^+\right) \tag{9}$$

In computing $I(0^-)$, the irradiance immediately below the sea surface, reflection losses for each of the two components of $I(0^+)$ are computed separately (Sathyendranath and Platt, 1988). The same holds for refraction, which modifies the angular distribution.

To specify the underwater light field given $I(0^-)$, we need only the vertical attenuation coefficients for direct and sky light, $K_d(z, \lambda)$ and $K_s(z, \lambda)$. Again, we are required to treat the direct and diffuse components separately because the attenuation depends on angular distribution, and a difference in angular distribution is the essential distinction between the two components. The attenuation coefficients include the effects of absorption and scattering, both dependent on wavelength. We use the approximations

$$K_d\left(z,\lambda\right) = \left\{ a\left(z,\lambda\right) + b_b\left(z,\lambda\right) \right\} \left(cos\theta_d \right)^{-1} \tag{10}$$

and

$$K_s\left(z,\lambda\right) = \left\{ a\left(z,\lambda\right) + b_b\left(z,\lambda\right) \right\} \left\langle cos\theta_s \right\rangle^{-1} \tag{11}$$

where $a(z, \lambda)$ is the volume absorption coefficient and $b_b(z, \lambda)$ is the volume backscattering coefficient. The quantity θ_d is the zenith angle for the direct component after refraction at the sea surface. The factor $\langle cos\theta_s \rangle$ is the mean path length of the diffuse component in traversing unit vertical distance.

It is of the greatest significance for the problem under discussion that the volume absorption coefficient is a strong function of the concentration of plant pigments $C(z)$. The relevant component of $a(z, \lambda)$ is $a_c(z, \lambda)$ where

$$a_c\left(z,\lambda\right) = a_c^*\left(\lambda\right) C\left(z\right) \tag{12}$$

We write $C(z)$ here rather than $B(z)$ with the implication that degraded pigments are included as well as photosynthetically active ones. Here, a_c^* is the absorption coefficient of phytoplankton per unit concentration of C

(generally taken to be the sum of chlorophyll a and phaeopigments). For the open-ocean waters, the absorption coefficient a can now be written as $a(\lambda)=a_w(\lambda)+a_c(\lambda)$, where a_w is the constant absorption by pure seawater. The backscattering coefficient b_b can be similarly expressed as the sum of the contributions of pure seawater and phytoplankton (Sathyendranath and Platt, 1988). The irradiance $I(z, \lambda)$ at any depth z can now be found from

$$I\left(z,\lambda,\theta\right)=I_d\left(z,\lambda,\theta_d\right)+I_s\left(z,\lambda,\theta_s\right) \tag{13}$$

where

$$I_d\left(z,\lambda,\theta_d\right)=I_d\left(0^-,\lambda,\theta_d\right)e^{-\int_0^z K_d\left(z',\lambda\right)dz'} \tag{14}$$

and

$$I_s\left(z,\lambda,\theta_s\right)=I_d\left(0^-,\lambda,\theta_s\right)e^{-\int_0^z K_s\left(z',\lambda\right)dz'} \tag{15}$$

Since the pigment concentration C in the near-surface waters is one of the variables most easily obtained from ocean colour, such parametrisations have the advantage of being easily applied in the remote sensing context. Morel (1988) has proposed a similar approach to computation of the underwater light field.

4. The Local Algorithm of Photosynthesis

We have seen that the normalised primary production can be expressed in the form

$$P^B(z)=p\left(I(z);a^B(z),P_m^B(z)\right) \tag{16}$$

where

$$a^B=\left.\frac{\partial P^B}{\partial I}\right|_{I\to 0} \tag{17}$$

and

$$P_m^B=\lim_{I\to\infty}P^B(I) \tag{18}$$

To evaluate (16) we need to specify the functional form $p(I)$. A convenient choice, that performs well, is the Smith equation (Smith, 1936).

In conventional, nonspectral form it is

$$P^B(I) = P_m^B\left(I \Big/ I_k\right)\left(1 + \left(I \Big/ I_k\right)^2\right)^{-\frac{1}{2}} =$$

$$= Ia^B\left(1 + \left(\frac{Ia^B}{P_m^B}\right)^2\right)^{-\frac{1}{2}} \tag{19}$$

where I_k is a derived parameter with the definition

$$I_k = \frac{P_m^B}{a^B} \tag{20}$$

When the angular and spectral distributions of available light are taken into account, we have to recognise that a^B is known to depend on both the angular distribution and wavelength of available light whereas P_m^B is believed to depend directly on neither. We find (Platt and Sathyendranath, 1988; Sathyendranath and Platt, 1989a; Sathyendranath and Platt, 1990)

$$P(z) = B(z) P^B(z) = B(z) \Pi(z) \left\{1 + \left(\Pi(z) \Big/ P_m^B(z)\right)^2\right\}^{-\frac{1}{2}} \tag{21}$$

where

$$\Pi(z) = \left(cos\theta_d\right)^{-1} \int a^B(\lambda, z) I_d\left(z, \lambda, \theta_d\right) d\lambda +$$

$$+ \left\langle cos\theta_s \right\rangle^{-1} \int a^B(\lambda, z) I_s\left(z, \lambda, \theta_s\right) d\lambda \tag{22}$$

Equations (21) and (22) constitute the most general statement of the formalism for computation of primary production from available light. Note that (22) contains two terms, one each for the direct and diffuse components of available light. In the absence of information on the angular distribution of I_s, the quantity $\langle cos\theta_s\rangle^{-1}$ may be taken to be 1.2 (Sathyendranath and Platt, 1989a). The function $a^B(\lambda)$ is the action spectrum of photosynthesis: note that $a^B(\lambda)$ is weighted by the wavelength distribution of the available light.

When integrated over depth, (21) and (22) constitute the local algorithm for calculation of primary production form available light. To do the integration, it is first required to specify the shape of the pigment biomass profile $B(z)$. In some applications, $B(z)$ may be known in detail, for example from *in vivo* fluorometry. Generally, however, this will not be the case (especially if the biomass information is obtained from satellites), and $B(z)$ will have to be parametrised. A useful form has been found to be (Platt *et al.*, 1988)

$$B(z) = B_0 + h\left(2\pi\sigma^2\right)^{-\frac{1}{2}} e^{-\left(\left(z-z_m\right)^2 \big/ \left(2\sigma^2\right)\right)} \tag{23}$$

where $B(z)$ is modelled as a background concentration B_0 and a superimposed Gaussian with parameters z_m (depth of maximum), σ (related to thickness of maximum) and h (related to amplitude).

It is only on exceedingly rare occasions that one has information on all the photosynthetic parameters essential for defining the local algorithm: the spectral form of a^B and its depth dependence, and the depth dependence of P_m^B. It therefore becomes necessary to incorporate simplifying assumptions. In the absence of information on the spectral form of local a^B, we may take it to have a universal shape but variable amplitude (Sathyendranath *et al.*, 1989a). Another simplifying assumption that we have used with satisfactory results is that both P_m^B and a^B are depth independent (Platt and Sathyendranath, 1988). Note that, in the surface waters, where available light reaches saturating levels, $P(z)$ is more sensitive to P_m^B than to a, and in deeper waters, where light-limiting conditions prevail, the inverse holds true. We have tested the performance of the local algorithm against field measurements for a variety of oceanographic regimes and found very encouraging results (Platt and Sathyendranath, 1988). Particularly simple algorithms are available when the biomass can be treated as uniform with depth (Platt *et al.* 1990, Platt and Sathyendranath, 1991): they offer considerable economy of computing time when primary production has to be calculated on many pixels.

5. Estimating Production at Large Horizontal Scale

Estimation of primary production at large horizontal scale involves application of the local algorithm at many points in the relevant spatial domain and summing the results. Although it may be possible to have excellent knowledge of the parameters of the local algorithm for a given station (one that had been studied in detail by ship), it is very unlikely to be true that the parameters are well known at each and every grid point in the domain of interest.

Recall that the minimum number of parameters required for application of the local algorithm is six (two photosynthesis parameters and four to describe the vertical pigment profile). We now make the explicit assumption that the estimation of primary production at large horizontal scale will not be made without using remotely-sensed data to specify the biomass field. The number of parameters for the pigment profile can then be reduced by one (Platt and Sathyendranath, 1988; Sathyendranath and Platt, 1989b).

The next step is to partition the domain into subdomains within which the five parameters of the local algorithm can be taken to be given constants. If the required number of such subdomains is very small compared to the total number of grid points at which primary production will be calculated, the computation will have been reduced to manageable proportions. Implied in this simplification is the existence of "dynamic biogeography" of the relevant parameters (physiological and vertical profile). Biogeography implies that the ocean can be subdivided into a series of broad biogeochemical provinces with respect to vertical structure and physiological performance. The qualifier "dynamic" suggests that the boundaries of the zones, and the magnitudes of the parameters within them, may change with season. Our data for the North Sargasso Sea suggest that it could be classified as a province, with well-defined seasonal patterns in both a^B and P_m^B (Platt $et\ al.$, 1992).

Since full calculations of primary production make heavy demands on computer time, it becomes desirable to introduce schemes for optimising the computation procedure. Some such schemes have been discussed by Platt and Sathyendranath (1991), in the context of incorporation of primary production models into general circulation models of the oceans.

6. Prospects and Problems

The last decade has seen considerable advances in the study of primary production at sea: (1) relatively simple models are now available for computation of spectral light penetration in the sea (Smith and Baker, 1978; Morel, 1988; Sathyendranath and Platt, 1988); (2) we have a much better understanding of the specific absorption coefficient of phytoplankton, and the nature and causes of its variability, based on theoretical studies and measurements on cultures and natural phytoplankton populations (Prieur and Sathyendranath, 1981; Morel and Bricaud, 1983; Kiefer and SooHoo, 1982; Kishino $et\ al.$, 1985; Lewis $et\ al.$, 1983; Bidigare $et\ al.$, 1987; Mitchell and Kiefer, 1988a and references therein, 1988b; Hoepffner and Sathyendranath, 1991); (3) techniques have been developed that make it possible to measure the action spectrum of natural phytoplankton populations (Lewis $et\ al.$, 1985); (4) as logical next step to these developments, it has been possible to extend the photosynthesis-light model to include spectral and angular effects (Platt and Sathyendranath, 1988; Sathyendranath and Platt, 1989a); (5) with the advent of the remote sensing technology, it is becoming possible to take a realistic look at the large-scale picture of variability in primary production (Platt and Sathyendranath, 1988; Platt $et\ al.$, 1991). But problem areas still remain, and in the following paragraphs, we look at some of these, and the prospects for their solution in the near future.

6.1 REMOTE SENSING SATELLITES

The Coastal Zone Color Scanner (CZCS), launched in 1978, functioned until 1986, thereby serving the scientific community for a much longer span than that for which it was designed. The vast amount of data collected by this satellite is still being processed. Unfortunately, no replacement for the CZCS was ready at the time it ceased to function. A new satellite sensor for ocean colour measurements, the Sea-viewing Wide Field-of-view Sensor (SeaWiFS), is expected to be launched in 1993. The SeaWiFS is expected to be followed by the Japanese Ocean Colour and Temperature Scanner (OCTS), about 1995 at the earliest.

Thus, there will be a gap in the ocean colour data record from 1986 to 1993. But on the brighter side, the SeaWiFS project has benefitted from the CZCS experience, and should have better radiometric and spectral resolution. The SeaWiFS is therefore expected to provide improved algorithms for chlorophyll estimation. The Japanese OCTS designs indicate compatibility with, and further improvements over, the SeaWiFS.

6.2 ESTIMATION OF SURFACE IRRADIANCE

One of the crucial elements in the model of biomass-light-photosynthesis is the estimation of surface irradiance. The results of Sathyendranath *et al.* (1989a) show that failure to incorporate the spectral quality of surface light could lead to non-negligible errors in estimated primary production. The quantity and spectral quality of surface irradiance is dependent on optical properties of aerosols (eg. Bird, 1984; Brine and Iqbal, 1983) and clouds (Paltridge and Platt, 1977).

At present, there are two approaches to estimating surface irradiance. Platt and Sathyendranath (1988) have used a spectral clear sky model (Bird, 1984) assuming constant aerosol concentrations, coupled with a simple parametrisation of the cloud effect. Their parametrisation (based on Paltridge and Platt, 1976) accounts roughly for changes in the magnitudes of direct and diffuse irradiance at the sea surface in the presence of clouds, but does not account for any change in the spectral quality of the diffuse light.

Application of the model also requires independent estimates of cloud cover. The method perfected by Gautier and co-workers (Gautier and Katsaros, 1984) uses geostationary satellite data to estimate surface irradiance under cloudy skies, but offers no spectral decomposition (but see Pinker and Ewing, 1985 who compute surface irradiance in three broad bands in the visible). Clearly, an improvement in the atmospheric transmission models is desirable.

One hopes that increased modelling efforts, and better estimates of aerosol distribution from remote sensing techniques (Deepak, 1982) would lead to improvements in this area in the near future.

6.3 PHYTOPLANKTON ABSORPTION

In spite of the recent advances in our understanding of the variability in the specific absorption curves of phytoplankton, some problems still remain with the measurements on natural samples: there appears to be no simple way of decoupling phytoplankton absorption from scattering.

Also, it is very difficult to separate absorption by live phytoplankton from that by covarying detrital particles. The latter problem is not relevant to computations of light transmission underwater, but could lead to underestimates of the quantum efficiency of primary production.

6.4 BIOGEOGRAPHY OF RELEVANT PARAMETERS

Extension of the local algorithms to large scales depends on the dynamic "biogeography" of the parameters of the chlorophyll profile and of the P-I curve. Only very rudimentary information is now available on the distribution of these parameters. But many data already exist in various archives which could be used for improving our biogeography. A large amount of relevant data is also expected to be collected in the near future, particularly under the umbrella of the Joint Global Ocean Flux Study (JGOFS). The coming years, therefore, should see progressive improvements in the biogeography of primary production-related parameters.

6.5 MODELLING THE COASTAL WATERS

The open-ocean waters present the easiest case for optical modelling, since the optical properties of these waters are determined primarily by changing concentrations   of phytoplankton signal from these other substances (Sathyendranath and Morel, 1983), mainly because of the limitations in spectral resolution. The SeaWiFS is expected to fare better in this respect. Ideally, this would be matched by more measurements and modelling efforts on the optical characteristics of coastal waters (see Prieur and Sathyendranath, 1981; Sathyendranath *et al.*, 1987; 1989b; Gordon *et al.*, 1988) that would lead to further improvements in the interpretation of the satellite data.

6.6 THE UNDERWATER ANGULAR LIGHT DISTRIBUTION

The simple spectral transmission models that have been developed fall into two categories: those based on direct correlations between the apparent optical property K and phytoplankton concentration (Smith and Baker, 1978; Morel, 1988), and those that compute K via the inherent properties and the angular distribution function μ (Sathyendranath and Platt, 1988). In the latter case, the separation of the roles of absorption and scattering from that of the angular distribution lends some flexibility to the model. But, on the other hand, application of such a model requires the underwater angular light distribution to be specified explicitly. Sathyendranath and Platt (1988) use a conservative estimate of the distribution function (μ), and neglect its depth-dependence. It has been shown (Kirk, 1983) that the depth dependence of μ could be very important, particularly in turbid coastal waters. Sathyendranath and Platt (1991) have suggested a scheme to compute the depth dependence of μ in turbid waters. This is an area where further modelling and observational efforts are required.

6.7 SECOND-ORDER EFFECTS

The photosynthesis-light model discussed here neglects all second order effects such as nutrient supply, temperature and light history. However, such effects are accounted for, to some extent, indirectly, through their effect on the parameters of the P-I curve. Wroblewski *et al.* (1988) have been able to show that the chlorophyll distribution patterns can be modelled realistically using a model that accounts for input of nitrate into the photic zone. This suggests the possibility of an inter-relationship between the biogeography of parameters, and the large-scale nutrient distribution.

6.8 NEW PRODUCTION

For many applications, for example in the study of the role of oceanic primary production in the global cycle, the variable of interest is not just the total primary production, but also the exportable part of the production, or, in other words, the "new" production. Since there is no electromagnetic signal amenable to remote sensing that is directly related to new production, one has to resort to indirect methods. Two methods have been used so far to compute new production, combining ocean colour and sea-surface temperature data from satellites (Dugdale *et al.*, 1989; Sathyendranath *et al.*, 1991). In both methods, sea-surface temperature data are used to compute the supply of nitrate to the euphotic zone. The empirical relationships used in the computation of new production however, are simpler in the approach of Sathyendranath *et al.* (1991).

TABLE 1   Glossary of Mathematical Notation.

Notation	Quantity and Description	Typical Units
α	Initial slope of P-I curve. Defined as $\partial P/\partial I\|_{I\to 0}$.	mg C h^{-1} (W m^{-2})$^{-1}$
α^B	Initial slope of P^B-I curve. Defined as $\partial P^B/\partial I\|_{I\to 0}$.	mg C (mg Chl a)$^{-1}$ h^{-1} (W m^{-2})$^{-1}$
θ	Zenith angle.	Dimensionless.
θ_d	Zenith angle of direct sunlight in water.	Dimensionless.
$\langle\theta_s\rangle$	Mean $\cos\theta$ for the diffuse sky light. The average is computed taking into account the angular distribution of diffuse light underwater.	Dimensionless.
λ	Wave length of light.	nm
μ	Mean path length of light rays per unit vertical distance in water. For a ray with zenith angle θ, $\mu = 1/\cos\theta$.	Dimensionless.
σ	Standard deviation of Gaussian curve describing chlorophyll profile.	m
ϕ	Azimuth angle of light.	Dimensionless.
a	Total absorption coefficient.	m^{-1}
a_c	Partial absorption coefficient asssociated with phytoplankton. $a_c = a_c^* C$.	m^{-1}

(Contd. . . . .)

(. . . . .Contd)

a_c^*	Specific absorption coefficient of phytoplankton. Partial absorption coefficient of phytoplankton per unit pigment concentration. $a_c^* = a_c/C$.	$m^{-1}(mg\ m^{-3})^{-1}$
a_w	Partial absorption coefficient due to pure sea water.	m^{-1}
b_b	Backscattering coefficient.	m^{-1}
B	Biomass. Usually, concentration of chlorophyll a.	$mg\ m^{-3}$
B_0	Background biomass. Parameter of the chlorophyll profile.	$mg\ m^{-3}$
C	Biomass and associated detritus. Usually, concentration of chlorophyll a plus phaeophytin.	$mg\ m^{-3}$
h	This parameter of the chlorophyll profile defines the area within the biomass peak. Height of the peak above the background is given by $h/(\sqrt{2\pi}\sigma)$.	$mg\ m^{-2}$
$I(0^-)$	Irradiance just above the water. Irradiance has notation E in most current literature in optical oceanography. We are following the common notation used in biological oceanography literature. Since, in oceanography, z is positive downwards, we have used the negative superscript for the value of irradiance above water.	$W\ m^{-2}$
$I(0^+)$	Irradiance just below water.	$W\ m^{-2}$
I_d	Diffuse component of irradiance.	$W\ m^{-2}$
I_k	Adaptation parameter of the $P\text{-}I$ curve. $I_k \equiv P_m^B/\alpha^B \equiv P_m/\alpha$.	$W\ m^{-2}$
I_s	Sunlight (direct) component of irradiance.	$W\ m^{-2}$
K_d	Attenuation coefficient for direct irradiance. $K_d = (1/I_d)dI_d/dz$.	m^{-1}
K_s	Attenuation coefficient for diffuse irradiance. $K_s = (1/I_s)dI_s/dz$.	m^{-1}
P	Primary production rate.	$mg\ C\ m^{-3}\ h^{-1}$
P^B	Specific production. Primary production rate normalised to biomass. $P^B = P/B$.	$mg\ C\ (mg\ Chl\ a)^{-1}\ h^{-1}$
P_m^B	Assimilation number. Specific production at saturating light. In the absence of photoinhibition, $P_m^B = P^B|_{I\to\infty}$	$mg\ C\ (mg\ Chl\ a)^{-1}\ h^{-1}$
P_m	Primary production (maximum) at saturating light. $P_m = B\ P_m^B$.	$mg\ C\ m^{-3}\ h^{-1}$
z	Depth.	m
z_m	Depth of chlorophyll maximum, a parameter of chlorophyll profile.	m

References

Bidigare, R. R., Smith, R C., Baker, K. S., and Marra, J. (1987) 'Oceanic primary production estimates from measurements of spectral irradiance and pigment concentrations', Global Biogeochemical Cycles 1, 171-186.

Bird, R. E. (1984) 'A simple, solar spectral model for direct-normal and diffuse horizontal irradiance', Solar Energy 32, 461-471.

Brine, D. T. and Iwbal, M. (1983) 'Diffuse and global spectral irradiance under cloudless skies', Solar Energy 30, 447-453.

Deepak, A. (1982) 'Aerosol remote sensing', in F. J. Vernberg and F. P. Diemer (eds.), Processes in Marine Remote Sensing, university of S. Carolina, Columbia, S.C., pp. 175-193.

Dugdale, R. C., Morel, A., Bricaud, A., and Wilkerson, F. P. (1989) 'Modeling new production in upwelling centres: a case study of modeling new production from remotely sensed temperature and colour', Journal of Geophysical Research 94, 18119-18132.

Gautier, C. and Katsaros, K. B. (1984) 'Insolation during STREX: comparisons between surface measurements and satellite estimates', Journal of Geophysical Research 89, 11779-11788.

Hoepffner, N. and Sathyendranath, S. (1991) 'Effect of pigment composition on absorption properties of phytoplankton', Marine Ecology Progress Series 73, 11-23.

Kirk, J. T. O. (1983) 'Light and photosynthesis in aquatic ecosystems', Cambridge University Press, New York.

Kishino, M., Sugihara, S., and Okami, N. (1985) 'Estimation of the spectral absorption coefficients of phytoplankton in the sea', Bulletin of Marine Science 37, 634-642.

Lewis, M. R., Warnock, R. E., Irwin, B., and Platt, T. (1985) 'Measuring photosynthetic action spectra of natural phytoplankton populations', Journal of Phycology 21, 310-315.

Mitchell, B. G. and Kiefer, D. A. (1988a) 'Chlorophyll a specific absorption and fluorescence excitation spectra for light-limited phytoplankton', Deep-Sea Research 35, 639-663.

Mitchell, B. G. and Kiefer, D. A. (1988b) 'Variability in pigment specific particulate fluorescence and absorption spectra in northeastern Pacific Ocean', Deep-Sea Research 35, 665-689.

Morel, A. (1988) 'Optical modeling of the upper ocean in relation to its biogenous matter content (case I waters), Journal of Geophysical Research 93, 10749-10768.

Morel, A. and Bricaud, A. (1981) 'Theoretical results concerning light absorption in a discrete medium and application to specific absorption of phytoplankton', Deep-Sea Research 28, 1375-1393.

Paltridge, G.W. and Platt, C.M.R. (1976) 'Radiative processes in meteorology and climatology', Elsevier, New York.

Pinker, R. T. and Ewing, J. A. (1985) 'Modeling surface solar radiation: formulation and validation', Journal of Climate and Applied Meteorology 24, 289-401.

Platt, T., Caverhill, C., and Sathyendranath, S. (1991) 'Basin-scale estimates of oceanic primary production by remote sensing: the North Atlantic', Journal of Geophysical Research 96, 15147-15159.

Platt, T., Denman, K. L., and Jassby, A. D. (1977) 'Modelling the productivity of phytoplankton', in E.D. Goldberg, I.N. McCave, J. J. O'Brien and J. H. Steele (eds.), The sea: ideas and observations on progress in the study of the seas 6, pp. 807-856.

Platt, T. and Sathyendranath, S. (1988) 'Oceanic primary production: estimation by remote sensing at local and regional scales', Science 241, 1613-1620.

Platt, T. and Sathyendranath, S. (1991) 'Biological production models as elements of coupled, atmosphere-ocean models for climate research', Journal of Geophysical Research 96, 2585-2592.

Platt, T., Sathyendranath, S., Caverhill, C. M., and Lewis, M. R. (1988) 'Ocean primary production and available light: further algorithms for remote sensing', Deep-Sea Research 35, 855-879.

Platt, T., Sathyendranath, S., and Ravidran, P. (1990) 'Primary production by phytoplankton: analytic solutions for daily rates per unit area of water surface', Proceedings Royal Society of London B 241, 101-111.

Platt, T., Sathyendranath, S., Ulloa, O., Harrison, W. G., Hoepffner, N., and Goes, J. (1992) 'Nutrient control of phytoplankton photosynthesis in the Western North Atlantic, Nature 356, 229-231.

Prieur, L. and Sathyendrananth, S. (1981) 'An optical classification of coastal and oceanic waters based on the specific spectral absorption curves of phytoplankton pigments, dissolved organic matter, and other particulate materials', Limnology and Oceanography 26, 671-689.

Sathyendranath, S., Prieur, L., and Morel, A. (1987) 'An evaluation of the problems of chlorophyll retrieval from ocean colour, for case 2 waters', Advances in Space Research 7, (2)27-(2)30.

Sathyendranath, S. and Platt, T. (1988) 'The spectral irradiance field at the surface and in the interior of the ocean: a model for applications in oceanography and remote sensing', Journal of Geophysical Research 93, 9270-9280.

Sathyendranath, S. and Platt, T. (1989a) 'Computation of aquatic primary production', Limnology and Oceanography 34, 188-198.

Sathyendranath, S. and Platt, T. (1989b) 'Remote sensing of ocean chlorophyll: consequence of nonuniform pigment profile', Applied Optics 28, 490-495.

Sathyendranath, S. and Platt, T. (1990) 'The light field in the ocean: its modification and exploitation by the pelagic biota', in P. J. Herring, A. K. Campbell, M. Whitfield and L. Maddock (eds.), Light and life in the sea, Cambridge University Press, Cambridge, pp. 3-18.

Sathyendranath, S. and Platt, T. (1991) 'Angular distribution of the submarine light field: modification by multiple scattering', Proceedings Royal Society of London Ser. A. 433, 287-297.

Sathyendranath, S., Platt, T., Caverhill, C. M., Warnock, R. E., and Lewis, M. R. (1989a) 'Remote sensing of oceanic primary production: computations using a pectral model', Deep-Sea Research 36, 431-453.

Sathyendranath, S., Platt, T., Horne, E. P. W., Harrison, W. G., Ulloa, O., Outerbridge, R., and Hoepffner, N. (1991) 'Estimation of new production in the ocean by compound remote sensing', Nature 353, 129-133.

Sathyendranath, S., Prieur, L., and Morel, A. (1989b) 'A three component model of ocean colour and its application to remote sensing of phytoplankton pigments in coastal waters', International Journal of Remote Sensing 10, 1373-1394.

Smith, E. L. (1936) 'Photosynthesis in relation to light and carbon dioxide', Proceedings National Academy of Sciences 22, 504-511.

Smith, R. C. and Baker, K. S. (1987) 'The bio-optical state of ocean waters and remote sensing', Limnology Oceanography 23, 247-259.

Wroblewski, J. S., Sarmiento, J. L., and Flierl, G. R. (1988) 'An ocean basin scale model of plankton dynamics for the climatological oceanographic conditions in May', Global Biogeochemical Cycles 2, 199-218.

CZCS DATA PROCESSING ALGORITHMS

B. STURM
Institute for Remote Sensing Applications
Joint Research Center, Commission of the European Communities
21020 Ispra (VA), Italy

ABSTRACT. A review of the CZCS atmospheric correction algorithms is given. The standard CZCS algorithm with prefixed Angstrom exponent (AE) and the recently proposed algorithm (Bricaud, Morel 1987) using a water optical model for case 1 waters and allowing a pixel-by-pixel evaluation of AE and pigment are discussed. Results of a study of sensitivity loss of CZCS are presented and compared with results of other investigations.

1. Formulation of the Problem

Remote sensing of ocean color can help to evaluate quantitatively the concentration of suspended matter (phytoplankton pigments, particulate dissolved matter, inorganic suspended matter) from measurements of the spectral composition of the light up welling from the ocean. The Coastal Zone Color Scanner (CZCS) launched on Nimbus 7 in October and working till June 1986 was the first satellite sensor dedicated to that issue (Hovis *et al.*, 1980).

The main problems in the interpretation of CZCS data are:

- to correct the measured total spectral radiance for atmospheric stray light and light reflected from the ocean surface (sky- and sun-glint);

- to evaluate the concentrations from the water leaving reflectance;

- to make these evaluation on a pixel by pixel basis for a great number of scenes with great number of pixels: CZCS has acquired about 60000 scenes with about 10^6 pixels to be processed for each of them.

For mastering this problem a simple but still sufficiently accurate optical model of the atmosphere-ocean system is needed. The single scattering approximation of the radiative transfer equation was suggested by Gordon (1978) to be sufficiently accurate.

In terms of radiance its formulation is:

$$L_t\left(\lambda,\mu,\mu_0,\phi\right) = L_a\left(\lambda,\mu,\mu_0,\phi\right) + L_r\left(\lambda,\mu,\mu_0,\phi\right) +$$

$$+ T_r\left(\lambda,\mu\right) + T_a\left(\lambda,\mu\right) t_{OZ}(\lambda,\mu) L_W\left(\lambda,\mu,\mu_0\right) \quad (1)$$

V. Barale and P.M. Schlittenhardt (eds.),
Ocean Colour: Theory and Applications in a Decade of CZCS Experience, 95–116.
© 1993 ECSC, EEC, EAEC, Brussels and Luxembourg. Printed in the Netherlands.

where:
L_t is the measured total radiance;
L_r, L_a are the radiances arriving into the sensor from Rayleigh and Mie scattering without ever having penetrated the ocean surface;
L_w is the water leaving radiance containing information about the suspended matter in the water;
$T_r(\lambda,\mu) = \exp(-\tau_r(\lambda)/2/\mu)$ is the Rayleigh and aerosol diffuse transmittance (Tanre *et al.*, 1979);
$T_a(\lambda,\mu) = \exp(-\tau_a(\lambda)/6/\mu)$ is the aerosol diffuse transmittance (Tanre *et al.*, 1979);
$t_{oz}(\lambda,\mu) = \exp(-\tau_{oz}(\lambda)/\mu)$ is the ozone transmittance.

The angular dependence of the terms is described by μ and μ_0, the cosines of the viewing angles θ and θ_0 and the azimuth angle ϕ between the planes sensor-nadir-pixel and pixel-zenith-sun (see figure 1).

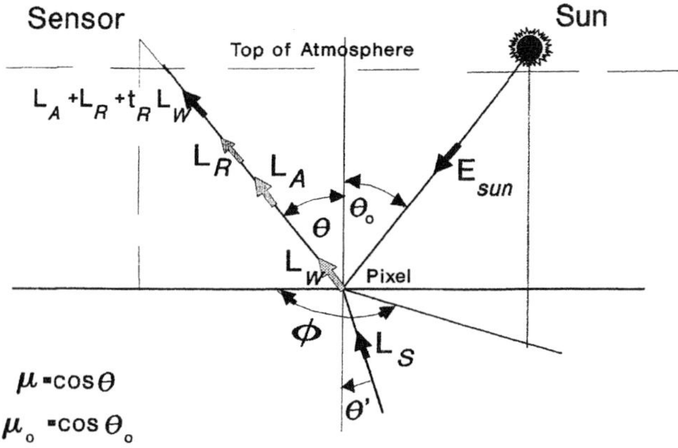

Figure 1.

This equation has to be solved for water leaving radiance $L_w(\lambda,\mu,\mu_0)$ which depends on λ, μ, and μ_0 and which can be expressed by the sub surface reflectance $R_s(\lambda)$ (which, in turn, depends only on wavelength λ):

$$L_W\left(\lambda,\mu,\mu_0\right) = \frac{\left(1-\rho(\mu)\right)\left(1-\bar{\rho}\right)}{n^2}$$

(2)

$$\frac{E_0(\lambda)\,\mu_0\,T_{r0}\left(\lambda,\mu_0\right)\,T_{a0}\left(\lambda,\mu_0\right)\,t_{oz}\left(\lambda,\mu_0\right)}{Q}\,R_s(\lambda)$$

Here $\rho(\mu)$ is the Fresnel reflectance of a calm water surface for the viewing angle θ given by:

$$\rho(\mu) = 1 - 2\,\mu\,y\,n\left[\frac{1}{(+n\,y)^2} + \frac{1}{(\mu\,n+y)^2}\right]$$

with:

$$y = \frac{1}{n}\left(n^2 + \mu^2 - 1\right)^{1/2}$$

n being the refractive index of sea water $n = 1.341$ (Austin 1974) and $\bar{\rho}$ being the reflectance of a calm water surface for diffuse radiation. Q is the irradiance to radiance conversion factor, which for constant radiance distribution is equal to π. The term $E_0(\lambda)$ is the extraterrestrial sun irradiance at the Julian Day of the overpass:

$$E_0(\lambda) = E_s(\lambda)\left(1 + 0.0167\,\cos\left(\frac{2\pi}{365}\,(Day-3)\right)\right)^2$$

where $E_s(\lambda)$ is the solar extraterrestrial irradiance at mean earth-sun distances. For the CZCS channels, $E_s(\lambda)$ is taken as:

$$E_s(\lambda) = 186.42,\,185.34,\,184.76,\,151.52\ mW\ \ cm^{-2}\mu^{-1}$$

(R. Austin's evaluation of Labs, Neckel data, NET meeting Jan. 1982).

Since in the numerical evaluations, to be described later, use is made of an optical water model for case 1 waters which is formulated in terms of sub-surface reflectance, and since reflectance, being a ratio, is independent of illumination and viewing geometry, it is convenient to transform eq.(1) in terms of reflectance. By multiplying by π and dividing by

$$\mu_0\,E_0(\lambda)\,exp\left(-\tau_{oz}(\lambda)\left(\frac{1}{\mu}+\frac{1}{\mu_0}\right)\right)$$

we pass from radiances to bidirectional reflectances

$$R_i = \frac{\pi\,.L_i}{\mu_0\,E_0}\,exp\left(-\tau_{oz}\left(\frac{1}{\mu}+\frac{1}{\mu_0}\right)\right)$$

Calling the R_i for $i=t$, a and r the total, aerosol and Rayleigh-reflectance respectively, we obtain from eq. (1):

$$R_t\left(\lambda,\mu,\mu_0,\phi\right) = R_a\left(\lambda,\mu,\mu_0,\phi\right) + R_r\left(\lambda,\mu,\mu_0,\phi\right) + \tag{3}$$

$$+ T_{2r}\left(\lambda,\mu,\mu_0\right) T_{2a}\left(\lambda,\mu,\mu_0\right)\frac{\left(1-\rho(\mu)\right)\left(1-\bar{\rho}\right)\pi}{n^2 Q} + R_s(\lambda)$$

in which T_{2r} and T_{2a} are defined as

$$T_{2x}\left(\lambda,\mu,\mu_0\right) = exp\left(-\tau_x(\lambda)\bigg/ X\left(\frac{1}{\mu}+\frac{1}{\mu_0}\right)\right)$$

where for $x = a$, $X = 6$ and for $x = r$, $X = 2$. These are the "two-way" diffuse transmittances for aerosol and Rayleigh attenuation. In analogy we introduce a "two-way" ozon extinction

$$T_{2oz}\left(\lambda,\mu,\mu_0\right) = exp\left(-\tau_{oz}(\lambda)\bigg/ X\left(\frac{1}{\mu}+\frac{1}{\mu_0}\right)\right)$$

2. Algorithms for Solution

We introduce now the simplifications $T_{2a} \approx 1$ and $Q = \pi$. The first is justified by the fact that even for values of $\tau a(670)$ of the order of 0.4, the error in the ratio of R_s remains less than some percent.

On the other side, taking account of T_{2a} by evaluating τa from the aerosol path radiance, is in principle possible, however it would greatly complicate the calculations, since the evaluation of aerosol radiance is evaluated by an iterative method as explained above. The second is of no importance since a constant factor does not influence the ratio from which the pigment is evaluated.

With the abbreviation

$$q(\mu) = \frac{\left(1-\rho(\mu)\right)\left(1-\bar{\rho}\right)}{n^2}$$

eq.(3) becomes

$$R_t\left(\lambda,\mu,\mu_0,\phi\right) = R_a\left(\lambda,\mu,\mu_0,\phi\right) + R_r\left(\lambda,\mu,\mu_0,\phi\right) + \tag{4}$$

$$+ T_{2r}\left(\lambda,\mu,\mu_0\right)\; q\left(\mu\right) + R_s\left(\lambda\right)$$

and we can formulate our problem in the following terms. Given N measured values of the Earth's bidirectional reflectance $R_t(\mu,\mu_0,\phi,\lambda_i)$, for $i=1...N$ discrete narrow wavelength intervals with center wavelength λ_i , we want to calculate N sub surface reflectances $R_S(\lambda_i)$ at the pixel from the equation system:

$$R_t\left(\mu,\mu_0,\phi,\lambda_i\right) = R_r\left(\mu,\mu_0,\phi,\lambda_i\right) + R_a\left(\mu,\mu_0,\phi,\lambda_i\right) +$$

$$+ T_{2r}\left(\mu,\mu_0,\lambda_i\right)\; q\left(\mu,\lambda_i\right) R_s\left(\lambda_i\right) \tag{5}$$

where R_r is the <u>known</u> bidirectional Rayleigh reflectance, and where R_a is the <u>unknown</u> bidirectional Aerosol reflectance. Subsequently, we subtract R_r from R_t, divide the equation by qT_{2r} and call

$$R_t^* = \frac{R_t - R_r}{qT_{2r}}$$

the reduced Rayleigh corrected total reflectance, while

$$R_a^* = \frac{R_a}{qT_{2r}}$$

becomes the reduced aerosol reflectance.

The division by $qT2r$ transforms the above-atmosphere values of the reflectances to sub-surface, *i.e* R are those underwater reflectances that give rise to the corresponding measured terms at sensor level. With this abbreviations we can formulate our problem at sub-surface level in a very simple way:

$$R_t^*\left(\lambda_i\right) = R_a^*\left(\lambda_i\right) + R_s\left(\lambda_i\right) \tag{6}$$

where for brevity we have retained only λ as variable. This is a system of N equations for $2.N$ unknowns: $R_s(\lambda_i)$, $R_a(\lambda_i)$, $i=1...N$.

According to Angstroem's law the wavelength dependence of the optical thickness of aerosols R_a can be expressed as

$$R_a\left(\lambda_i\right) = R_a\left(\lambda_N\right)\left(\lambda_N/\lambda_i\right)^v \tag{7}$$

v is the Angstroem exponent (AE). For R_a^* we obtain

$$R_a^*\left(\lambda_i\right) = R_a^*\left(\lambda_N\right)\frac{T_{2r}\left(\lambda_N\right)}{T_{2r}\left(\lambda_i\right)}\left(\lambda_N/\lambda_i\right)^v$$

$$= R_a^*\left(\lambda_N\right) S\left(\lambda_i, v\right) \tag{8}$$

with

$$S\left(\lambda_i\right) = \frac{T_{2r}\left(\lambda_N\right)}{T_{2r}\left(\lambda_i\right)}\left(\lambda_i/\lambda_N\right)^v =$$

$$= \left(\lambda_N/\lambda_i\right)^v \exp\left(\frac{\tau_r\left(\lambda_i\right) - \tau_r\left(\lambda_N\right)}{2}\left(\frac{1}{\mu_0} + \frac{1}{\mu}\right)\right)$$

Since $S(\lambda_N, v) = 1$, introducing this into the above equation gives:

$$R_t^*\left(\lambda_i\right) = R_a^*\left(\lambda_N\right) S\left(\lambda_i, v\right) + R_s\left(\lambda_i\right) \quad \text{for } i = 1 \text{ to } N-1 \tag{9}$$

and

$$R_t^*\left(\lambda_N\right) = R_a^*\left(\lambda_N\right) + R_s\left(\lambda_N\right)$$

which is a system of N equations for $N+2$ unknowns. $R_s(\lambda_i)$ $i=1...N$, $R_a(\lambda_N)$ and v. Substituting $R_a^*(\lambda_N)$ by $R_t^*(\lambda_N)$ - $R_s(\lambda_N)$ in N-1 equations we can reduce it to N-1 equations for $N+1$ unknowns: $R_s(\lambda_i)$ $i=1...N$ and v.

3. The Standard Algorithm

In the case that the wavelength channels of the sensor are positioned so that at two wavelength say λ_{N-1} and λ_N the condition $R_s\left(\lambda\right) = 0$ is met (which for λ > 700 nm is the case for nearly all oceanic waters, except for strong cocolithophore blooms or near to river plumes) a solution is possible: v can be evaluated from

$$S\left(\lambda_{N-1},v\right) = \frac{R_t^*\left(\lambda_{N-1}\right)}{R_t^*\left(\lambda_N\right)}$$

and the nonzero $R_s(\lambda)$ from

$$R_s\left(\lambda_i\right) = R_t^*\left(\lambda_i\right) - R_t^*\left(\lambda_N\right) S\left(\lambda_i,v\right), \quad i=1...N-2$$

In the case of CZCS with $N=4$ this condition is not fulfilled. Therefore the algorithms for CZCS atmospheric correction must add to the equation system additional information:

1. assume that v is known or calculated from an independent procedure (Gordon and Clark, 1981);

2. use an empirical relation between $R_s(\lambda_4)$ and the ratio $R_s(\lambda_1)/R_s(\lambda_3)$ or $R_s(\lambda_2)/R_s(\lambda_3)$ to add and equation (Smith and Wilson, 1981).

This gives a nonlinear system of four equations for four unknowns:

$$R_s\left(\lambda_i\right) = R_t^*\left(\lambda_i\right) - \left(R_t^*\left(\lambda_4\right) - R_s\left(\lambda_4\right)\right) S\left(\lambda_i,v\right), \quad i=1, 2, 3$$

and

$$R_s\left(\lambda_4\right) = F\left(R_s\left(\lambda_1\right) \Big/ R_s\left(\lambda_3\right) \right) \tag{10}$$

Formulations of the function F have been given by Austin and Petzold (1981) for oceanic (case 1) waters and by Viollier and Sturm (1984) for turbid coastal waters. This is the standard method for CZCS atmospheric corrections: the equations are solved by a iteration procedure: one starts with $R_s(\lambda_4) = 0$, evaluating of $R_s(\lambda_i)$ for $i=1,2,3$ followed by a calculation of a new $R_s(\lambda_4)$ which gives a new set of $R_s(\lambda)$ until convergence.

4. Algorithm using a Water Optical Model

Bricaud and Morel (1987) introduced a method which makes use of an optical water model that is based on the well founded hypothesis that in case 1 waters the phytoplankton produced by photosynthesis is the primary source of the optically active material. This assumption makes it possible to express all $R_s(\lambda_i)$ ($i=1..4$) as functions of pigment concentration C (see figure2). In this

102

water optical model C itself is a unique function of the ratio $X = R_s(\lambda_1)/R_s(\lambda_3)$ and we can express the $R_s(\lambda)$ as:

$$R_s\left(\lambda_i\right) = exp\left(a_i + b_i\, lnX + c_i\left(lnX\right)^2 + d_i\left(lnX\right)^3\right)$$

(numerical values of the constants a_i, b_i, c_i, d_i are given in the Appendix).

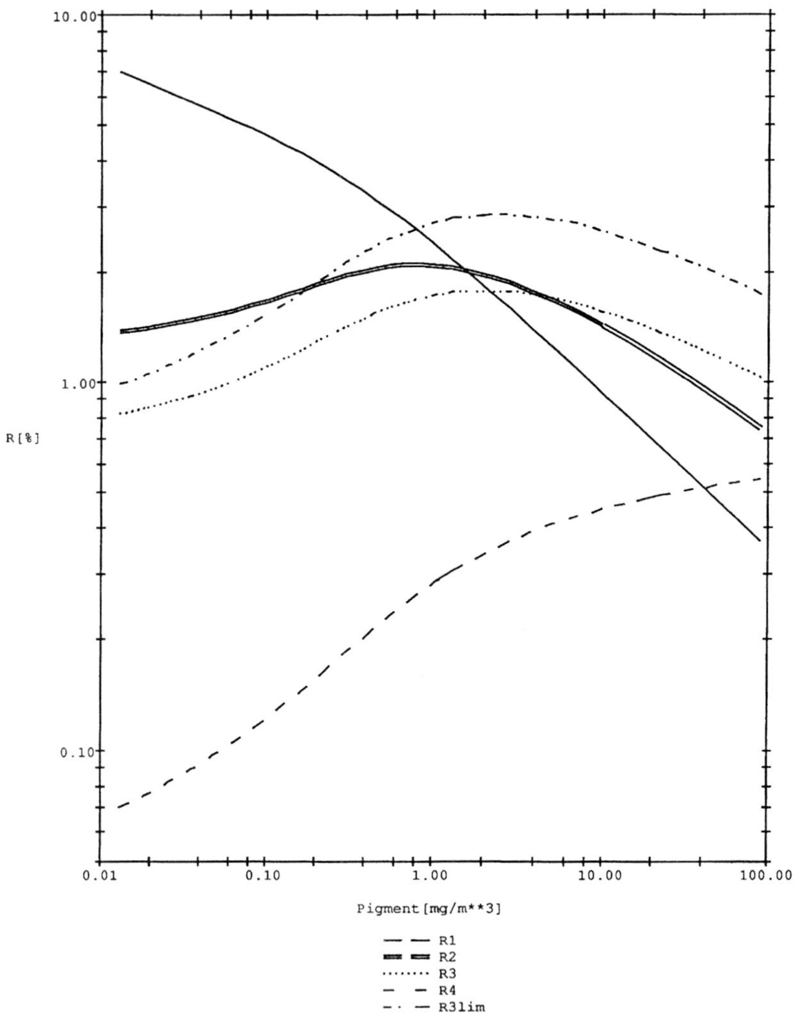

Figure 2. Reflectance vs pigment for Case 1 water.

It is easy to see that with this assumptions it is possible to reduce the system (10) into a system of two nonlinear equations for the two unknowns $X = R_s(\lambda_1)/R_s(\lambda_3)$ and v, as:

$$X = \frac{R_t^*\left(\lambda_1\right) - \left[R_t^*\left(\lambda_4\right) - F(X)\right] S\left(\lambda_1, v\right)}{R_t^*\left(\lambda_3\right) - \left[R_t^*\left(\lambda_4\right) - F(X)\right] S\left(\lambda_3, v\right)}$$

$$G(X) = R_t^*\left(\lambda_4\right) - \left[R_t^*\left(\lambda_4\right) - F(X)\right] S\left(\lambda_2, v\right)$$

where F, and G are known functions of X:

$$F(X) = R_s\left(\lambda_4\right) = exp\left(a_4 + b_4 \ln X + c_4 (\ln X)^2 + d_4 (\ln X)^3\right)$$

$$G(X) = R_s\left(\lambda_2\right) = exp\left(a_2 + b_2 \ln X + c_2 (\ln X)^2 + d_2 (\ln X)^3\right)$$

This solution is obtained again by interation. The first equation is solved with a fixed v by the substitution method:

$$X_i = \frac{R_t^*\left(\lambda_1\right) - \left[R_t^*\left(\lambda_4\right) - F(X_{i-1})\right] S\left(\lambda_1, v\right)}{R_t^*\left(\lambda_3\right) - \left[R_t^*\left(\lambda_4\right) - F(X_{i-1})\right] S\left(\lambda_3, v\right)}$$

and a new guess for v is then obtained from solving the second equation for v:

$$S\left(\lambda_2, v\right) = \frac{R_t^*\left(\lambda_2\right) - G(X)}{R_t^*\left(\lambda_4\right) - F(X)}$$

Note that in order to calculate X and v in this system independently, measured R_t^* from channel 2 (and not 3) must be used, because otherwise, in the v-iterations, $R_s(\lambda_3)$ in X will be evaluated from the model value instead from the measured quantity $R_t^*(\lambda_3)-[R_t^*(\lambda_4)-F(X)] S(\lambda_3, v)$.

Equally well the problem can be solved using as model parameter $Y = R_s(\lambda_2)/R_s(\lambda_3)$.

104

In this case v has to be evaluated from

$$S\left(\lambda_1, v\right) = \frac{R_t^*\left(\lambda_1\right) - H(Y)}{R_t^*\left(\lambda_4\right) - F(Y)}$$

where $H(Y)$ is the corresponding case 1 water model expression for $R_s(\lambda_1)$.

In the practical application of the method on a pixel-by-pixel basis one needs criteria to decide whether a pixel belongs to case 1 and if so, when to use the ratio X and when to use Y. The first criterion has been given by Bricaud and Morel (1987) : a pixel is a case 1 pixel if its reflectance $R_s(\lambda_3)$ is less than a upper limit R_{lim} given as function of X or Y (see Appendix).

The criterion when to use X or Y has been suggested by André and Morel (1991) : according to them Y should be used at high pigment concentrations in case 1 water when the sub surface reflectance $R_s(\lambda_1)$ is so low that it approaches the detection limit of sub surface reflectance corresponding to one digital count. In the case of highest gain of CZCS this value is of the order of 0.002 at the begin of its lifetime (no sensitivity loss) and about 0.003 at the end. As can be seen from figure 2 such low values of $R_s(\lambda_1)$ correspond to pigment concentrations hardly reached by case 1 water. Therefore we use in our programs C-values of the order of 2 - 3 mg as switching limit.

Pixels which in both cases do not fulfill the $R_s(\lambda_3)$ limit condition are declared case 2 pixels and are processed by the standard method (see eq. (10)) with a constant Angstroem coefficient. For the iteration on channel 4 we use the relation:

$$R_s\left(\lambda_4\right) = R_s\left(\lambda_1\right) \left[R_s\left(\lambda_1\right) \Big/ R_s\left(\lambda_3\right) \right]^{-2.48} \qquad \text{(Bricaud, Morel 1987)}$$

when $R_s(\lambda_1)$ is significant and the relation:

$$R_s\left(\lambda_4\right) = R_s\left(\lambda_3\right) \left[R_s\left(\lambda_2\right) \Big/ R_s\left(\lambda_3\right) \right]^{-2.0} \qquad \text{(Viollier, Sturm 1984)}$$

otherwise.

5. Calculation of the Rayleigh Path Radiance

We have assumed that the Rayleigh path radiance L_R and respectively the Rayleigh atmospheric reflectance can be calculated from first principles i.e only knowing the geometric relations between sensor, pixel and sun and the Rayleigh optical thickness of the atmosphere. Three methods to calculate L_R are available:

(1) Gordon's approximation to the single scattering formula (Gordon *et al.*, 1983);

(2) Sobolev's approximate multiple scattering formula (Sobolev 1975);

(3) Gordon's exact multiple scattering tables taking account also of polarization of Rayleigh scattering and surface reflection by the ocean (Gordon *et al.*,1988).

The derivation of the single scattering formula starts from the RTE

$$\mu \, \frac{dL\left(\tau,\mu,\phi\right)}{d\tau} = L\left(\tau,\mu,\phi\right) - \frac{1}{4\pi} \int_{0}^{2\pi} d\phi' p\left(\mu,\phi,\mu',\phi'\right) L\left(\tau',\mu',\phi'\right) -$$

$$- \frac{E_0}{4\pi} \, p\left(\mu,\phi,\mu_0,\phi_0\right) \, exp\left(\tau/\mu_0\right)$$

with the Rayleigh scattering phase function

$$p\left(\mu,\phi,\mu_0,\phi_0\right) = \frac{3}{4}\left(1+cos\left(\psi_-\right)^2\right)$$

at the scattering angle

$$\psi_- = \pi - acos\left[\mu\,\mu_0 + \sqrt{\left(1-\mu^2\right)\left(1-\mu_0^2\right)}\,cos\left(\phi-\phi_0\right)\right]$$

By neglecting the second term on the right side of the equation that describes the secondary scattering, a inhomogeneous differential equation for $L(\tau)$ is obtained:

$$\mu \, \frac{dL\left(\tau,\mu,\phi\right)}{d\tau} = L\left(\tau,\mu,\phi\right) - \frac{E_0}{4\pi} \, p\left(\mu,\phi,\mu_0,\phi_0\right) \, exp\left(\tau/\mu_0\right)$$

which has the general solution:

$$L = \frac{E_0}{4\pi\mu} p\left(\psi_-\right) \, exp\left(\tau/\mu\right)\left[\frac{\mu\,\mu_0}{\mu+\mu_0} \, exp\left(-\tau\left(1/\mu+1/\mu_0\right)\right)+C_1\right]$$

Assuming a black ocean, *i.e* $L(\tau_r, \mu, \mu_0) = 0$ at $\tau = \tau_r$ the integration constant C_1 can be evaluated:

$$C_1 = -\frac{\mu \cdot \mu_0}{\mu + \mu_0} \; exp\left(-\tau_r\left(1\Big/\mu + 1\Big/\mu_0\right)\right)$$

where τ_r is the total optical thickness of the Rayleigh atmosphere. The complete solution is then:

$$L\left(\tau, \mu, \mu_0, \phi - \phi_0\right) = \frac{E_0}{4\pi\mu} p\left(\psi_-\right) \; exp\left(\tau\Big/\mu\right) \frac{\mu \cdot \mu_0}{\mu + \mu_0}$$

$$\left[exp\left(-\tau\left(1\Big/\mu + 1\Big/\mu_0\right)\right) - exp\left(-\tau_r\left(1\Big/\mu + 1\Big/\mu_0\right)\right)\right]$$

In this form the path radiance equation was also derived by Kriebel and Quenzel (1976) by integrating the scattered direct sun light along the viewing path. For satellite observation (top of the atmosphere *i.e.* $\tau = 0$) we obtain:

$$L_r = L\left(0, \mu, \mu_0, \phi - \phi_0\right) = \frac{E_0}{4\pi\mu} p\left(\psi_-\right) \frac{\mu \, \mu_0}{\mu + \mu_0}$$

$$\left[1 - exp\left(-\tau_r\left(1\Big/\mu + 1\Big/\mu_0\right)\right)\right]$$

For $\tau_r \ll 1$ the exponential function can be approximated by $1 - \tau_r\,(1/\mu + 1/\mu_0)$ and we obtain Gordon's "single scattering" formula

$$L_r = \frac{E_0}{4\pi\mu} p\left(\psi_-\right) \tau_r \qquad\qquad \text{(Gordon et. al. 1983)}$$

Sobolev's approximation of multiple scattering does not neglect completely the secondary scattering but uses in the secondary scattering term of the RTE azimuthal averaging and a simple scattering phase function

$$p\left(\psi_-\right) = 1 + x_1 \, cos\left(\psi_-\right)$$

with

$$x_1 = \frac{3}{2} \int_0^\pi p(\gamma) \, cos(\gamma) \, sin(\gamma) \, d\gamma$$

For the Rayleigh scattering phase function $x_1 = 0$ i.e the second order scattering is taken into account with an isotropic phase function. The detailed derivation of the solution is rather complex and can be found in Sobolev (1975). The resulting formula for the Rayleigh path radiance becomes:

$$L_r = \frac{\mu_0 E}{\pi} \left[1 - \frac{R(\mu, \tau_r) \, R(\mu_0, \tau_r)}{4 + 3 \, \tau_r} + \right.$$

$$\left. + \left(3 \, \mu \, \mu_0 - 2 (\mu + \mu_0) + p_r(\psi_-) \right) \frac{1 - exp\left(-\tau_r \left(\frac{1}{\mu} + \frac{1}{\mu_0} \right) \right)}{4(\mu + \mu_0)} \right]$$

with $R(\mu, \tau)$ given by

$$R(x, y) = 1 + 1.5 \, x + (1 - 1.5 \, x) \, exp\left(-\frac{x}{y} \right)$$

All the above formulae assume a black ocean surface, however we need to take into account two additional contributions to the Rayleigh path radiance (see figure 3a,b):

a) single scattering by molecules of direct sun light reflected at the ocean surface;

b) surface reflection of down welling diffusely Rayleigh-scattered light.

These two contributions are taken into account by a correction factor:

$$1 + \left(\rho(\mu) + \rho(\mu_0) \right) \frac{p_r(\psi_+)}{p_r(\psi_-)}$$

where $\rho(\mu)$ is the Fresnel reflectance at incidence angle with cosine μ, and where ψ_+ is the "reflected scattering angle"

$$\psi_+ = acos\left[\mu\,\mu_0 - \sqrt{\left(1-\mu^2\right)\left(1-\mu_0^2\right)}\,\cos\left(\phi-\phi_0\right)\right]$$

Note that $cos(\psi_+) = cos(\psi_-) + 2\,\mu\,\mu_0$.

For the calculation of Rayleigh path radiance according to the multiple scattering approximation we use the numerically derived coefficients $Im(\theta, \theta_0)$, $m = 0,1,2$ tabulated by Gordon *et al.*, (1988) to calculate the Rayleigh path radiance from

$$L_r\left(\theta, \theta_0, \phi-\phi_0\right) = \sum_{m=0}^{2} I_m\left(\theta, \theta_0\right)\cos\left(m\left(\phi-\phi_0\right)\right)$$

These coefficients were derived by an exact solution (including polarization by Rayleigh scattering and surface reflection) of the radiative transfer equation for a plane parallel atmosphere and flat ocean.

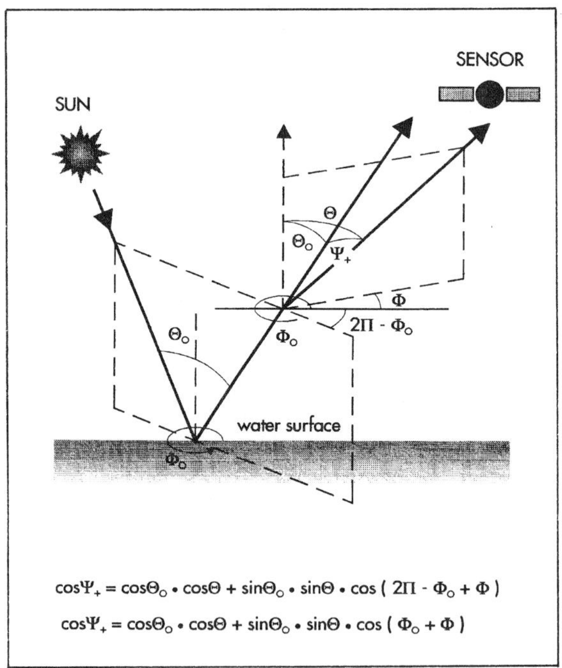

Figure 3. Scattering of Reflected sun to sensor.

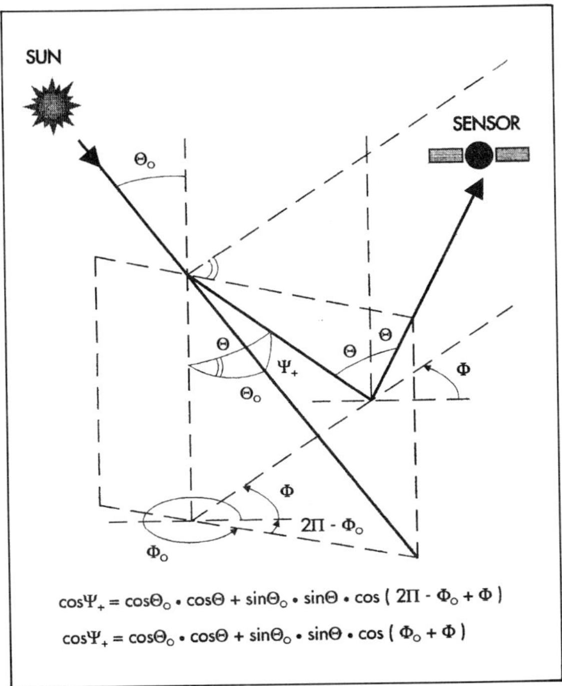

Figure 4. Reflection of scattered sun to sensor.

6. CZCS Calibration and Correction for Sensitivity Decay

The CZCS was equipped with an internal calibration system and the sensor had been calibrated before launch (Ball Aerospace Systems Division 1979). Rather early after the first CZCS data became available it appeared that the sensitivity of the sensor was decreasing (Austin 1982, Viollier 1982). Since the internal calibration system did not detect any decay of sensitivity it was concluded that one or more optical surfaces placed ahead of the internal calibration system had deteriorated.

Aircraft measurements to confirm and evaluate the sensitivity loss were performed by Hovis et al.,(1985). When it became evident that the sensitivity decay was continuous in time, the only way to solve the problem appeared to be the evaluation of correction factors by comparing measured and expected radiances over water with known reflectance properties were well known i. e. clear water areas.

The sensitivity decay of the CZCS has been evaluated by Hering et al., (1984) by normalization of measured total CZCS radiance over clear water to a standard illumination and viewing geometry. Gordon et al., (1984), Mueller (1984) and Sturm (1986) made similar evaluations by comparing measured radiances with expected radiances over clear water pixels. Such evaluations are sensitive to the the model used for the calculation atmospheric parameters and to the values used for the "clear water" radiance. In order to minimize systematic errors the method of simulation of the expected sensor radiances must be exactly the same as that used later for the atmospheric correction of the CZCS data.

The correction factors (CF) for use with the most recent version of the CZCS processing software at Ispra have been re-evaluated using the existing 'Clear-Water-pixel' catalog with data up to orbit no. 37834 (April 21,1986). The evaluations were made using a catalog of 660 clear water pixels from 59 orbits ranging from Nr.302 to 37834. Most of the data in this catalog are from clear water pixels from the Adriatic Sea, Ligurian Sea, West- and South-West African Atlantic and some data from the U.S. East coast Atlantic waters.

The correction factor is defined as ratio of true radiance to the measured radiance obtained from the preflight calibration slopes and intercepts.
If we call L_t the "true" radiance at the satellite aperture for the orbit N, then the sensitivity-loss compensation factor $F(\lambda,N)$ which accounts for sensitivity loss with time is obtained from

$$L_t(\lambda) = \left[CR(\lambda, N)\, a(\lambda) + b(\lambda) \right] F(\lambda, N)$$

The term in the brackets is the "measured" radiance $L_m(\lambda,N)$; then the compensation factor $F(\lambda,N)$ becomes:

$$F(\lambda, N) = \frac{L_t(\lambda)}{L_m(\lambda, N)}$$

The true sensor radiance is evaluated by a model that uses the same assumptions as the model used for the atmospheric correction. If we call the ocean water-leaving radiance $L_w(\lambda)$, we can express the true radiance as:

$$L_t(\lambda) = L_a(\lambda) + L_r(\lambda) + T_r(\lambda)\, T_a(\lambda)\, t_{oz}(\lambda)\, L_w(\lambda)$$

or in terms of reflectance and assuming $T_{za} = 1$:

$$R_t(\lambda) = R_r(\lambda) + R_a(\lambda) + T_{2r}(\lambda)\, q(\mu,\lambda)\, R_S(\lambda)$$

so that the "measured" reflectance becomes:

$$R_m(\lambda, N) = \frac{\pi \left[CR(\lambda, N)\, a(\lambda) + b(\lambda) \right]}{\mu_0 E_0(\lambda)\, exp\left(-\tau_{oz}(\lambda) \left(\frac{1}{\mu} + \frac{1}{\mu_0} \right) \right)}$$

As shown earlier, $R_a(\lambda)$ can be expressed as:

$$R_a\left(\lambda_4\right)\left(\lambda_4\big/\lambda\right)^{\nu} = \left[\, R_t\left(\lambda_4\right) - R_r\left(\lambda_4\right) - T_{2r}\left(\lambda_4\right) q R_S\left(\lambda_4\right)\,\right]$$

We assume, that $R_t(\lambda_4)=R_m(\lambda_4, N)$ i.e that the channel 4 of CZCS has no sensitivity loss. This has been confirmed by investigations of Hering *et al.*,(1985) who evaluating measured aperture radiances for channel 4 as function of orbit number. Figure 4 shows normalized (according to McGlamery, 1984) measured CZCS total radiances for our clear water pixel set for channels 1, 3 and 4: it is seen that in channel 4 no significant sensitivity loss can be detected.

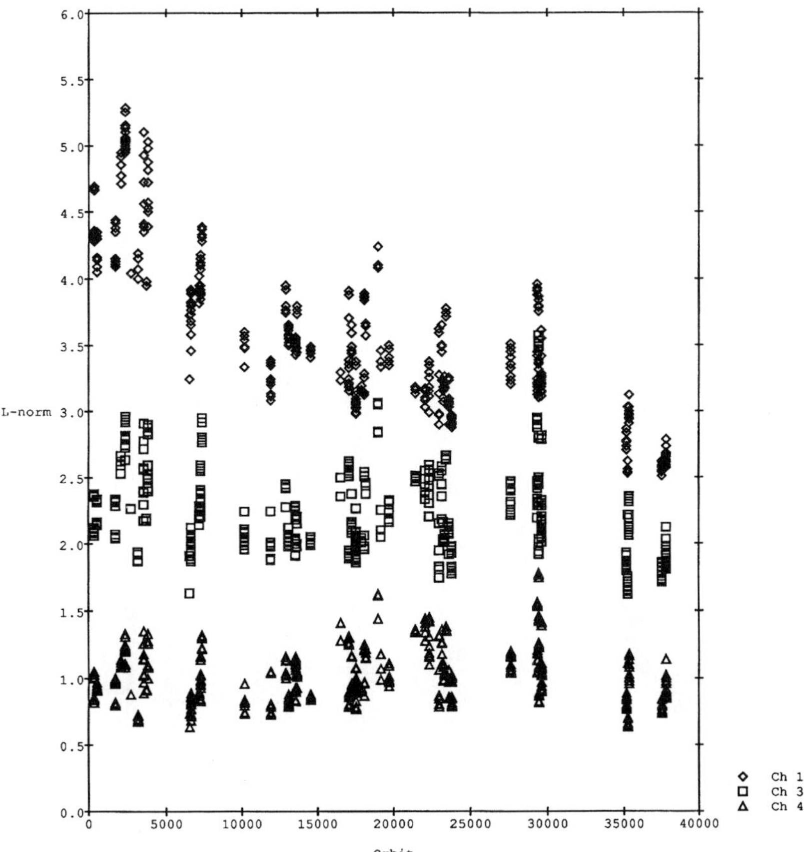

Figure 5. CZCS sensitivity decay.

With this assumption we can calculate the sensitivity loss correction factor from

$$F\left(\lambda, N\right) = \frac{R_t\left(\lambda\right)}{R_m\left(\lambda, N\right)}$$

We have to assume in this calculations values for the Angstroem exponent and $R_s(\lambda)$.

For calculating the clear water reflectances R_s we use the case 1 water model with a value of $R_s(\lambda_1)/R_s(\lambda_3)= 3.9$ which corresponds in to a pigment concentration $C = 0.113$ mg/m^3. A standard atmosphere with an Angstroem exponent of $\nu= 0.9$ is used for the aerosol path radiance calculation figure 5 the results are shown in comparison with those of other investigators. We find that the CF ar well represented by polynomials of the type:

$$F\left(\lambda, N\right) = f_0\left(\lambda\right) + f_1\left(\lambda\right)N + f_2\left(\lambda\right)N^2 + f_3\left(\lambda\right)N^3$$

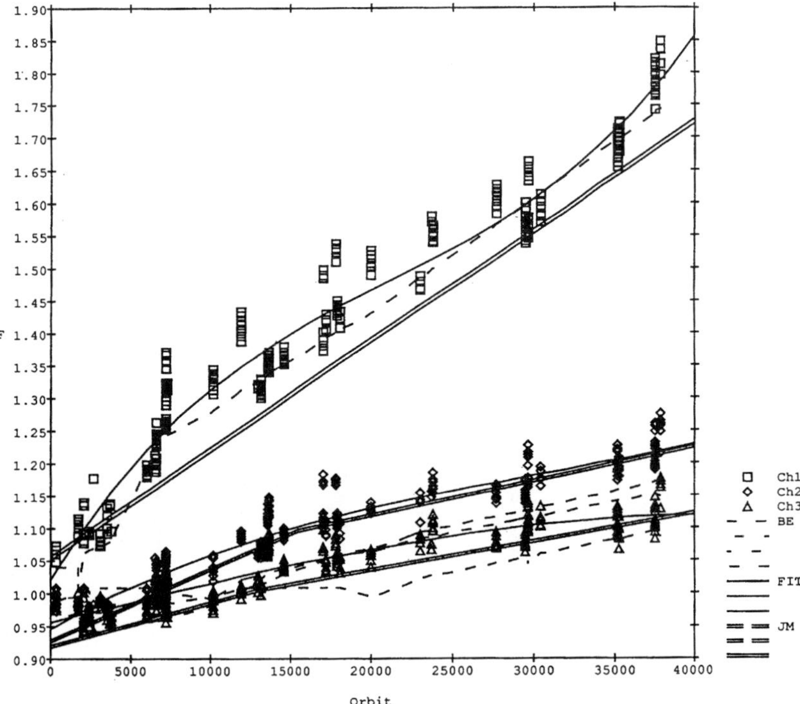

Figure 6. Sensitivity loss correction factors.

The curves labeled *BE* are the correction factors used by the Miami group (H. Fukushima, R.H. Evans, private communication), while JM labels the result s of J.L. Mueller (J.L. Mueller, private communication).

Besides the systematic biases due to the used optical model of the system ocean-atmosphere other, it is evident that statistical fluctuations are introduced by assumptions about the average radiance of clear water and average Angstroem exponent. Nevertheless it is encouraging that the evaluations of the various groups seem to agree within the statistical error limits. Some disagreement exists about the sensitivity loss of channel 4: the Miami data show a significant decay for this channel beyond orbit No. 20000.

7. Conclusions

The limited number (4) of sensitive channels of the CZCS make the atmospheric correction and the subsequent evaluation of sub surface reflectance difficult due to the presence of unknown parameters like the aerosol path radiance and its wavelength dependence (Angstroem exponent). The solution of the problem is possible using empirical relations between the values of the spectral reflectances which are well established for 1 waters. By introducing an optical model describing sub surface reflectance as function of one optically active parameter, the chlorophyll like pigment concentration, it is possible to evaluate both the pigment concentration and the Angstroem exponent in case 1 water.

Non case 1 water pixels must be processed by the standard method using a constant Angstroem exponent, which however can be obtained from the next nearest case 1 water region so giving a more reliable evaluation of aerosol path radiance. If more general optical models describing spectral reflectance as function of more parameters (pigment, suspended matter, dissolved organic matter and more spectral channels to measure ocean color will be available, a complete solution of the ocean color remote sensing problem will be possible.

8. Appendix: <u>Case 1 water model</u>

Morel's case 1 water model (Morel 1988, Morel private communication, André and Morel 1989, 1991) expresses the spectral sub surface reflectance $R_s(\lambda)$ as function of the chlorophyll like pigment concentration C.

When C itself is expressed in terms of reflectance ratios

$$X = R_S\left(\lambda_1\right) \Big/ R_S\left(\lambda_3\right) \quad R_S\left(\lambda_i\right)$$

can be expressed as functions of X:

$$R_S\left(\lambda_i\right) = exp\left(a_i + b_i \; lnX + c_i \left(lnX\right)^2 + d_i \left(lnX\right)^3\right)$$

R is given in percent!

The constants a_i, b_i, c_i and d_i for $i=1$ to 4 for the CZCS channels are given in the following Table

TABLE 1.

Channel i	a	b	c	d
1	0.571	1.04	-0.407	0.102
2	0.663	0.352	-0.488	0.115
3	0.571	0.039	-0.407	0.102
4	-1.060	-0.605	-0.130	0.026

The Pigment in mg/m^3 is given as function of X by:

$$C = exp \left(0.768 - 2.61 \ lnX + 0.791 \ (lnX)^2 - 0.388 \ (lnX)^3 \right)$$

and the value of reflectance in channel 3 (in percent) that represents the limit between case 1 and case 2 waters:

$$R_{lim} = exp \left(1.05 - 0.02 \ lnX - 0.429 \ (lnX)^2 + 0.094 \ (lnX)^3 \right)$$

The corresponding equations in terms of the ratio $Y = R_s(\lambda_2)/R_s(\lambda_3)$ can easily be derived from those using X.

References

André, J.M. and Morel, A. (1989) 'Simulated effects of barometric pressure and ozone content upon the estimate of marine phytoplankton form space', Journal of Geophysical Research 94 (C1), 1029-1037.

André, J.M., Morel, A. (1991) 'Atmospheric corrections and interpretation of marine radiances in CZCS imagery: revisited', Oceanologica Acta 14, 3-22.

Austin, R.W. (1974) 'The Remote Sensing of spectral Radiance from below the Ocean Surface', in N.G. Jerlov and E.S. Nielsen (eds.), Optical Aspects of Oceanography, Academic Press, pp. 317-344.

Austin, R.W and Petzold, T.J. (1981) 'The determination of the Diffuse Attenuation Coefficient of Sea Water using the Coastal Zone Color Scanner', in J.F.R.Gower(ed), Oceanography from Space, Plenum Press, New York, pp. 239-256.

Austin, R.W. (1982) 18. NET Meeting, NASA/GSFC, Jan 1982.

Ball Aerospace Systems Division, 1979, Development of the Coastal Zone Color Scanner for Nimbus 7, Vol.1,2 , NASA Final Report F78-11, Rev. A.

Bricaud, A. and Morel, A. (1987) 'Atmospheric corrections and interpretation of marine radiances in CZCS imagery: use   of a reflectance model', Oceanologica Acta 1987, No.SP, 33-49.

Gordon, H.R. (1978) 'Removal of atmospheric effects from satellite imagery of the oceans', Applied Optics 17(10), 1631-1636.

Gordon, H.R. and Clark, D.K. (1981) 'Clear water radiances for atmospheric correction of Coastal Zone Color Scanner imagery', Applied Optics 20, 4175-4180.

Gordon, H.R., Clark, D.K., Brown, J.W., Brown, O.B., Evans, R.H., Broenkow, W.W. (1983) 'Phytoplankton pigment concentrations in the Middle Atlantic Bight: comparison   of ship determinations and CZCS estimates',Applied Optics 22, 20-36.

Gordon, H.R., Brown, J.W., Evans, R.H. (1988) 'Exact Rayleigh scattering calculations for use with the Nimbus-7 Coastal Zone Color Scanner', Applied Optics 27 (5), 862-871.

Gordon, H.R., Brown, J.W., Brown, O.B., Evans, R.H., Clark, D.K. (1983) 'Nimbus 7 CZCS: reduction of its radiometric sensitivity with time', Applied Optics 22 (24), 3929-3931.

Hering, W.S., Austin, R.W., McGlamery, B.L. (1984) 'Extended Analysis of CZCS Radiometric Sensitivity Decay, Preliminary Report, SIO, 1985.

Hovis, W.A., Clark, D.K, Anderson, F., Austin, R.W., Wilson, W.H., Baker, E.T., Ball, D., Gordon, H.R., Mueller, J.L., ElSayed, S.Z., Sturm, B., Wrigley, R.C.B, Yentsch, C.S. (1980) 'Nimbus-7 coastal zone color scanner: System description and initial imagery', Science 210, 60-63.

Hovis, W.A., Knoll, J.S., Smith, G.R. (1985) 'Aircraft measurements for calibration of an orbiting spacecraft sensor', Applied Optics 24 (3), 407-410.

Kriebel, K.T., Quenzel, H., Scholze, W. (1976) 'On the Determination of the Atmospheric Air Light as a normally underestimated Perturbation Signal', Atmospheric Environment 10, 645-653.

McGlamery, B.L. (1984) 'CZCS Sensor Decay Study', NET Meeting, Aug.1984.

Mueller, J.L. (1984) 'Radiometric Sensitivity Decay of the Nimbus-7 Coastal Zone Color Scanner (CZCS) Determined Using Data from the Central NE Pacific Ocean', NET Meeting AUG. 1984.

116

Morel, A. (1988) 'Optical modeling of the upper ocean in relation to its biogeneous matter content (Case I waters)', Journal of Geophysical Research 93 (C9), 10749-10768.

Smith, R.C. and Wilson, W.H. (1981) 'Ship and Satellite Bio-Optical Research in the California Bight', in: J.F.R.Gower (ed.) Oceanography from Space, Plenum Press 1981, pp. 281-294.

Sobolev, V.V. (1975) 'Light Scattering in Planetary Atmospheres', Pergamon Press 1975, pp. 256.

Sturm, B. (1986) 'Correction of the sensor degradation of the Coastal Zone Color Scanner on Nimbus-7', in Europe from Space, ESA SP-258, ESA-EARSeL Proceedings, Lyngby, Denmark June 1986.

Tanre, D., Herman, M., Deschamps, P.Y., deLeffe, A. (1979) 'Atmospheric modeling for space measurements of ground reflectances, including bidirectional properties', Applied Optics 18 (21), 3587-3594.

Viollier, M. (1982) 'Radiometric calibration of the Coastal Zone Color Scanner on Nimbus 7: a proposed adjustment', Applied Optics 21, 1142-1145.

Viollier, M., Sturm, B. (1984) 'CZCS Data Analysis in Turbid Coastal Water', Journal of Geophysical Research 89 (D4), 4977-4985.

THE COASTAL ZONE COLOR SCANNER (CZCS) ALGORITHM. A CRITICAL REVIEW OF RESIDUAL PROBLEMS

J. J. SIMPSON
Satellite Oceanography Center
Scripps Institution of Oceanography
University of California at San Diego
La Jolla, CA 92093
USA

ABSTRACT. Computation of phytoplankton pigment concentrations from radiances received by the Coastal Zone Color Scanner (CZCS) is a complex process. Water-leaving radiances, a small portion of the total signal, control the value of pigment computed, but fully 80 to 90 % of the total radiance is due to Rayleigh and aerosol scattering. These latter effects must be accurately modeled if reliable pigment concentrations are to be retrieved from CZCS data. In addition, prior to pigment retrieval a series of pre-processing steps must be performed to: (1) correct for instrumentation problems (*e.g.* drop-outs, zero counts, electronic overshoot); and (2) eliminate clouds from the scene. This paper critically reviews all of the aforementioned factors involved in the valid extraction of phytoplankton pigment concentrations from CZCS data. Also, several possible new approaches for improving the basic closure of the radiative transfer model used in CZCS pigment extraction algorithms (i.e., more robust approaches to the aerosol radiance terms) are introduced. Finally, based upon the community-wide CZCS experience, a few guidelines are suggested for the calibration and timely distribution of ocean color data from future generation sensors.

1. Introduction

Accurate climate prediction can be achieved only with models which treat the ocean, atmosphere and land as a synchronously coupled system, *i.e.* each component simultaneously influences the other two in a complex network of energy, moisture, momentum and CO_2 feedback loops (Ramanthan, 1987). Climate changes, for example, are highly correlated with the amount of primary productivity and the type, state, density and space-time distribution of oceanic (terrestrial) plant biomass because each of these vegetation parameters directly affects the CO_2 flux between ocean (land) and atmosphere and the global radiation balance (*e.g.* Simpson, 1985; Blumel and Tonn, 1986). Thus, useful climate models and analyses require accurate assessments of both oceanic and terrestrial plant biomass and its rate of change on global scales. Satellites provide the only feasible way to collect these types of data on the spatial and temporal scales needed.

The Coastal Zone Color Scanner (CZCS) sensor measured total radiance at wavelengths in the visible region (centered at 443, 520, 550, and 670 nm with a 20 nm bandwidth in channels 1-4) and in the near infrared region (750

117

V. Barale and P.M. Schlittenhardt (eds.),
Ocean Colour: Theory and Applications in a Decade of CZCS Experience, 117–166.
© 1993 *ECSC, EEC, EAEC, Brussels and Luxembourg. Printed in the Netherlands.*

nm$\pm$50 nm in channel 5) of the electromagnetic spectrum. Unfortunately, a complicated algorithm is needed to convert these raw data to the water-leaving radiances from which phytoplankton-like pigment concentrations are computed.

Historically, most CZCS algorithms have approximated the total measured radiance L_t as the sum of the atmospheric scattering (Rayleigh and aerosol) and the attenuated water-leaving radiance

$$L_t(\lambda) = L_R(\lambda) + L_a(\lambda) + t(\lambda) L_W(\lambda) \tag{1}$$

where L_R, L_a, and L_W are the Rayleigh, aerosol and water-leaving radiances, respectively, and $t(\lambda)$ is the diffuse transmittance (attenuation) of the atmosphere between the ocean and the satellite (Gordon *et al.*, 1983a). Note carefully that Equation 1 ignores the possibility of combined Rayleigh and aerosol scattering. Moveover, equation 1 has two unknowns: L_a and L_W (L_R can be estimated theoretically with some degree of accuracy). To extract the water-leaving radiance (the signal) from atmospheric and instrumentation effects (the noise), any algorithm must use empirical and theoretical relations with many explicit and implicit assumptions. Even larger errors in derived pigment concentration are likely to occur because $L_W(\lambda)$ is usually only 10-20% of the total radiance and pigment content is proportional to a power of ratios of water-leaving radiances (*e.g.* Eckstein and Simpson, 1991a).

The purpose of this paper is to review the historical processing procedures used to extract oceanic plant pigment concentration from CZCS data and to identify unresolved, critical issues whose solutions could lead to improved pigment estimates from future ocean color sensors (*e.g.* the Sea-Viewing Wide Field of View Sensor (SeaWIFS), the Advanced Earth Observing Satellite (ADEOS) with core sensors like the Ocean Color and Temperature Sensor (OCTS), the Medium Range Imaging Spectrometer (MERIS), the Moderate Resolution Imaging Spectrometer (MODIS), and the Reflective Optics Imaging Spectrometer (ROSIS). Moreover, a scenario is developed for retrospectively reprocessing CZCS data with the hope of obtaining a significantly improved global ocean color data set for immediate use and for comparison with future global ocean color data sets. Finally, a few recommendations are made concerning the calibration and near-real time distribution of future ocean color data sets for subsequent use by the global oceanographic community.

2. Problem Identification in the Historical CZCS Pigment Algorithm

CZCS data have been used to study the biological dynamics of coastal upwelling systems off the west coast of the United States (*e.g.* Peláez and McGowan, 1986) and off northwest Africa (*e.g.* Shannon, 1985), plankton distributions in the mid-Atlantic Bight (*e.g.* Gordon *et al.*, 1983a), mesoscale eddy entrainment processes (*e.g.* Haury *et al.*, 1986; Simpson *et al.*, 1986), El Niño-Southern Oscillation events and their impact on commercial fisheries (*e.g.* Fiedler, 1984; Simpson, 1992a) and seasonal and inter-annual variability in major oceanic current systems (*e.g.* Strub *et al.*, 1990). Almost all of these applications have used level_3 pigment products derived from noise-contaminated CZCS data using pigment extraction algorithms based upon equivalent assumptions. A schematic diagram of the pigment extraction algorithm typically used to retrieve level_3 pigment concentrations from level_1 CZCS data prior to 1985 is shown in figure 1. A second schematic diagram, reflecting post 1985 knowledge about sensor performance and atmospheric corrections, is shown in figure 2. Below, both the contributions to sensor noise and the underlying assumptions historically used in CZCS pigment extraction (*e.g.* figures 1 and 2) are identified.

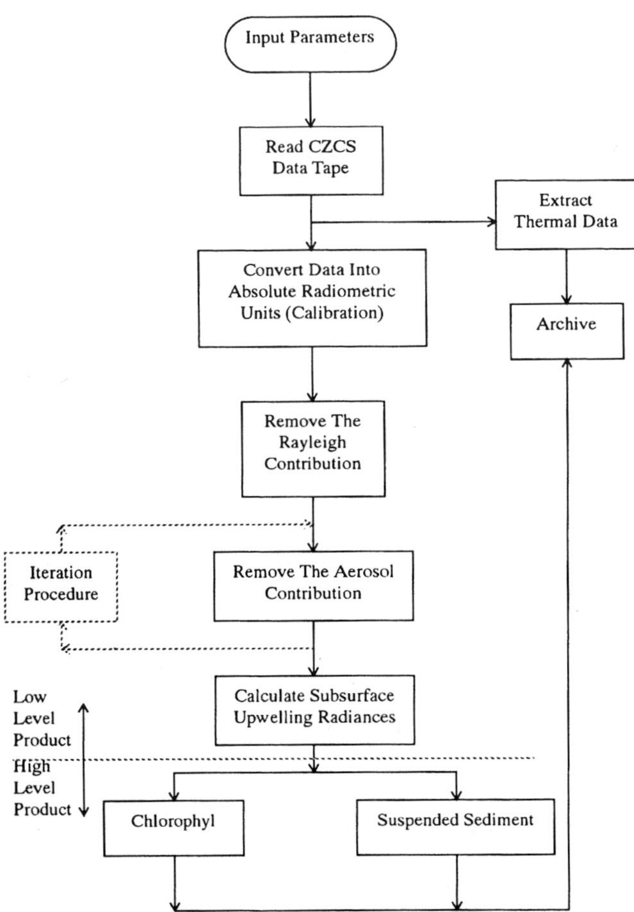

Figure 1. A schematic diagram of the plant pigment algorithm typically used to retrieve level_3 plant pigment and suspended sediment concentrations from level_1 CZCS data (from Walters, 1985). Note, not all of the errors in CZCS pigment processing discussed in the text were known at the time this figure was published by Walters. GOFS and GLOBEC concentrate on the high level product. This, however, requires accurate low level products.

120

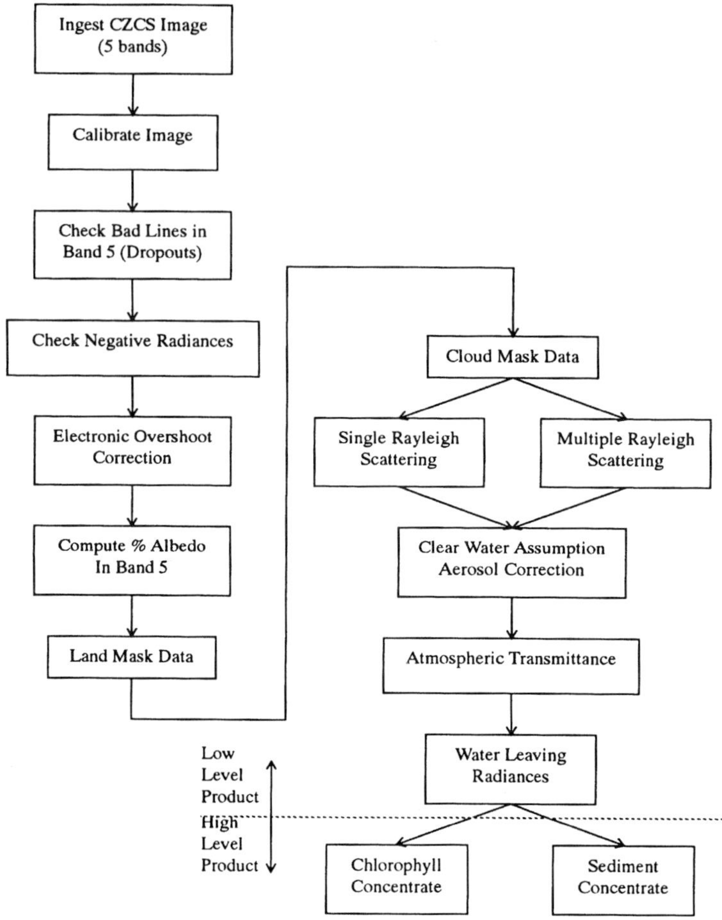

Figure 2. Analogous to figure 1, except incorporating post-1985 knowledge about CZCS sensor performance and atmospheric corrections. GOFS and GLOBEC concentrate on the high level product. This, however, requires accurate low level products.

2.1. INSTRUMENTATION PROBLEMS

2.1.1. *Sensor Drift.* Several instrumental problems developed over the lifetime of the CZCS sensor. Its sensitivity decreased with time, forcing an alteration in the initial calibration constants. Fiedler's (1986) correction formulae, which are similar to those of Mueller (1985) and Herring (1985), have been used in this study. These three corrections apply through mid-1984 and are improvements over Gordon *et al.*'s (1983b) equations, which are applicable only through mid-1982 because their correction for channel 1 breaks down.

2.1.2. Zero Count. Regression models generally were used to determine calibration coefficients for the various CZCS channels (*e.g.* Walters, 1985). For some of these channels (*e.g.* channel 5), the fitting equations have a negative slope. A zero count measurement, combined with a negative intercept for the calibration equation, will produce a negative radiance for that channel. Such data must be excluded from further analysis because negative radiances are physically meaningless.

2.1.3. Anomalously Low Values in Scan Lines. Anomalously low pixel values are routinely recorded at approximately every 16th line and may easily be corrected by linearly interpolating across adjacent lines in the count domain (Eckstein and Simpson, 1991b). These lines cannot be restored by a simple DC offset because the error is not always constant for a given line. The offset appears to fluctuate between one and two counts. These bad lines may be seen if channel 5 is displayed with the proper intensity and contrast settings. Operation of the sensor calibration lamp seems to be related to anomalously low recorded radiances (Mueller, 1989 personal communication). Note, interpolation must be done in the count domain to avoid digitization error.

2.1.4. Electronic Overshoot. Electronic overshoot occurs in channels 1-4 because of the slow response of the analog amplifiers when the sensor passes from bright clouds to open water (Mueller, 1988). Under these conditions, pixels east of the cloud edge will have unrealistically large radiances. Moreover, Eckstein and Simpson (1991b) have shown that electronic overshoot also occurs at land-ocean boundaries; the amount depends on the areal extent and brightness of land and clouds and on the crosstrack geometry of the individual CZCS pass.

2.2. LAND- AND CLOUD-MASKING

Land- and cloud-masking must be performed on a CZCS scene prior to other atmospheric corrections because: (1) the distribution (*e.g.* type, concentration) of terrestrial aerosols generally is quite different from that of marine aerosols; (2) undetected clouds can produce significant errors in derived water-leaving radiances; and (3) undetected clouds frequently inhibit the selection of pixels which might satisfy the Clear Water Assumption (see below). This latter issue is of special importance for the traditionally used CZCS pigment retrieval algorithms.

Historically, cloud-screening algorithms adopted for CZCS processing (*e.g.* Zion, 1983; Abbott, 1988; Strub *et al.*, 1990) have used a constant count threshold (T-model) in channel 5 to exclude clouds. For example, Mueller (1989, personal communication) uses a value of 18 counts to mask land and clouds, and lowers this to 14 counts at high solar zenith angles (*e.g.* for images of the Gulf of Alaska). An 18-count threshold translates to an albedo of $a \approx 1.6\%$ at small zenith angles. At this albedo threshold, the results of cloud masking are almost exactly the same whether or not the anomalously low radiance lines have been restored by interpolation. At lower thresholds, however, the T-model often produces supposedly uncloudy regions deep within clouds at the uncorrected bad lines (*e.g.* Eckstein and Simpson, 1991b). Fiedler (1986) use $\alpha \approx 1.3\%$ to distinguish clouds and land from ocean pixels. Eckstein and Simpson (1991b) have shown that threshold methods of cloud-screening generally produce unsatisfactory results when compared against more sophisticated techniques in a CZCS context.

Moreover, Simpson (1992b) has shown that cloud-masking does not necessarily exclude all land pixels from a scene. In fact, land-masking processes should be performed prior to cloud masking for optimal cloud masking in an arbitrary scene.

2.3. CLEAR WATER ASSUMPTION

The Clear Water Assumption was developed by Gordon and Clark (1981) to determine one unknown in Eq. 1, the water-leaving radiance L_W, at low pigment concentrations. Then, given a known Rayleigh radiance, Eq. 1 may be solved for L_a, the aerosol radiance:

$$L_t(\lambda) = L_R(\lambda) + L_a(\lambda) + t(\lambda) L_W(\lambda) \tag{1}$$

This assumption is critical to CZCS processing because historically it has been the only method of finding L_a.

For low pigment concentrations, the water-leaving radiance L_W is fairly constant for channels 2-4 (Gordon and Clark, 1981). In general, for a solar zenith angle θ_0, L_W is given by

$$L_W(\lambda) = t_0(\lambda) L_{WN} \cos\theta_0 \tag{2}$$

$$t_0(\lambda) = exp\left[\frac{-\left(\frac{1}{2}\tau_R(\lambda) + \tau_{OZ}(\lambda) + \left(1 - \omega_a(\lambda) F(\lambda)\right)\tau_a(\lambda)\right)}{\cos\theta_0}\right] \tag{3}$$

where L_{WN} is the normalized water-leaving radiance, t_0 is the diffuse transmittance of the atmosphere between the sun and the ocean surface, τ_R, τ_{OZ}, and τ_a are the Rayleigh, ozone, and aerosol optical depths respectively, ω_a is the aerosol single-scattering albedo, and F is the probability that a photon will be scattered through an angle $< 90°$ by aerosols. L_{WN} is almost independent of solar zenith angle:

$$L_{WN} = \left[\frac{\left(1 - \rho(\theta_0)\right) \cdot \left(1 - \rho(\theta)\right) R F_0}{n_r^2 Q}\right] \tag{4}$$

where $\rho(\theta)$ is the Fresnel reflectance of the sea surface at viewing angle θ, R is the reflectance of the ocean, n_r is the index of refraction of water, and F_0 is the instantaneous extraterrestrial solar radiance (Gordon and Clark, 1981; Robinson, 1985). Q accounts for the directional distribution of undersea radiation (Stewart, 1985). Typical empirically determined values of Q are 4.4 and 4.7, which is somewhat dependent on wavelength (Gordon and Clark, 1981). If the sea surface were a true Lambertian light source, Q would equal π. The Fresnel reflectance is

$$\rho(\theta) = \frac{1}{2}\left[\frac{tan^2\left(\theta-\theta_r\right)}{tan^2\left(\theta+\theta_r\right)} + \frac{sin^2\left(\theta-\theta_r\right)}{sin^2\left(\theta+\theta_r\right)}\right] \tag{5}$$

where the angle of refraction θ_r between water and air is determined via Snell's law:

$$sin\theta_r = \frac{sin\theta}{n_r(\lambda)} \tag{6}$$

and the refractive index of seawater $n_r(\lambda)$ is wavelength dependent (Neumann and Pierson, 1966). Gordon and Clark (1981) assume that L_{WN} has little directional dependence. When weighted by the response of the CZCS sensor, L_{WN} is 0.498, 0.30, and < 0.015 mW/cm^2 μm sr in channels 2, 3, and 4, respectively.

The Clear Water Assumption is difficult to use for channel 1 because L_W (443) is highly dependent on pigment concentration. If pigment content is already known (which is generally not the case), one can find L_W(443) from Fig. 2 of Gordon and Clark (*e.g.* Gordon *et al.*, 1983a). An uncertainty of only 0.05 mg/m^3, however, may produce errors as much as 0.17 mW/cm^2 μm sr for pigment greater than 0.15 mg/m^3 and 0.38 mW/cm^2 μm sr for pigments less than 0.10 mg/m^3 when this method is used (Eckstein and Simpson, 1991a). Alternatively, one can use Austin and Petzold's (1981) empirical formula relating L_W(443) to L_W(520) and L_W(550):

$$\frac{L_W(443)}{L_W(520)} = 0.900\left[\frac{L_W(443)}{L_W(550)}\right]^{0.742} \tag{7}$$

which translates to

$$L_W(443) = 0.665\, L_W(520) \cdot \left[\frac{L_W(520)}{L_W(550)}\right]^{2.876} \tag{8}$$

Austin and Petzold evaluated Eq. 7 over several orders of magnitude for the ratios on the left- and right-hand sides of the equation, yielding 0.985 as a coefficient of determination. Hence, without restrictions on pigment content, one can use Eq. 2 with L_{WN} equal to 0.498 and 0.30 to find $L_W(520)$ and $L_W(550)$, and then Eq. 8 to determine $L_W(443)$ with sufficient reliability. Note, from a radiative transfer point of view, the effect of the Clear Water Assumption on the retrieval of plant pigment concentrations from CZCS data is seen most clearly by close examination of equations 17-21 given below.

2.4. ATMOSPHERIC EFFECTS

The water-leaving radiance L_W (Eq. 1) is the radiance obtained once the Rayleigh and aerosol contributions (L_R and L_a) have been subtracted from L_t and corrected for atmospheric attenuation $t(\lambda)$. These effects depend on cross-track geometry and optical depths. Both $t(\lambda)$ and L_R have theoretically-derived expressions. Unfortunately, L_a must be empirically determined.

2.4.1. Atmospheric Attenuation. The atmosphere attenuates the water-leaving radiance $L_W(\lambda)$ by a factor of $t(\lambda)$, the diffuse transmittance of the atmosphere between the sea surface and the sensor:

$$t(\lambda) = exp\left[-\left(\frac{1}{2}\tau_R(\lambda) + \tau_{OZ}(\lambda)\right)\Big/ cos\theta\right] t_a(\lambda) \tag{9}$$

where θ is the zenith angle of a vector from the point on the sea surface (pixel) to the satellite (Gordon *et al.*, 1983a). The aerosol transmittance t_a is

$$t_a(\lambda) = exp\left[-\tau_a(\lambda)\left(1 - \omega_a(\lambda) F(\lambda)\right)\Big/ cos\theta\right] \tag{10}$$

Gordon *et al.* (1983a) state that t_a is only weakly dependent on aerosol optical depth because $[1-\omega_a(\lambda)F(\lambda)] \leq 1/6$. They also set t_a equal to unity because the entire algorithm will fail before t_a becomes large enough to significantly affect the product $t(\lambda)L_W(\lambda)$ in Eq. 1 (*e.g.* Eckstein and Simpson, 1991a).

2.4.2. Rayleigh Scattering: Single Scattering Model. Traditionally, single-scattering theory has been used to compute L_R, the Rayleigh component of the total radiance (*e.g.* Gordon *et al.*, 1980; Guan *et al.*, 1985; Mueller, 1985; Walters, 1985; Pelaez and McGowan, 1986). Multiple Rayleigh scattering computations of L_R (Gordon *et al.*, 1988a), however, can differ significantly from those produced by the single-scattering approximation. L_R^s is determined by

$$L_R^s = \frac{\tau_R(\lambda) F_0{}'(\lambda)}{4\pi cos\theta}\left\{P\left(\Psi_-\right) + \left[\rho(\theta) + \rho(\theta_0)\right] P\left(\Psi_+\right)\right\} \tag{11}$$

where the superscript s refers to the single-scattering approximation and $P(\Psi_\pm)$, the Rayleigh phase functions (Gordon *et al.*, 1983a), are

$$P\left(\Psi_\pm\right) = \frac{3}{4}\left[1 + \cos^2\Psi_\pm\right] \tag{12a}$$

$$\cos\Psi_\pm = \pm\cos\theta\cos\theta_0 - \sin\theta_0\sin\theta\cos\left(\emptyset - \emptyset_0\right) \tag{12b}$$

Equation 12b is explained more fully in Appendix II of Eckstein and Simpson (1991a). In Eq. 12, Ψ_- is the angle between a vector from the sun to the pixel and a vector from the pixel to the sun, Ψ_+ is the angle between the reflected sunlight and the pixel-satellite vector, and $\emptyset$ and $\emptyset_0$ are the azimuth angles of vectors from the pixel to the satellite and sun respectively (Sturm, 1981).

$F_0'(\lambda)$ is the instantaneous extraterrestrial solar radiance $F_0(\lambda)$ reduced by two trips through the ozone layer (Gordon *et al.*, 1983a):

$$F_0{}'(\lambda) = F_0(\lambda)\exp\left[-\tau_{OZ}(\lambda)\left(\frac{1}{\cos\theta} + \frac{1}{\cos\theta_0}\right)\right] \tag{13}$$

and

$$F_0(\lambda) = \overline{F_0}(\lambda)\left[1 + e\cos\left(\frac{2\pi\left(D-3\right)}{365}\right)\right]^2 \tag{14}$$

where $e = 0.0167$ is the eccentricity of earth's orbit and D is the Julian day. $\overline{F_0}(\lambda)$ is the mean extraterrestrial solar radiance for a given wavelength λ. Until 1988, all algorithms set $\overline{F_0}(\lambda)$ equal to 186.42, 185.34, 184.76, and 151.52 mW/cm^2 µm sr for channels 1-4 respectively. In 1988, Gordon *et al.* (1988a) published new values of 189.96, 187.02, 186.81, and 153.09 mW/cm^2 µm sr.

2.4.3. *Rayleigh Scattering: Multiple scattering model.* Gordon *et al.* (1988a) have developed a method of computing multiple Rayleigh scattering. Under this formulation, the Rayleigh radiance is

$$L_R{}^m(\lambda) = F_0{}'(\lambda)I\left(\lambda,\theta,\theta_0,\Delta\emptyset\right) \tag{15}$$

$$I\left(\lambda,\theta,\theta_0,\Delta\emptyset\right) = \sum_{k=0}^{2}I_k\left(\lambda,\theta,\theta_0\right)\cos\left(k\Delta\emptyset\right) \tag{16}$$

where the superscript m refers to multiple-scattering theory, $\Delta\emptyset = \emptyset - \emptyset_0$ is the azimuth angle difference, $I(\lambda,\theta,\theta_0,\Delta\emptyset)$ is the Rayleigh coefficient, and $I_k(\lambda,\theta,\theta_0)$ is the Fourier coefficient of the Rayleigh radiance. Gordon *et al.* (1988a) have computed tables of I_k for $0 <\theta< 90°$ and $0 \le \theta_0 \le 76°$ in two degree increments, using the exact radiative transfer equation and the assumption that the ocean is a totally absorbing flat surface.
From these, $I(\lambda,\theta,\theta_0,\Delta\emptyset)$ is determined for any angles θ and θ_0 using bilinear interpolation.

Gordon *et al.* (1988a) use Rayleigh optical depths τ_R of 0.237, 0.123, 0.098, and 0.044 for channels 1-4. The tables of $I_k(\lambda,\theta,\theta_0)$ may be used at any season and latitude because the Rayleigh optical depth varies by no more than $\pm 0.8\%$ (based on Zion, 1983).

The ozone optical depth τ_{OZ} implicit in $F_0{}'(\lambda)$ changes with season and geographic location. From Zion's tables, $\tau_{OZ}(\lambda_i)$ can vary by as much as 50%, affecting L_R by up to $\sim 5\%$ (*e.g.* Eckstein and Simpson, 1991a). By comparing their algorithm against other Rayleigh radiance models, Gordon *et al.* (1988a) conclude that the multiple-scattering Rayleigh radiance $L_R{}^m$ is accurate to within $\sim 0.1\%$. Eckstein and Simpson (1991a) have raised several questions about the multiple-scattering Rayleigh radiances derived by Gordon *et al.* (1988a).

2.4.4. Aerosol Scattering. The aerosol component of the total radiance is much more difficult to compute than either $L_R(\lambda)$ or $t(\lambda)$ in Equation 1. Theoretically, an expression similar to Eq. 11 can be derived. The relevant optical parameters (*e.g.* aerosol optical depth, phase function, and albedo), however, are dependent on the aerosol type and concentration at any location and time.

Gordon *et al.* (1983a) developed a method of relating aerosol radiance in channel 4 ($\lambda = 670$ nm) to aerosol radiance in the other CZCS channels. Using the Clear Water Assumption, they note that the water-leaving radiance in channel 4 is essentially zero at low pigment concentrations ($< \sim 0.25$ mg/m^3). Then the aerosol radiance at the sensor in channel 4 is

$$L_a\left(670\right) = L_t\left(670\right) - L_R\left(670\right) \tag{17}$$

where $L_t(670)$ is measured by the sensor and $L_R(670)$ is computed from either a single or multiple Rayleigh scattering model. $L_a(670)$ is related to the aerosol radiances in the other CZCS channels via

$$L_a\left(\lambda_i\right) = S\left(\lambda_i, 670\right) \cdot L_a\left(670\right) \tag{18}$$

where

$$S\left(\lambda_i, 670\right) = \varepsilon\left(\lambda_i, 670\right)\frac{F_0{}'\left(\lambda_i\right)}{F_0{}'\left(670\right)} \tag{19}$$

and the epsilon factor $\varepsilon(\lambda_i, 670)$ is usually written as

$$\varepsilon\left(\lambda_i, 670\right) \approx \left[\frac{\lambda_i}{670}\right]^{n\left(\lambda_i\right)} \tag{20}$$

The exponent $n(\lambda_i)$ is called Angstrom's coefficient. Angstrom's coefficient is not well understood theoretically and may be positive or negative (Robinson, 1985). The water-leaving radiances in channels 2 and 3 do not vary greatly with wavelength and can be estimated from Gordon and Clark's work for Clear Waters (Eq. 2). Then one can extract the aerosol component of the signal and compute $\varepsilon(\lambda_i, 670)$, or Angstrom's coefficient. Gordon *et al.* (1983a) apply constant values of $(\lambda_i, 670)$ to an entire image. The practice of using constant values of ε for an entire image or set of images has been used extensively in the preparation of large-scale CZCS data sets (*e.g.* Abbott, 1988; Feldman *et al.*, 1989). Unfortunately, the use of constant ε values generally isn't appropriate for reasons cited below.

The water-leaving radiance in channel 1 is highly dependent on pigment content and cannot be accurately estimated from Gordon and Clark's relations. Gordon *et al.* (1983a) circumvent this problem by assuming

$$n\left(443\right) = \frac{1}{2}\left[n\left(520\right) + n\left(550\right)\right] \tag{21}$$

Eckstein and Simpson (1991a) note that a more robust method of estimating ε is to first use Eq. 8 to determine $L_W(443)$ and thus $L_a(443)$. Then $\varepsilon(443,670)$ and $n(443)$ may be determined from Eq. 18-21.

2.5. PIGMENT CONCENTRATION

Gordon *et al.* (1983a) compute pigment content from CZCS data using the general relation

$$P'_{ij} = X_{ij}\left[\frac{L_W\left(\lambda_i\right)}{L_W\left(\lambda_j\right)}\right]^{Y_{ij}} \tag{22}$$

where P' is the pigment concentration of an optically stratified ocean of depth z with *in situ* pigment content $P(z)$. The subscripts i and j in Eq. 22 refer to CZCS channels. From empirical determinations of X_{ij} and Y_{ij},

$$P'_{13} = 1.130\left[\frac{L_W\left(443\right)}{L_W\left(550\right)}\right]^{-1.71} \tag{23a}$$

$$P'_{23} = 3.327 \left[\frac{L_W(520)}{L_W(550)}\right]^{-2.44} \tag{23b}$$

where X_{ij} and Y_{ij} are valid over limited concentration ranges. Empirically, $P' = P'_{13}$ if $P'_{13} < 1.5$ mg/m^3 or if $P'_{13} > 1.5$ mg/m^3 but $P'_{23} < 1.5$ mg/m^3, and $P' = P'_{23}$ if $P'_{13} > 1.5$ mg/m^3 and $P'_{23} > 1.5$ mg/m^3. Gordon *et al.* (1983a) found no statistical difference between P' and *in situ* measured surface phytoplankton pigment concentration. Eckstein and Simpson (1991a), however, found significant differences in the values of plant pigment concentration retrieved from an arbitrary CZCS scene (*e.g.* figure 3); these differences depended greatly upon the nature of the pigment extraction algorithm used (*e.g.* single vs. multiple Rayleigh scattering).

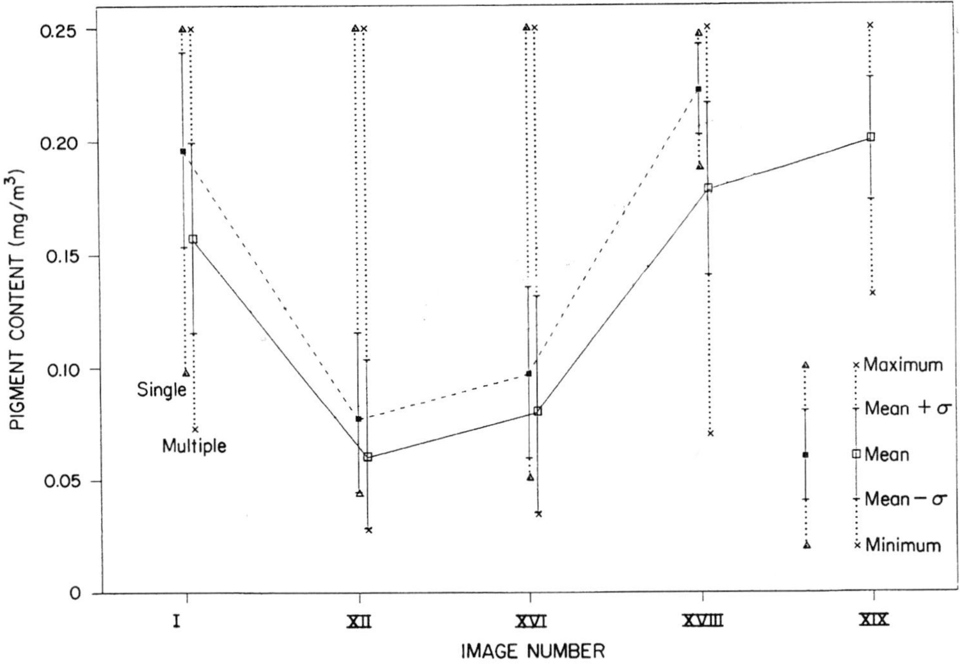

Figure 3. Plant concentrations (mg m^{-3}) retrieved from five CZCS images. Both single- and multiple-scattering approximations were used to compute Rayleigh radiance. Clear water areas are 7×7 pixel areas that pass the criteria for low pigment regions defined in the text. The symbol σ stands for standard deviation about the mean (from Eckstein and Simpson, 1991a).

3. Progress in Problem Areas

3.1. INSTRUMENTATION PROBLEMS

3.1.1. *Sensor Drift.* Degradation of sensor and sensor electronics is a fact of life for any instrument placed in orbit on board a satellite. The only practical solution to the general problem of sensor drift is to maintain an ongoing series of calibration experiments throughout the lifetime of the sensor. Depending upon the type of sensor, such calibration experiments may need to be performed 2 to 4 times per year; more often if non-linearities begin to develop in the sensor or sensor electronics. Moreover, such calibration experiments should be performed by several different groups located in different geographical areas. Such a procedure will help to minimize regional influences on accurately modeling sensor performance as well as prevent community dependence on the continued good performance of a single group.

3.1.2. *Negative radiances.* Negative radiances are invalid and must be excluded from analysis. Consider for example, channel 5. Raw digital counts $N(750)$ in channel 5 are converted to total radiances $L_t(750)$, in mW/cm^2 µm sr, using

$$L_t\left(750\right) = A\left(750\right) N\left(750\right) + B\left(750\right) \tag{24}$$

where $N(750)$ is the number of digital counts at 750 nm. The calibration constants A (slope) and B (intercept) are 9.416×10^{-2} mW/cm^2 µm sr per count and -2.825×10^{-2} mW/cm^2 µm sr respectively (Walters, 1985). The negative intercept value yields negative radiance when $N(750)$ is zero; a condition which is physically unrealistic. Negative radiances typically result from instrumentation noise.

For example, noise fluctuations may reduce very low radiances even further; to the point that the calibration of the pixels to the valid range of albedo is not possible. Also, least count digitization errors may invalidate the pixels with extremely low radiances. Careful screening of all calibrated pixels must be performed to insure that such pixels are set to zero radiance and are excluded from further analysis (Eckstein and Simpson, 1991b).

3.1.3. *Anomalously Low Values in Scan Lines.* Anomalously low values in scan lines must be removed. Failure to remove these data can result in compromised cloud screening and/or invalid pigment concentrations. Eckstein and Simpson (1991a) used manual visual enhancement of the CZCS scene to determine the location(s) of the scan lines with anomalously low values. While this procedure may be useful for processing a small number of CZCS scenes, it clearly is inadequate for processing global ocean color data sets.

More recent research indicates that a variety of statistical filtering procedures designed to eliminate isolated noise segments from data may prove useful in the above context. Definitions and examples of five such procedures are given in figure 4; their application to a CZCS scene is shown in figures 5 and 6. These attempts at isolated noise removal should be viewed as

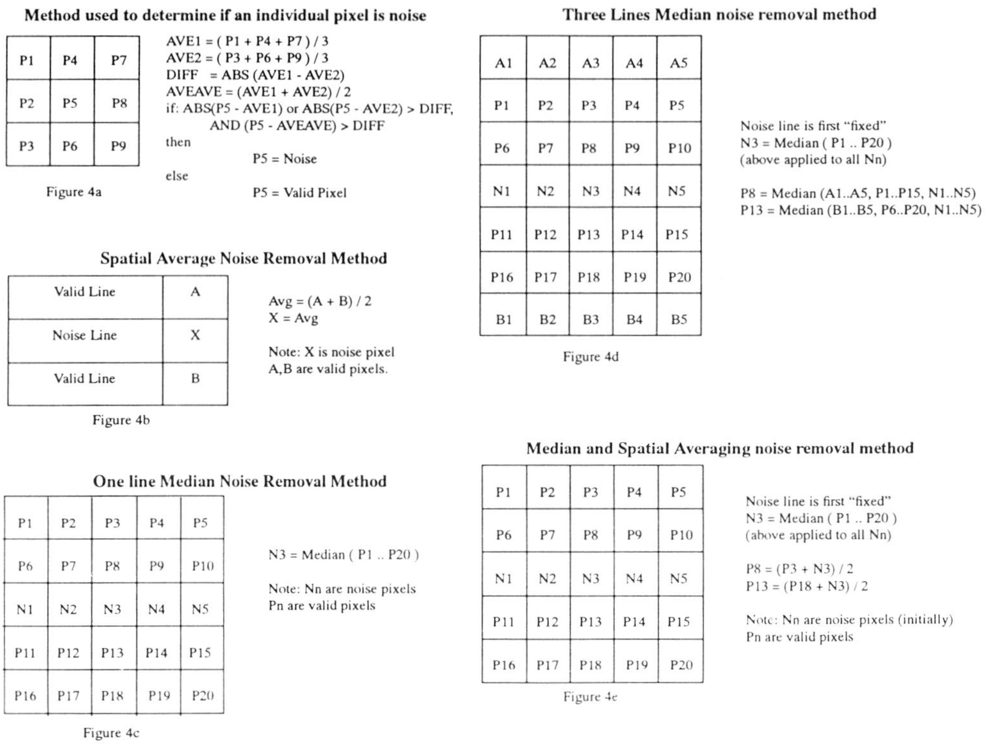

Method used to determine if an individual pixel is noise

P1	P4	P7
P2	P5	P8
P3	P6	P9

AVE1 = (P1 + P4 + P7) / 3
AVE2 = (P3 + P6 + P9) / 3
DIFF   = ABS (AVE1 - AVE2)
AVEAVE = (AVE1 + AVE2) / 2
if: ABS(P5 - AVE1) or ABS(P5 - AVE2) > DIFF,
         AND (P5 - AVEAVE) > DIFF
then
         P5 = Noise
else
         P5 = Valid Pixel

Figure 4a

Three Lines Median noise removal method

A1	A2	A3	A4	A5
P1	P2	P3	P4	P5
P6	P7	P8	P9	P10
N1	N2	N3	N4	N5
P11	P12	P13	P14	P15
P16	P17	P18	P19	P20
B1	B2	B3	B4	B5

Noise line is first "fixed"
N3 = Median (P1 .. P20)
(above applied to all Nn)

P8 = Median (A1..A5, P1..P15, N1..N5)
P13 = Median (B1..B5, P6..P20, N1..N5)

Figure 4d

Spatial Average Noise Removal Method

Valid Line	A
Noise Line	X
Valid Line	B

Avg = (A + B) / 2
X = Avg

Note: X is noise pixel
A,B are valid pixels.

Figure 4b

One line Median Noise Removal Method

P1	P2	P3	P4	P5
P6	P7	P8	P9	P10
N1	N2	N3	N4	N5
P11	P12	P13	P14	P15
P16	P17	P18	P19	P20

N3 = Median (P1 .. P20)

Note: Nn are noise pixels
Pn are valid pixels

Figure 4c

Median and Spatial Averaging noise removal method

P1	P2	P3	P4	P5
P6	P7	P8	P9	P10
N1	N2	N3	N4	N5
P11	P12	P13	P14	P15
P16	P17	P18	P19	P20

Noise line is first "fixed"
N3 = Median (P1 .. P20)
(above applied to all Nn)

P8 = (P3 + N3) / 2
P13 = (P18 + N3) / 2

Note: Nn are noise pixels (initially)
Pn are valid pixels

Figure 4e

Figure 4. a) Isolated noise detection procedure. Various methods of noise removal as defined in the different panels (b-e).

preliminary in nature. These initial results, however, indicate that a fully automated, statistically robust procedure for removing anomalously low (or high) values from scan lines can be developed. Such a procedure would greatly benefit automated efforts to reprocess the global and/or regional CZCS data and may prove useful for future ocean color sensors such as SeaWIFS and OCTS should similar instrumentation errors develop with these newer sensors.

3.1.4. Electronic Overshoot. The effects of electronic overshoot decrease with increasing wavelength (*i.e.* with increasing channel number); channel 5 is virtually unaffected by electronic overshoot (Mueller, 1988). Channel 5 radiances, however, provide information on those pixels in channels 1-4 which have been corrupted by electronic overshoot.

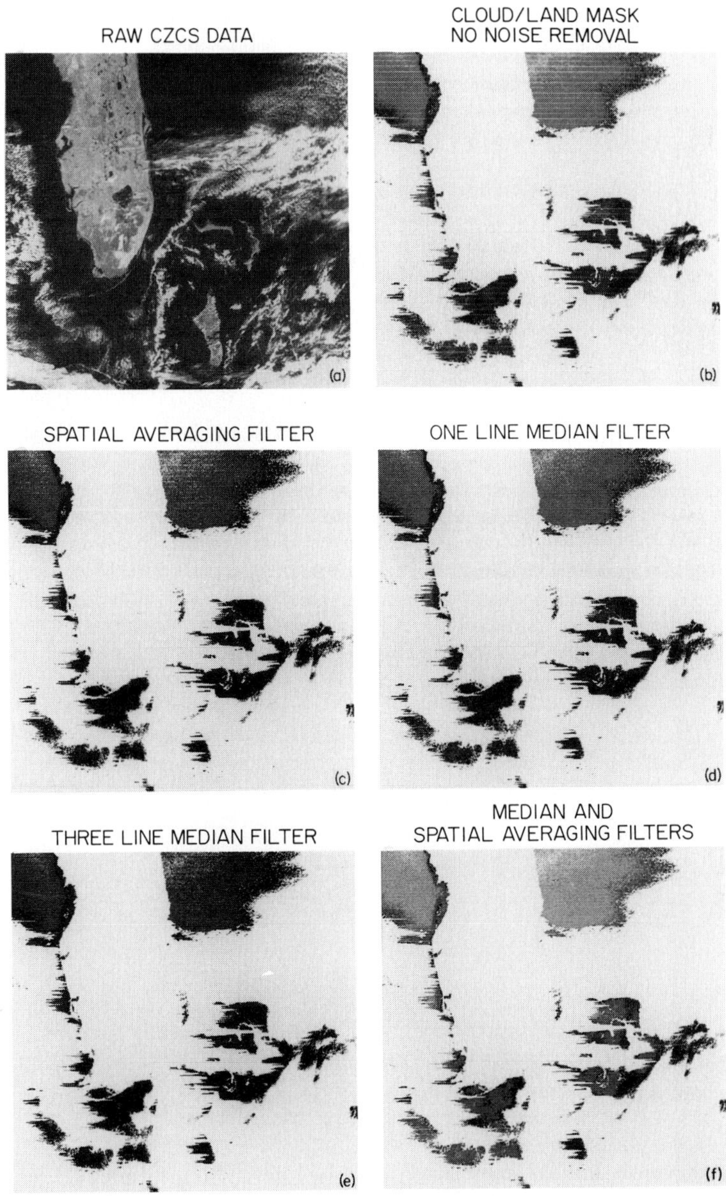

Figure 5. a) Raw CZCS scene off the coast of Florida; b) land- and cloud-masked version of (a) with no isolated noise removal; Land- and cloud-masked versions of (a) with different forms of noise removal: (c) spatial averaging filter; (d) one line median filter; (e) three line median filter; and (f) median and spatial averaging filters.

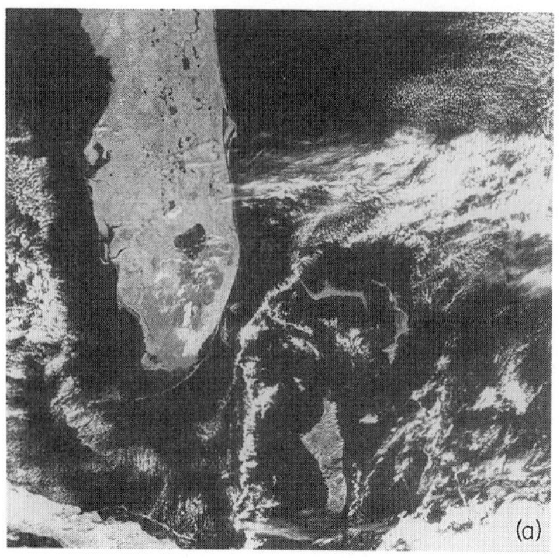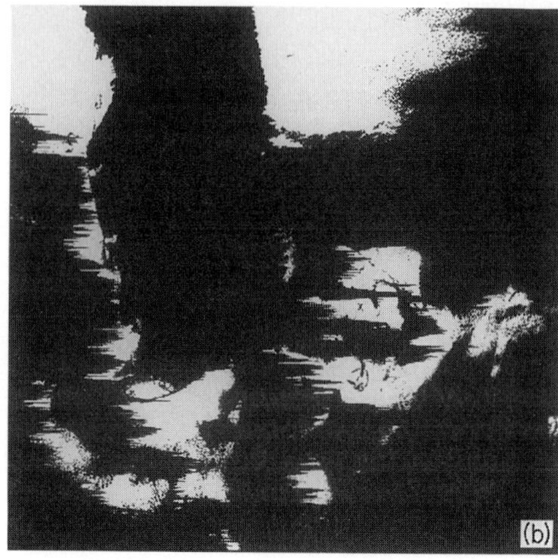

Figure 6. a) Raw CZCS image off the coast of Florida; b) Land- and cloud-masked version of (a). Isolated noise removal performed prior to land- and cloud-masking. Black areas are pixels screened by land- and cloud-masking.

Mueller's equation for electronic overshoot is

$$Y_c = 3.9\left[\pm 9.7\right] + 30.8\left[\pm 3.3\right]\left\{ ln\left[\sum_{i=1}^{5}\overline{B}_i\, e^{-0.32i}\right] + InG\right\} \tag{25}$$

$$\overline{B}_i = \frac{1}{10}\sum_{j=0}^{9} B\left(y_e - 10i + j\right) \qquad i = 1, ..., 5 \tag{26}$$

$$B = \begin{cases} L_t(750) - L_0(750)\Big/ G & L_t(750) > L_0(750)\Big/ G \\ 0 & L_t(750) \le L_0(750)\Big/ G \end{cases} \tag{27}$$

where Y_c is the number of corrupted pixels east of the cloud edge; y_e is the sample number of the edge of the cloud; B is the brightness of the cloud above the overshoot threshold; G is the gain fac tor or relative radiometric intensity, *i.e.* the radiance change per count relative to the radiance change per count at the lowest gain setting; and $L_t(750)$ is the total radiance at the sensor in

133

nnel 5. In Eq. 26, $B(y_e - 10i + j)$ refers to the brightness at sample number $(y_e - 10i + j)$. For sensor gains 0, 1, 2, and 3, G is 1.00, 1.25, 1.50, and 2.10 respectively. (Note, gains reported here are given in Scripps Satellite Oceanography Center ZIP format; their correspondence to NASA Goddard Space Flight Center's CRT format is given in the Appendix of Eckstein and Simpson (1991b)). As a worst case, Y_c is

$$Y_c = 13.6 + 34.1 \left\{ In \left[\sum_{i=1}^{5} \overline{B}_i \, e^{-0.32i} \right] + lnG \right\} \tag{28}$$

In Eq. 27, $L_0 = 2.45$ mW/cm^2 µm sr. This is the radiance threshold at 750 nm (Mueller, 1988). From Eq. 27, we conclude that pixels with radiances above this threshold indicate the presence of electric overshoot in channels 1-4 because larger radiances in channel 5 are associated with larger radiances in channels 1-4.

At low brightness sums, the logarithmic coefficient in Eq. 28 may be negative. This can produce negative values for Y_c, but the physics of the problem demand that Y_c always be positive. This discrepancy occurs because of the failure of the logarithmic equation to determine the origin, *i.e.* the point at which $Y_c = 0$. If negative Y_c values are set to the threshold of 13.6, the resultant image may have many short line segments of corrupted data resulting from only one or two bright pixels. Eckstein and Simpson (1991b) chose to set Y_c equal to zero whenever it assumes a negative value. Mueller (personal communication) concurs with this constraint, which indicates that the threshold L_0 is too low.

Electronic overshoot may contaminate a large portion of an image; the amount depends on the areal extent and brightness of land and clouds and on the cross-track geometry of the individual CZCS pass. Eckstein and Simpson (1991b) show that electronic overshoot due to land masses is likely to be a more serious problem for western boundary currents than for eastern boundary currents; a result consistent with CZCS amplifier behavior and Nimbus cross-track geometry. In fact, all their examples show that ocean data east of land masses were corrupted by overshoot, whereas, only two of their images contained any zero count measurements.

Finally, it is important to recognize that problems associated with electronic overshoot are likely to manifest themselves in future high gain ocean color sensors (*e.g.* SeaWIFS, OCTS, MERIS, ROSIS, MODIS). In the case of CZCS, the equations used to model electronic overshoot (equations 25-28) were limited in accuracy because the number of data points (hence statistical degrees of freedom) was relatively small. Future ocean color missions, such as SeaWIFS and OCTS, should develop careful protocols for tracking amplifier response characteristics at land-ocean and cloud-ocean boundaries so that statistically improved models of electronic overshoot can be used to rigorously eliminate contaminated data from further analysis.

3.2. LAND- AND CLOUD-MASKING

3.2.1. *Conversion of Digital Counts to Percent Albedo.* Prior to land-masking and cloud-screening, pixels with positive radiances in CZCS channel 5, which

have successfully passed the criteria discussed in III.A, should be converted to percent albedo a using

$$a = \frac{L_t\,(750)}{t(\theta).L_i} \times 100\%$$

(29)

where $t(\theta)$ is the diffuse transmittance between the reflective surface and the sensor, L_t is the total radiance at the sensor, L_i is the light incident on the reflector (Fiedler, 1988, personal communication), and θ is the satellite zenith angle with respect to the reflector. If aerosol scattering is negligible,

$$t(\theta) = exp\left[-\left(\frac{1}{2}\tau_R + \tau_{OZ}\right)\cdot\frac{1}{cos\theta}\right]$$

(30)

where τ_R and τ_{OZ} are the Rayleigh and ozone optical depths respectively (Gordon $et\ al.$, 1983). The incident light L_i is

$$L_i = F_0\,.t\left(\theta_0\right)$$

(31)

$$t\left(\theta_0\right) = exp\left[-\left(\frac{1}{2}\tau_R + \tau_{OZ}\right)\cdot\frac{1}{cos\theta_0}\right]$$

(32)

where $t(\theta_0)$ is the diffuse transmittance of the atmosphere between the sun and the reflector, θ_0 is the solar zenith angle, and F_0 is the solar radiance:

$$F_0 = \overline{F_0}\cdot\left[1 + e\,cos\left(2\pi(D-3)\Big/365\right)\right]^2$$

(33)

e and D are defined as in Eq. 14, and $F_0 = 127.50$ mW/cm^2 μm sr is the mean extraterrestrial solar radiance at 750 nm (Sturm, 1981; Fiedler, 1986).

3.2.2. *Land masking.* Precision land-masking should be done prior to cloud-screening CZCS data for two reasons: (1) cloud-screening algorithms which use high vs. low levels of brightness to discriminate land from clouds invariably will allow some land pixels to pass undetected because landform relief often shades some pixels in an image (*e.g.* Simpson and Humphrey, 1990); and (2) the performance of unsupervised classification schemes (*e.g.* cloud detection) is significantly enhanced if the statistical dimensionality of the data set is minimized prior to the classification (*e.g.* Gallaudet and Simpson, 1991). Land-masking prior to cloud-masking satisfactorily handles both of these issues.

Simpson (1992b) developed procedures for land-masking both simply- and multiply- connected regions using polygon fills and morphological transformations. Moreover, this study also shows that errors in Geographical Information System (GIS)-derived coastlines, often used in landmasking, can be corrected using morphological transformations. Application of these techniques to a multiply-connected region such as the Gulf of California (figure 7) illustrates their use with Advanced Very High Resolution Radiometer (AVHRR) data. These same methods can be applied to CZCS data without modification. It is important to emphasize that rigorous land-masking also simplifies subsequent pigment retrieval from CZCS data because most pixels contaminated by large concentrations of terrestrial aerosols are eliminated from further consideration.

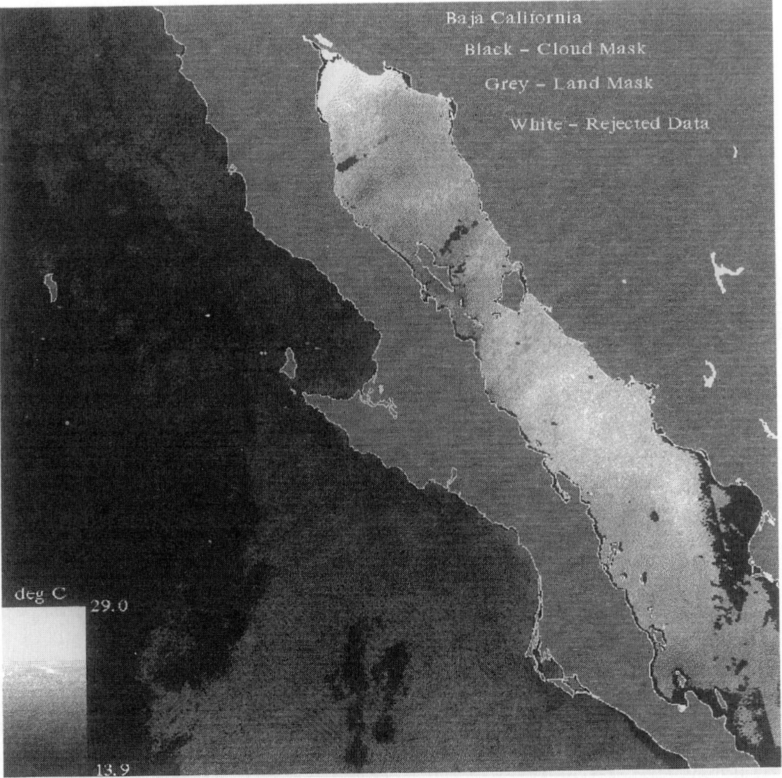

Figure 7. Land- and cloud-masked AVHRR channel 4 image showing the Pacific Ocean, Baja California, the Gulf of California, and part of the Mexican mainland. Landmasks are based on polygon fill operations and morphological transformations. Set operations are given by Simpson (1992a). Black implies cloud over open ocean, gray is land and coastlines are clearly marked (from Simpson, 1992b). These same operations can be used with CZCS data without modification.

3.2.3. *Cloud Masking.* A typical CZCS scene of an oceanic coastal area includes land, ocean, and clouds of various types. Thus, an accurate segmentation (*i.e.*, separation of clouds, land, and ocean in the CZCS scene) must be performed prior to meaningfully computing quantities of geophysical interest (*e.g.*, the water-leaving radiances, plant pigment concentration, sediment concentration) from satellite data. Traditionally, threshold models of albedo have been used to screen clouds from CZCS data using channel 5. Eckstein and Simpson (1991b) have shown that threshold models of albedo produce relatively poor cloud masks for CZCS data and that the value of the constant threshold used in such models is subjectively determined. To circumvent these problems associated with traditional cloud screening techniques, Eckstein and Simpson (1991b) preprocessed channel 5 CZCS data as outlined above and then successfully applied the Local Dynamical Threshold Non-Linear Rayleigh (LDTNLR) algorithm developed for AVHRR data by Simpson and Humphrey (1990) to CZCS data. The LDTNLR model for AVHRR data (channel 2, $\lambda = 910 \pm 190$ nm) may be applied to CZCS data (channel 5, $\lambda = 750 \pm 50$ nm) because both sensors have similar cross-track geometry and these bands have overlapping wavelengths. Below, a brief description of the LDTNLR algorithm as developed for AVHRR data is presented; details are given by Simpson and Humphrey (1990). Application of a slightly modified version (*i.e.* omit use of thermal channel filters) of the LDTNLR algorithm for use with CZCS data is shown as an example.

3.2.4. *Cloud Masking : The Local Dynamic Threshold Non-Linear Rayleigh (LDTNLR) Algorithm.* A combination of visible and infrared data is used. Albedo is defined as the ratio of the effective radiance as seen by channel 2 in the spectral band (0.725 - 1.10 µm) to the effective radiance of the radiometer viewing reflected sunlight (NOAA, 1985). An empirical model of albedo as a function of solar zenith angle θ_s is computed. Note, θ_0 (CZCS literature) corresponds to θ_s (AVHRR literature). This requires that the pertinent sun-pixel-satellite angles be computed for each pixel in the image. Next, a subset of albedo values is made by sampling AVHRR channel 2 along the image diagonals and at equally spaced grid locations throughout the image. This grid is proportional to image size. Hereinafter, the population of data values subsampled along the image diagonals and at the grid points is called the initially subsampled data set. The initially subsampled data set provides a large number of data values which includes the range of sun-pixel-satellite angles associated with that image. This minimizes errors which might result from an incomplete representation of the range of sun-pixel-satellite angles in a given image. Then an iterative procedure is used to develop a model of cloud-free albedo as a function of solar zenith angle ($a_1 = F_1(cos\ \theta_s)$). This procedure requires an initialization constant a_0 which is objectively and automatically determined. Solar zenith angle (or its complement, solar altitude) was chosen as the independent variable for this empirical model because observations (Payne, 1972; Simpson and Paulson, 1979) have shown that albedo can be parameterized well by solar altitude for a given value of atmospheric transmittance.

Dynamic local rejection of cloud-contaminated pixels is now done using this empirical model. A localized albedo threshold is computed from the empirical model ($a_1 = F_1(cos\ \theta_s)$) on a pixel-by-pixel basis. Thus each pixel in the image has its own rejection criterion defined by its particular cross-track scan geometry within the image. Then a nonlinear statistical model of cloud-free albedo as a function of Rayleigh scattering cross-section ($a_2 = F_2(R)$) is developed.

Pixel rejection based upon albedo from channel 2 uses a combination of these empirical and theoretical criteria. The empirical criteria require that each pixel in the image (channel 2) must satisfy the condition $0.0 < a \leq a_1$. The theoretical criteria require that each pixel in the image (channel 2) must satisfy the condition $a_2 - \sigma \leq a < a_2 + \sigma$, where σ is the standard deviation of the cloudfree data set obtained from the F_1 criteria. A pixel is retained as valid (uncontaminated) if its channel 2 albedo satisfies either of these constraints.

Clouds in the shadow of other clouds may have low radiances similar to those of the sea surface. Hence use of channel 2 alone to mask clouds may be insufficient for some images. For the AVHRR case, these shadow areas, however, can be identified with channel 4 data. All channel 4 pixels whose corresponding channel 2 albedos satisfy the above LDTNLR criteria are then highpass filtered to exclude pixels whose temperatures are below -2.0 °C. This minimum value of temperature is consistent with the minimum global SST (U. S. Navy, 1981). A small negative SST value is possible because of the effect of the saline coefficient of contraction on the freezing point of salt water (Neumann and Pierson, 1966). Optimization of the final cloud-masked SST is achieved by filtering this data set with a band-pass filter whose width is 4 standard deviations about the mean of the high-passed SST image. These band 4 criteria have little overall effect (typically, $< 1\%$) on total pixels masked (Simpson and Humphrey, 1990). Because band 6 on the CZCS sensor failed soon after launch, these last two SST filters are not invoked by the LDTNLR algorithm when cloud-screening CZCS data.

The cloud-screening ability of the LDTNLR algorithm was compared in detail with several other methods currently used with AVHRR data. Random worldwide AVHRR images (*i.e.*, California Current, Kuroshio Current, Mediterranean Sea, Falkland Current, the North Atlantic Ocean and the Indian Ocean) were selected for these intercomparisons. Results are presented in Simpson and Humphrey (1990).

The LDTNLR method offers a totally automated, statistically reproducible procedure to cloud-screen daytime AVHRR data which eliminates subjective inputs and subsequent performance evaluation related to questions concerning subjective inputs. The LDTNLR method uses empirical and theoretical criteria for pixel rejection, computed on a pixel-by-pixel basis because traditionally used static radiance thresholds do not take into account the natural variability in the reflectance of cloud-free pixels having different sun zenith angles in the cross-track scan geometry of a particular image. The LDTNLR method, unlike the spatial coherence method of Coakley and Bretherton (1982), does not require at least one completely cloud-filled region in the image. Moreover, the LDTNLR algorithm also discriminates on geometric criteria, computed pixel-by-pixel, based on the details of the cross-track geometry of a given image (*i.e.*, computed from a detailed knowledge of the particular orbit of a given image). These geometric criteria enable the LDTNLR algorithm to determine pixels with poor illumination, poor viewing angle, or pixels with high Fresnel reflection. Note, pixels with a high Fresnel reflection represent a major source of contamination in remotely-sensed ocean color data. Finally, the performance (*i.e.*, speed) of the algorithm (measured on a Hewlett-Packard 9000/375 workstation) is such that it has been used in near-real time support of *in situ* studies by the Scripps Satellite Oceanography Center for over two years. Thus, adverse computational constraints often associated with image segmentation will not affect the construction of large time-space series of CZCS-derived products. This is an important consideration, especially within the context of Global Change studies (*e.g.* GOFS, GLOBEC).

3.2.5. *Cloud Masking : Examples.*   Figure 8 shows the preprocessing steps required prior to cloud-screening. Figure 8a is unprocessed CZCS channel 5 data for an area west of Baja California (Image XII), taken at 19:07:40 GMT on May 12, 1979. A bright cloud exists to the northwest. This image contained significant zero count measurements, occurring as two line segments near Punta Eugenia and Cabo San Quintin (Fig. 8b).

These pixels comprise only 0.03% of the total image. Electronic overshoot eliminates 1.54% of the pixels (Fig. 8c). Overshoot exists east of some land masses as well as east of bright clouds. Land masses, even in the absence of clouds, can produce extensive image degradation due to electronic overshoot. This point is not discussed by Mueller (1988). Figure 8d is percent albedo. The vertical boundary separating darker and lighter gray shades at the left of the image is not an albedo structure but an artifact of the coarseness of the display device.

Figure 9 shows the result of cloud-screening Image XII. For comparison purposes, Fig. 9a shows unprocessed channel 5 data (*i.e.* the same as Fig. 8a). T-model cloud-masks using 1.3% and 1.6% albedo thresholds (Figs. 9b and 9c respectively) produce almost identical results: the lower threshold excludes 6.50% while the higher one excludes only 5.69% of the pixels. The black areas in Fig. 9d are the pixels screened only by the 1.3% threshold. As expected, the lower threshold extends the cloud boundaries. It also often increases land mass by about one pixel along coastlines. Including electronic overshoot and zero counts, the 1.3% T-model masks a total of 8.06% of the image; the 1.6% threshold masks 7.25%. In most cases, there is little difference between a 1.3% and 1.6% threshold. Only two of the nineteen images studied by Eckstein and Simpson (1991b) showed significant deviation (10.97 and 8.49% respectively).

Figure 9e is the cloud mask obtained from the LDTNLR method. This method excludes 13.42% more of the image than the T-model. Figure 9f is the difference between the 1.6% T-model and LDTNLR-model, where black areas are pixels masked only by the LDTNLR-model. Fine structure, such as parallel filaments southeast of Guadalupe Island (near the center of the image), suggest that the LDTNLR-model is correctly masking clouds undetected by the T-model. The linear cloud boundary near the center indicates sun glint contamination. Including electronic overshoot and zeros, 20.67% of the image is rejected.

3.3. ATMOSPHERIC EFFECTS

3.3.1 *Rayleigh Scattering.*   A comparison between the radiances obtained with single- and multiple-Rayleigh scattering with two of the three variables (θ, θ_0, and $\Delta\varnothing$) held constant is shown in figure 10. The computations were performed for winter at mid-latitudes (Julian day $D = 20$). Figure 10a shows the satellite zenith angle θ dependence. The large difference between L_R^m and

$L_R{}^s$ at high values of θ is caused by a breakdown in the single-scattering approximation (Gordon *et al.*, 1988). For $\theta_0 = 20°$, $\Delta\varnothing = 30°$, $L_R{}^m < L_R{}^s$ when $\theta < {\sim}8°$. The dependence of Rayleigh radiance on solar zenith angle θ_0 is illustrated by Fig. 10b, where $\theta = 10°$ and $\Delta\varnothing = 30°$. Again, except for $\theta_0 < {\sim}15$, $L_R{}^m < L_R{}^s$. Figure 10c gives the dependence of Rayleigh scattering on azimuth angle difference, $\Delta\varnothing$. Multiple-scattering is greater than single-scattering radiance for ${\sim}70° < \Delta\varnothing < {\sim}300°$. The precise cross-over points are wavelength

IMAGE XII

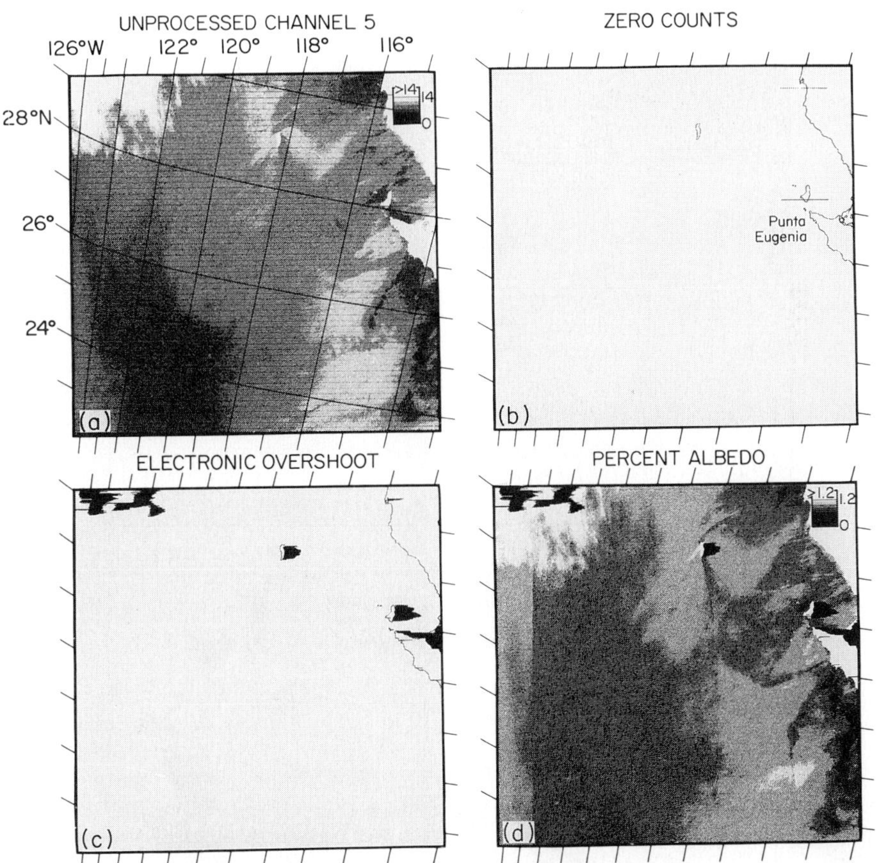

Figure 8. Pre-processing of Image XII prior to cloud-screening. (a) Unprocessed channel 5 brightness counts. Dark to light shades correspond to low to high counts. (b) Zero counts measured in channel 5. Zeros form two horizontal black lines near Punta Eugenia and Cabo San Quintin. (c) Electronic overshoot contamination (black areas). Overshoot occurs east of <u>both</u> clouds and land. In (b), (c), the coastline is drawn in black. (d) Percent albedo of reconstructed channel 5. Dark to light shades correspond to low to high albedos. Black areas are electronic overshoot and zeros (from Eckstein and Simpson, 1991b).

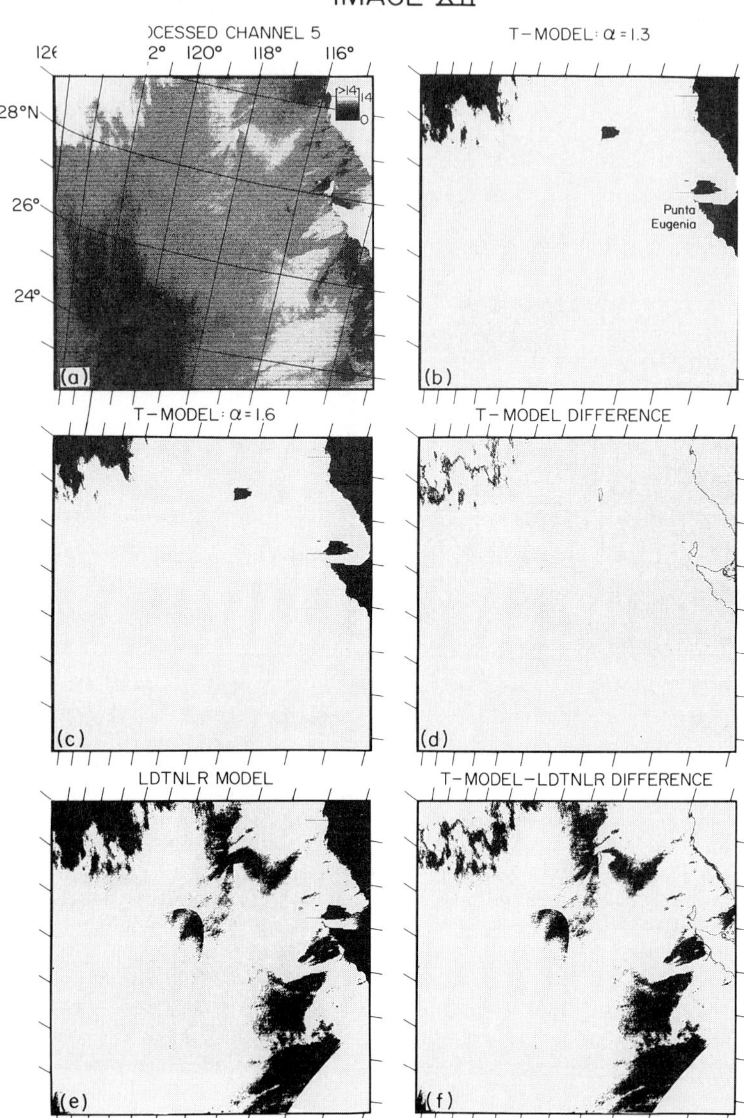

Figure 9. Cloud masking procedure for Image XII. (a) Unprocessed channel 5 brightness counts. Dark-to-light shades correspond to low-to-high counts (same as figure 8). Panels (b), (c), and (e) are cloud masks for a threshold (T-model) with albedo of 1.3%, T-model with albedo 1.6%, and LDTNLR model. Black areas are clouds. (d) Difference between the 1.3% and 1.6% T-model cloud masks. Black areas are pixels screened by only the 1.3% T-model. (f) Difference between the 1.6% T-model and the LDTNLR cloud masks. Black areas are pixels screened by only the LDTNLR-model. In panels (d) and (f), the coastline is shown in black (from Eckstein and Simpson, 1991b).

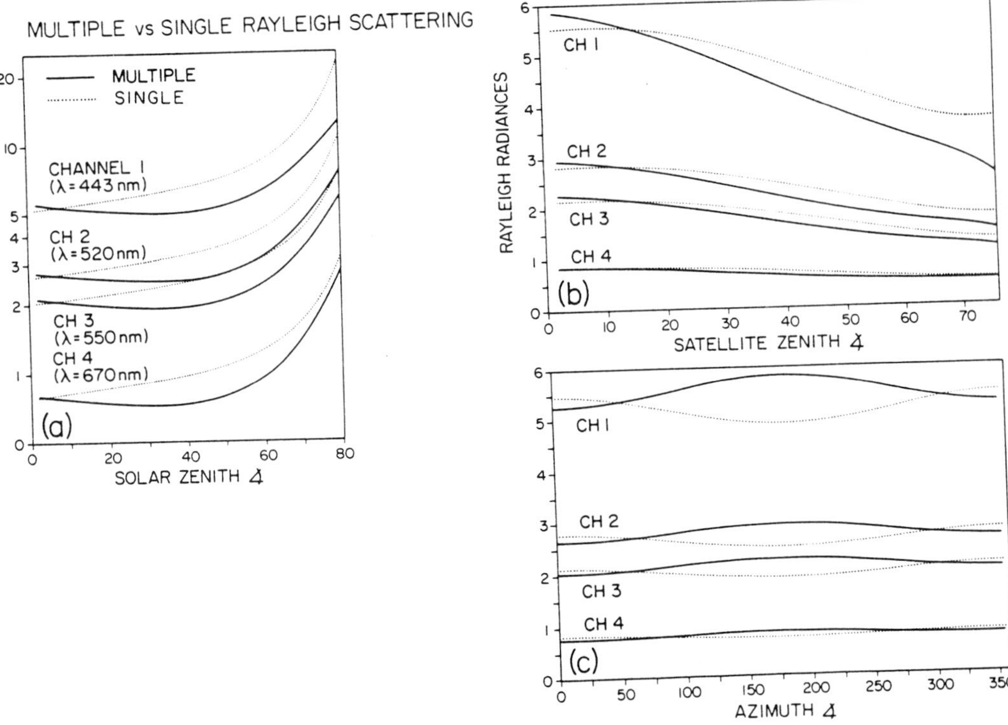

Figure 10. Comparison of multiple-scattering (solid lines) and single-scattering (dotted lines) Rayleigh radiance, in mW/cm^2 μm sr. Atmospheric parameters are from winter at mid-latitude (25-55°, Julian day = 20). (a) Rayleigh radiance as a function of solar zenith angle. The satellite zenith angle is constant at 10°; the azimuth difference is fixed at 30°. The vertical axis is a logarithmic scale. (b) Rayleigh radiance as a function of satellite zenith angle. The solar zenith angle is constant at 20°; the azimuth difference is fixed at 30°. (c) Rayleigh radiance as a function of azimuth angle difference. The satellite zenith angle is constant at 10°; the solar zenith angle is fixed at 20. Figure is from Eckstein and Simpson (1991a).

dependent. These diagrams (figures 10a-c) clearly show that both $L_R{}^m$ and $L_R{}^s$ are complex functions of satellite cross-track geometry and that $L_R{}^m$ can differ by over 80% from $L_R{}^s$.

Eckstein and Simpson (1991a) observed other differences between Rayleigh radiances computed using single- and multiple-scattering theory (see their figures 3 and 4). They found that multiple-scattering theory has a fundamental disagreement with the physics of the single-scattering formulation. With $L_R{}^s$, a minimum occurs when $\Delta\varnothing = 180°$, and a maximum occurs when $\Delta\varnothing = 0°$. In $L_R{}^m$, this relationship is reversed: the minimum is found when $\Delta\emptyset = 0°$. Hence, multiple-scattering theory produces larger radiances when the sun and satellite are further apart. Radiation traverses a shorter path when the azimuth angle difference is smaller (-90° < $\Delta\emptyset$ < 90°), and more light reaches the sensor. At large azimuth angle differences (90° < $\Delta\emptyset$ < 270°), radiation encounters more atmospheric particles and hence is more attenuated. However, if $\Delta\emptyset$ is redefined as the difference between the reflected sun vector and the satellite vector, $L_R{}^m$ has the same $\Delta\emptyset$ functionality as $L_R{}^s$. Gordon *et al.* (1983b, 1988a) have not been consistent in the definition of $\Psi^i_\pm$, and perhaps the azimuth angle dependence in the 1988 article is misstated (see Appendix II of Eckstein and Simpson, 1991a for a more detailed discussion of this point). This inference is consistent with a notation error (J. Mueller, private communication to H. Gordon) briefly mentioned in Gordon and Castaño (1987). If this is the case, then $L_R{}^m$ is simply $L_R{}^s$ plus a zenith angle dependent constant. In addition, $L_R{}^m$ theory significantly increased the correlation between $L_a(670)$ and ε. In effect, this increased correlation can decrease the reliability of pigment concentrations retrieved from CZCS data using multiple-Rayleigh scattering compared to single-Rayleigh scattering.

3.3.2. *Aerosol Scattering : Methodology.* The Gordon *et al.* (1983b) method of estimating aerosol radiances and $\varepsilon(\lambda_i, 750)$ using the Clear Water Assumption was evaluated by Eckstein and Simpson (1991a) with several images representative of different oceanic regimes. Only a brief review of their results are presented here.

Open ocean pixels, that is, pixels unaffected by clouds, land, or electronic overshoot, were located using the method developed by Eckstein and Simpson (1991b) and the automated cloud-screening process of Simpson and Humphrey (1990).

Then, the CZCS image was examined in 7×7 pixel areas for Clear Water. This size was selected based on several investigators work. Fiedler (1986) uses 7×7 areas, while Gordon *et al.* (1983b) use 5×5 areas to minimize the effects of noise. The larger size was chosen because it may provide a larger signal-to-noise ratio. Images were scanned west-to-east in blocks of seven lines. If a 7×7 area did not pass the Clear Water test, the next area selected for examination was started one sample east of the previous area. If the previous area contained Clear Water, the next area's western boundary was adjacent to the previous area's eastern edge. This method of search produced statistically independent determi-nations of aerosol radiance, $\varepsilon(\lambda_i, 670)$, and pigment concentration from as much of the image as possible. Images were created of the resultant values of $\varepsilon(\lambda_i, 670)$.

Because there is no completely objective method for locating low pigment waters, several tests were required for an area to qualify as Clear Water. These checks were invoked in the following order. If any condition were true, the area was rejected:

a. Number of open ocean pixels < 40
b. $L_a(670) \leq 0$
c. $L_t - L_R \leq 0$ in channels 2 and 3
d. $L_a \leq 0$ in channels 2 and 3
e. $L_W(443) \leq 0$
f. $P' \geq 0.25$ mg/m^3

The last condition is the test for low pigment content because pigment concentration is not known a priori. Note that this screening process is not entirely objective because the Clear Water Assumption is used to determine water-leaving radiances in channels 2-4. Conditions b-e therefore are checks for physical consistency because a ratio would otherwise be meaningless. This set of Clear Water tests was performed using both single- and multiple-scattering Rayleigh formulations.

3.3.3. *Aerosol Scattering : Example.* The Clear Water areas in Image XVI, a view of the Pacific Ocean off the coast of Point Conception captured on Aug. 29, 1979, are shown in figure 11. Approximately 34% of the image is corrupted by clouds, land, and electronic overshoot (gray areas in Fig. 11a). Using L_R^s, ~50% of the image passed the Clear Water test (Fig. 11b); L_R^m theory passed 57% (Fig. 11c). Figure 11 illustrates the difference in the Clear Water areas: gray represents like areas (~93%), white is Clear Water only in the L_R^m derivation (~7%), and black is Clear Water only in the L_R^s method (~0.04%).

Aerosol radiance at 670 nm in mW/cm^2 μm sr (Fig. 12a), and $\varepsilon(\lambda_i,670)$ for $i=1,2,3$ (Figs. 12b, 12c, and 12d respectively) from single Rayleigh scattering suggest a possible inverse relationship between $L_a(670)$ and ε. High aerosol radiances are usually found near low ε values, and the inverse is also true. Figures 12b, 12c, and 12d were mapped so that similar gray shades correspond to similar ε values.

The results of multiple-scattering theory are shown in Fig. 13, which indicates a stronger correlation between $L_a(670)$ shown in figure 13a and $\varepsilon(\lambda_i,670)$ for $i=1,2,3$ (Figs. 13b, 13c, and 13d respectively). Both L_R^s and L_R^m formulations show high aerosol/low ε values occurring in the same regions, although the magnitudes have changed. The same gray shades in Figs. 13b, 13c, and 13d represent the same ε values.

Pigment concentrations were also computed. Figures 14a and 14c are pigment content, in mg/m^3, from L_R^s and L_R^m theory respectively, using the locally varying epsilon factors; gray shades correspond to the same concentrations. Both methods show the same areas of low and high pigment. Pigment content also was computed using mean ε values for the entire image, and the differences $\Delta P'$ from the locally computed pigments

$$\Delta P' = P'\left(\varepsilon\right) - P'\left(\bar{\varepsilon}\right) \tag{34}$$

are shown in Figs. 14b (single scattering, $\varepsilon = 1.239, 1.190$, and 1.038) and 14d (multiple scattering, $\varepsilon = 2.567, 1.832, 1.495$ for channels 1, 2, and 3). The greatest differences occur at lower $P'(\varepsilon)$ values. In Figs. 14b and 14d, gray shades represent the same values, and regions rejected by the Clear Water test are solid gray outlined in white.

3.3.4. *Aerosol Scattering : Residual problems with atmospheric aerosol correction for CZCS.* The correlation between $L_a(670)$ and ε (Eqs. 17-21) as seen in Image XVI (figures 11-14) and four other images studied by Eckstein and Simpson (1991a) show that the assumption of a constant ε for a given image is invalid in most cases. Gordon *et al.* (1983b) state that this assumption is valid only when aerosol type is constant for a given image.

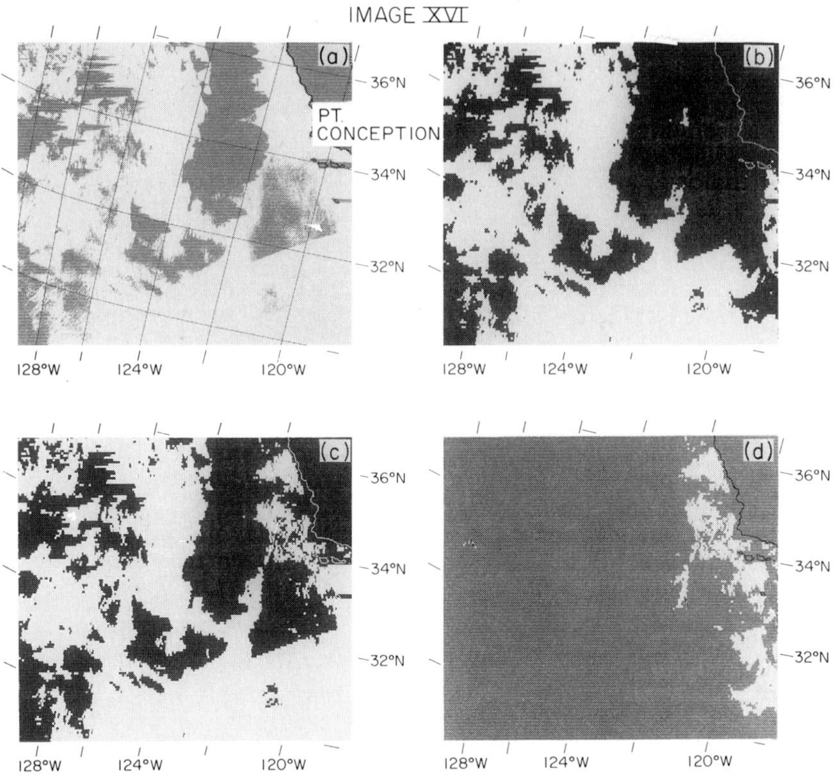

Figure 11. Clear water areas in Image XVI. (a) Image geography: coastline and latitude/longitude lines are in black. White areas are open ocean; gray is clouds, land, and electronic overshoot. (b) Clear Water areas from single-Rayleigh scattering theory. Black areas are rejected from Clear Water processing. White areas are Clear Water. Coastline is also in white. (c) Clear Water areas from multiple-Rayleigh scattering theory. Black areas are rejected from Clear Water processing. White areas are Clear Water. Coastline is also in white. (d) Difference between single- and multiple-scattering Clear Water masks. Gray represents like areas (92.91%). White is Clear Water only in multiple-scattering theory (7.05%); black is clear (0.04%) only in single-scattering theory. Figure is from Eckstein and Simpson (1991a).

Thus, one must either assume that ε is not constant or risk significant errors in pigment content because no information about aerosol type is available from the CZCS data. The correlations between $L_a(670)$ and ε (see Table 5 of Eckstein and Simpson, 1991a) suggest that either very few images have constant aerosol type, or that the ratio method of determining aerosol radiance in channels 1-3 also is dependent on concentration for a given aerosol type.

It has been suggested that changing aerosol type can be evaluated in certain cases (Gordon *et al.*, 1983b). They recommend that each pixel serve as a Clear Water calibrator in low pigment waters. However, earlier in their

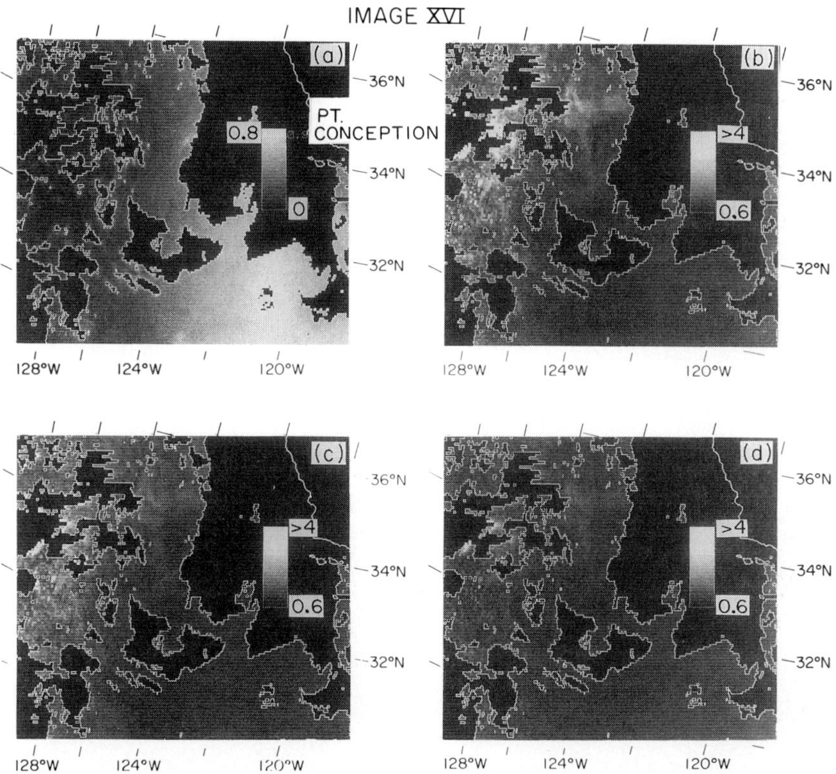

Figure 12. Image XVI: Atmospheric parameters from single-Rayleigh scattering theory. Black represents areas masked from the Clear Water algorithm. Coastline and Clear Water boundaries are outlined in white. (a) Aerosol radiance in channel 4, in mW/cm^2 μm sr. (b) Epsilon factor in channel 1. (c) Epsilon factor in channel 2. (d) Epsilon factor in channel 3. Figure is from Eckstein and Simpson (1991a).

article 5×5 calibration areas were used to reduce noise effects. A more significant objection to this method is that pigment content is the unknown one wishes to measure, and that few investigators are interested in pigment content when pigment concentration is negligible.

It has also been suggested that aerosol radiance can be estimated when two distinct aerosol types exist, either with the "more turbid atmosphere" over Clear Water areas or over no Clear Water areas (Gordon *et al.*, 1983b). In the first case, Gordon *et al.* (1983b) recommend using the ε values for the more turbid atmosphere, adding the caveat that the errors for the other aerosol type may be insignificant. This approach presupposes that one already has considerable information about the aerosol distribution, which requires data sources other than CZCS imagery. For the second case, in which ε can only be determined for the less turbid atmosphere, Gordon *et al.* (1983b) recommend systematically guessing ε or n in the more turbid atmosphere until the

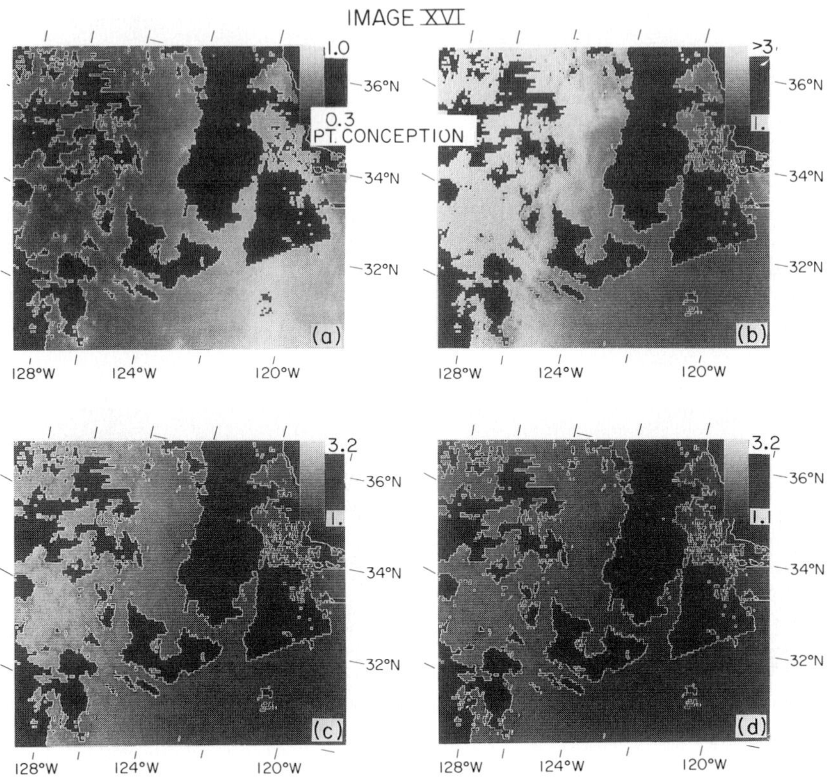

Figure 13. Image XVI: Atmospheric parameters from multiple-Rayleigh scattering theory. Black represents areas masked from the Clear Water algorithm. Coastline and Clear Water boundaries are outlined in white. (a) Aerosol radiance in channel 4, in mW/cm^2 µm sr. (b) Epsilon factor in channel 1. (c) Epsilon factor in channel 2. (d) Epsilon factor in channel 3.Figure is from Eckstein and Simpson (1991a).

boundary between the two aerosol regions disappears in all bands. Unfortunately, one cannot determine if a boundary is due to variations in ε or to actual changes in water-leaving and/or aerosol radiance because there is no *a priori* information about aerosol type or distribution. Hence, Eckstein and Simpson (1991a) concluded that these approaches are not valid when the only data source is CZCS imagery.

The results of Eckstein and Simpson (1991a) show that the greatest source of error in conventional CZCS pigment extraction algorithms based upon the Clear Water Assumption is the notion that the contributions of molecular and aerosol scattering to the total radiance observed at the sensor can be computed separately; this conclusion is consistent with the earlier work of Deschamps *et al.* (1983). The results of Eckstein and Simpson (1991a) clearly show that ε generally is not a constant across even small subsections of a

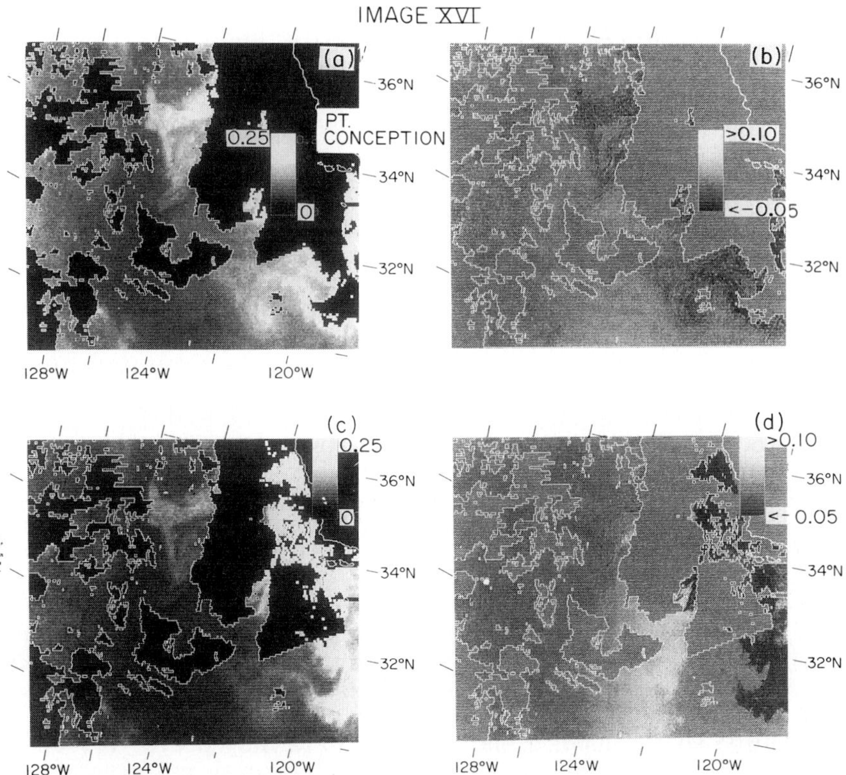

Figure 14. Image XVI: Pigment concentrations and pigment differences, in mg/m³. Coastline and Clear Water boundaries are outlined in white. (a) Pigment concentration computed form local determinations of epsilon factors using single-Rayleigh scattering. (b) Pigment concentration differences: pigment from local determinations of epsilon factors minus pigment from mean epsilon factors using single-scattering. (c) Same as (a) using multiple-scattering. (d) Same as (b) using multiple-scattering. Figure is from Eckstein and Simpson (1991a).

parent image and thus, ε ideally should be computed at each pixel in the image. This conclusion is very consistent with the independent analysis of Gordon and Castaño (1987) who suggested that one choose an "apparent" value of ε, called ε', for the atmospheric parameter correction (their Eq. 18) and that ε' bears little resemblance to the correct and theoretically meaningful ε. Moreover, Gordon and Castaño (1987) conclude that because traditional CZCS pigment algorithms assume that the optical properties of the aerosol are independent of position in the image, it is highly desirable to estimate the value of ε' at each pixel in the image.

The results of Eckstein and Simpson's (1991a) aerosol radiance analysis, particularly the correlation coefficients, raise serious questions about the validity of the Clear Water Assumption and aerosol ratios $S(\lambda_i,670)$ and

$\varepsilon(\lambda_i,670)$ in determining aerosol radiance. The fundamental problem is the undetermined nature of Eq. 1, (*i.e.* L_a and L_W are unknown):

$$L_t(\lambda) = L_R(\lambda) + L_a(\lambda) + t(\lambda) L_W(\lambda) \tag{1}$$

The Clear Water Assumption provides an estimate of L_W in channels 2-4 and thus determines one of the two unknowns for low pigment waters, but channel 1 is not so easily resolved. The CZCS images examined by Eckstein and Simpson (1991a) indicate that ε is usually dependent on both aerosol type and aerosol concentration, so that constant ε factors are not applicable to an entire image. If constants are used, pigment concentration may change significantly. Because ε can only be estimated in low pigment waters, the area of interest - the pigment-rich coastal regions - cannot be accurately quantified using only CZCS data. Also, the failure to find Clear Water in some of the images studied by Eckstein and Simpson (*e.g.* Images II and III) suggests that Eq. 21 is incorrect. This result strongly suggests that one should be suspicious of pigment extrapolation algorithms which rely on linear interpolation across a range of optical wavelengths (*e.g.* one possible scenario being suggested for pigment retrieval from SeaWIFS data). These conclusions are supported by the findings of other CZCS researchers (*e.g.* Khosraviani and Cracknell, 1987). These inconsistencies must be addressed before reliable pigment concentrations can be extracted from an arbitrary CZCS image. It is probably necessary to either compute ε at each pixel in the image or perhaps for each statistically homogeneous cluster of pixels within the image. Also, the use of constant ε values over a sequence of images which spans seasons must be strictly avoided because both aerosol type and concentration at a given location change with season. Finally, CZCS data may prove very useful for determining variability in the spatial and temporal patterns of aerosol distributions.

4. Implications for Historically-Processed Time-Space Series of Level 3 CZCS Products

The extraction of chlorophyll-like pigment concentrations from CZCS level 1 data is a formidable task, complicated by instrumentation problems, time varying calibrations, the presence of clouds, fog and residual glitter in the scene, and a wide variety of atmospheric effects (*e.g.* atmospheric attenuation, Rayleigh and aerosol scattering). Hence, many scientists have preferred to use level 3 products (*e.g.* distributions of chlorophyll-like pigment; attenuation coefficient) provided by large, specially funded processing projects (*e.g.* NASA Goddard Global Ocean Color Data Set, Feldman et.al, 1989; West Coast Time Series Project, Abbott, 1988) to address their specific scientific questions (*e.g.* role of phytoplankton in biogeochemical cycles) rather than process the CZCS level 1 data themselves.

Many of the above cited problems (*i.e.* Sections II and III herein) associated with the retrieval of chlorophyll-like pigment (and other level 3 products) from the CZCS level 1 data, however, have not been properly handled in the various iterations of these large-scale processing efforts. In particular, the failure to properly cloud screen data from the CZCS scene prior to atmospheric correction, the use of a constant set of ε values to implement an atmospheric correction, and neglect of the combined Rayleigh-aerosol scattering radiance $(L_{Ra}(\lambda))$ are three primary sources of residual error in the two major level 3 CZCS data sets (*i.e.* the NASA Goddard Global Ocean Color Data Set; the West Coast Time Series Project's CZCS Data Set) used by the oceanographic community. Unfortunately, these three particular types of errors can lead to some of the most serious errors in the final estimate of pigment concentration (*e.g.* Eckstein and Simpson, 1991a, b). Moreover, such globally processed data sets generally have not provided the end user with an associated numerical estimate of the error (or alternately confidence interval) in the final derived product.

These considerations all may help to explain some of the inconsistencies between CZCS derived pigments and *in situ* pigments observed by other investigators who have used large CZCS data sets processed with the traditional algorithm. For example, Fargion (1989) compared the seasonal cycle of phytoplankton abundance determined from West Coast Time Series data (Abbott, 1988) with *in situ* chlorophyll-a data taken by the California Cooperative Oceanic Fisheries Investigations (CalCOFI). She found that for many regions of the California Current System the seasonal cycle of phytoplankton abundance determined from the CZCS data was about 180° out of phase with that determined from the in situ data. Abbott (1988) carefully points out many of the limitations in the preliminary West Coast Time Series processing of CZCS pigment. Eckstein and Simpson (1991a) conclude that the use of the same image-wide constant ε values for all images, incomplete removal of pixels contaminated by electronic overshoot from either land or clouds (Mueller, 1988; Eckstein and Simpson, 1991b), inadequate cloud-screening (Eckstein and Simpson, 1991b; Simpson and Humphrey, 1990), and neglect of the Rayleigh-aerosol scattering term are the major deficiencies in the preliminary processing of the West Coast Time Series data which have contributed to the contradictory results reported by Fargion.

After ten years of experience with CZCS data the community has learned a great deal about what really is required to retrieve the best possible chlorophyll-like pigment fields from level 1 CZCS data. It now may be appropriate to consider reprocessing the global CZCS data set in light of this collective experience.

5. Importance of Case 2 Waters

Jerlov and Steeman Nielsen (1974) and Jerlov (1976) provide a scientific basis for both ocean optics and the interpretation of remotely-sensed ocean color. On the large-scale, the ocean can be divided into two main optical cases, generally referred to as Case 1 and Case 2 waters respectively (Morel and Prieur, 1977). Case 1 waters are dominated by biological constituents (*e.g.* photosynthetic related pigments from both living algal cells and associated decay products) and other dissolved organic matter produced by algae and their decay products. Case 1 water can be found in blue oligotrophic waters (pigment concentration ≤ 0.1 mg m^{-3}), in biologically active waters (pigment concentration ≈ 1.0 mg m^{-3}) and in the green euphotic waters (pigment concentration ≥ 10 mg m^{-3}) of upwelling areas. Case 2 waters generally contain all the components of Case 1 water but accompanied by terrigenous material associated with coastal runoff, suspended sediments, and yellow substance. Yellow substance generally represent a complex mixture of dissolved organic matter of terrestrial origin in some stage of decay. Case 2 waters often are found in enclosed European seas (*e.g.* the Baltic) but also can be found in the Mediterranean and North Seas, as well as in Asian waters. Case 2 waters can contain also all of the above substances plus high concentrations of suspended solid associated with high volume river flow (*e.g.* the Mississippi delta of the Gulf of Mexico, the Yellow and East China Seas).

Case 2 waters constitute a relatively small portion of the total oceanic surface area. These regions, however, are scientifically, environmentally and economically extremely important because these areas often coincide with the locations of major oceanic fisheries and human recreational zones (*e.g.* Simpson, 1992a). Moreover, these areas often receive a disproportionately high influx of pollutants from land-based industrial and agricultural sources.

Most of our experiences with CZCS algorithm development and validation is based solely on Case 1 waters. In general, the CZCS pigment retrieval algorithms have not been systematically applied and tested in Case 2 waters and in the few instances where such efforts have been made (*e.g.* Fukushima and Ishizaka, 1991) the traditional CZCS pigment extraction algorithms have proved inadequate. Given the importance of the oceanic areas containing Case 2 waters, however, it is clear that careful plans need to be made to design new pigment retrieval algorithms capable of handling the full suite of components found in these more optically complex waters. This will be especially important for future narrow-banded, multi-channel ocean color sensors such as SeaWIFS, OCTS, MERIS, MODIS, and ROSIS.

6. Other Approaches to Closure of the Basic CZCS Constraint Equation

6.1. THE RAYLEIGH-AEROSOL SCATTERING INTERACTION

The fundamental problem of pigment extraction from CZCS data is the undetermined and incomplete nature of the basic constraint equation used historically in CZCS processing (Eq. 1).

The greatest source of error in conventional CZCS pigment extraction algorithms based upon the Clear Water Assumption (*e.g.* Gordon *et al.*, 1983a; Walters, 1985) is the assumption that the contribution of the molecular ($L_R(\lambda)$) and the aerosol ($L_a(\lambda)$) scattering to the total radiance observed at the sensor ($L_t(\lambda)$) can be computed separately (Eckstein and Simpson, 1991a); their results also are consistent with the earlier work of Deschamps *et al.* (1983). In order to compute the atmospheric correction more accurately Eq. 1 must be replaced by an equation of the form:

$$L_t(\lambda) = L_R(\lambda) + L_a(\lambda) + \underline{L_{Ra}(\lambda)} + \underline{L_g(\lambda)} + t(\lambda)L_W(\lambda) \tag{35}$$

All non-underlined terms in Eq. 35 are identical to those in Eq. 1. $L_{Ra}(\lambda)$ is the Rayleigh-Aerosol scattering interaction term. It allows for the possibility that photons can scatter from both aerosols (*i.e.* $L_a(\lambda)$) and from air molecules (*i.e.* $L_R(\lambda)$). $L_g(\lambda)$ is a glint radiance.

Glint is associated with regions of high Fresnel reflection and generally results in amplifier saturation at the sensor. Thus, pixels with glint ($L_g(\lambda) > 0$) must be eliminated from the scene. The LDTNLR cloud-screening algorithm (Simpson and Humphrey, 1990) uses geometric criteria based on scene-specific cross-track geometry to eliminate glint contaminated pixels from a scene on a pixel by pixel basis. Thus, the recommended approach for cloud screening CZCS data cited herein (III.B.) incorporates a rigorous method for handling the $L_g(\lambda)$ term in Eq. 35. $L_g(\lambda)$, however, will not necessarily be removed completely from the CZCS scene by the traditionally used simple constant threshold methods of cloud-masking which operate on channel 5 data. This is another reason for not using such methods to cloud-screen remotely-sensed visible data.

The cross-scattering term $L_{Ra}(\lambda)$ requires that the multiple Rayleigh scattering formulation (*e.g.* Eqs. 15 and 16) be used and that the radiative transfer equation be solved simultaneously for both the Rayleigh and aerosol scattering.

Since about 1989, several ongoing efforts have sought alternative closure schemes for a CZCS pigment extraction algorithm; some using Eq. 1 and others using an equation similar in form to Eq. 35 as the basic constraint equation. Much of this work is presently in an exploratory stage. If successful, however, it could provide a basis for more accurately reprocessing the CZCS global ocean color data set. Because of the preliminary nature of this work, only a brief overview of such efforts is given below. The reader should note carefully that the notation in sections VI.B - VI.E. is that of the original author. Therefore, it may differ somewhat from that used previously in this paper.

6.2. SCALAR TRANSFER USING SUPERPOSITION

Mukai (1990) models the atmospheric correction as a scalar transfer process, excluding polarization, using a superposition method (Mukai and Ueno, 1978) in an atmosphere-rough surface model. He uses a scalar as opposed to a vector transfer model because the algorithm requires intensity values not degree of polarization. A schematic of the atmospheric correction algorithm implemented by Mukai (1990) is given in figure 15. Aerosol optical thickness, t_a, is measured with lidar by the National Institute for Environmental Studies of Japan. (Note, lidar measurements do not provide information about the aerosol phase function; these must be determined from atmospheric models of radiative transfer processes).

These data are used to characterize the seasonal variation in aerosol type and concentration in the oceanic regimes around Japan. Total optical thickness of the atmosphere, t_1, and the mixing ratio, F_g, are derived for winter and summer model atmospheres using Lowtran-6 (*e.g.* Mukai *et al.*, 1990). The Lowtran-6 code allows for different aerosol types (*e.g.* rural, maritime, navy maritime, etc.). In the solution of the radiative transfer equation, multiple reflection at the ruffled sea surface, as well as multiple atmospheric scattering, is accommodated using the superposition method. Preliminary examples of the algorithm's performance are given by Mukai (1990) using images of the Sea of Japan. His results show considerably more structure in CZCS-derived plankton distributions compared to results obtained from the traditional CZCS algorithm.

152

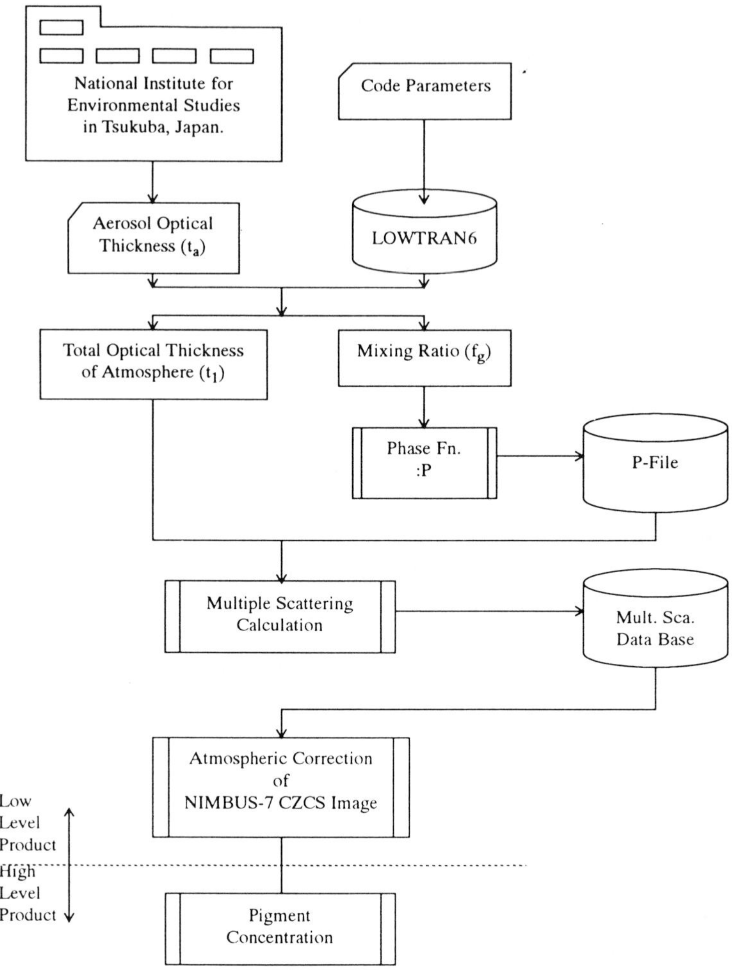

Figure 15. Schematic diagram of the CZCS atmospheric correction procedure used by Mukai (from Mukai, 1990). GOFS and GLOBEC concentrate on the high level product. This, however, requires accurate low level products.

Mukai (1990) circumvents four problems associated with traditional CZCS processing: (1) the Clear Water Assumption; (2) single-scattering approximation; (3) separation of molecular and aerosol scattering processes; and (4) the flat sea surface approximation. These benefits, however, are obtained by: (1) using models to compute atmospheric optical thickness and mixing ratios from which phase functions are constructed; and (2) direct measurements of atmospheric aerosols. As part of the process a climatology of parameters related to atmospheric correction is stored on disc for future use.

6.3. THE ATMOTOOLS INTERATIVE PROCEDURE FOR ATMOSPHERIC CORRECTION

Zege and Prikhach (1991) carefully examined the Clear Water Assumption (Gordon and Clark, 1981), as well as proposed calibration procedures for remote sensing of the ocean in the visible region of the spectrum (Gordon, 1984). They correctly noted that these procedures use an averaged optical atmosphere model and retain only the linear term in the Taylor series expansion of the optical thickness. Under these constraints, the radiative transfer equations have a well known solution for single-scattering (*e.g.* Sobolev, 1963).

Zege *et al.* (1991), like Mukai (1990), have combined a set of models with atmospheric observations (*e.g.* aerosol database) in order to achieve an alternate approach to the atmospheric closure problem of CZCS. This combination of models and observations is called ATMOTOOLS. Two different ways of retrieving plant pigment concentration from remotely-sensed visible data are proposed by Zege and Prikhach (1991). Method 1 assumes that independent spectral information about the atmosphere (*e.g.* total spectral transparency or data from lidar soundings) is available for pigment extraction from a CZCS scene. Such information is most likely to be available only in coastal domains. When available, however, this information is combined with radiative transfer calculations in ATMOTOOLS (*e.g.* multiple-scattering, real atmospheric stratification, rough ocean surface) to provide closure for the pigment extraction algorithm. When no scene specific aerosol calibration data are available (*i.e.* no simultaneous ground truth measurements of atmospheric aerosols) a reference area in the satellite pass is chosen. Note, this reference area is not a Clear Water Area, rather it is any oceanic area for which there is at least seasonal knowledge of the spectral radiance coefficient of light reflected by sea water. Again, radiative transfer models from ATMOTOOLS are used to provide closure for the CZCS pigment extraction algorithm. It is important to emphasize that the alternative procedures developed by Zege *et al.* (1991), like those of Mukai (1990), perform closure by merging ancillary information (*e.g.* optical models of ocean surfaces, atmospheric aerosols, sea water optical characteristics) with data directly received by the CZCS sensor. Both sets of results point to the conclusion that the CZCS data is insufficient by itself to properly close the basic constraint equation (Eq. 35) on pigment extraction from the remote-sensed water-leaving radiances; this result is consistent with the work of Deschamps *et al.* (1983), Khosraviani and Cracknell (1987) and Eckstein and Simpson (1991a).

6.4. REFLECTANCE MODEL FOR CZCS PIGMENT EXTRACTION

Bricaud and Morel (1987) and Andre and Morel (1991) use a reflectance model developed for Case 1 waters to extract plant-pigment concentration from CZCS data. These authors note that in visible remote sensing of the ocean (*e.g.* CZCS) no *a priori* or independently acquired information concerning aerosol is available. Therefore, the water-leaving radiances and the aerosol radiances must be extracted simultaneously from the total radiance measured from the sensor. Decoupling of the terms is accomplished by external constraints. Bricaud and Morel (1987) and Andre and Morel (1991) use ocean

reflectance models, coupled with empirical relations for interlinking the various water-leaving radiances, as the required external constraints. An iterative scheme, originally proposed by Smith and Wilson (1981), and based on an empirical relation between water-leaving radiances at three wavelengths is used to estimate the aerosol contribution.
The general form of this relation is

$$r_{1,4} = A \left(r_{1,3} \right)^B \tag{36}$$

where $r_{i,j} = R(\lambda_i)/R(\lambda_j)$, R being the diffuse reflectance of the ocean just below the surface. In this scenario for closure, pigment concentration is estimated with a similar relation

$$C = D \left(r_{i,j} \right)^E \tag{37}$$

where C is pigment concentration (mg m^{-3}), $i = 1$ or 2, and $j = 3$ for CZCS spectral channels. The constants D and E, like the constants A and B in Eq. 36, are determined statistically from field data.

6.5. CLASSIFICATION AND ATMOSPHERIC MODELING APPROACH TO CLOSURE

The basis for this approach to closure (figure 16) is a combination of four techniques: (1) rigorous pre-processing of the CZCS data to eliminate all spurious pixels; (2) classification of the residual pixels into a set of clusters such that within cluster scatter is minimized and the between cluster scatter is maximized using unsupervised classification procedures; (3) determine aerosol type and concentration on a cluster by cluster basis using models and supervised classification techniques; and (4) solve the radiative transfer equations (Eq. 35) on a cluster by cluster basis to determine water-leaving radiances and derived quantities (e.g. pigment concentration, sediment concentration). This is the method of CZCS closure currently under investigation by Simpson and his group at Scripps. In essence, this approach uses pre-processing techniques to reduce the statistical dimensionality of the data set prior to the unsupervised clustering. Then, spectral differencing and the Principal Component Transformation are performed on the residual CZCS pixels; again, as a means of further reducing the statistical dimensionality of the data prior to classification (figure 16). The unsupervised clustering of the residual valid pixels solves an eigenvalue problem of the form

$$\left(S_W + S_B \right) G^T - \Lambda G^T = 0 \tag{38}$$

where S, the scatter matrix of the image of residual pixels, is the sum of two matrices, S_W, the within cluster scatter matrix defined by

$$S_W = \sum_{k=1}^{K} S^{(k)} = \sum_{i=1}^{n_k} \left(y_i^{(k)} - m^{(k)} \right) \left(y_i^{(k)} - m^{(k)} \right)^T \tag{39}$$

and S_B, the between cluster scatter matrix defined by

$$S_B = \sum_{k=1}^{K} n_k \left(m^{(k)} - m \right)\left(m^{(k)} - m \right)^T$$
$$= \sum_{k=1}^{K} n_k\, m^{(k)} m^{(k)\,T} - N m m^T \tag{40}$$

The transformation G^T is constructed from the spectrally differenced CZCS data (channels 1-4) in such a way as to optimize the solution space of Eq. 38; that is, to minimize the variance in S_W subject to the constraint of conserving the total variance and thereby maximizing the variance in S_B. In the notation of Eqs. 37, 38, and 39, $S^{(k)}$ is the scatter matrix of the kth cluster in the set of K clusters, $y_i^{(k)}$ is the ith vector in the kth cluster, $m^{(k)}$ is the mean vector of the kth cluster, m is the mean vector for the entire set of K clusters, n_k is the number of vectors in the kth cluster and N is the total number of vectors in the set of K clusters.

Supervised classification, based upon a variety of statistical and physical measures (*e.g.* brightness, texture) is used to assign each cluster an aerosol type. Direct measurements (*e.g.* lidar) and climatology also can be incorporated into the supervised classification procedure. Again note that lidar measurements provide no information on the scattering phase functions; these must be computed from atmospheric models of radiative transfer processes. Then, the radiative transfer equations (multiple-Rayleigh scattering, aerosol scattering, roughened sea surface, etc.) are solved on a cluster by cluster basis. Finally, plant pigment concentration is determined on a pixel by pixel basis.

The second step of this procedure performs image segmentation using a split-and-merge clustering procedure on the PC transform of the differenced image, Y. This results in a segmented image, Y', in which the natural spectral classes in the original image are separated into distinct groups. The method of clustering that is employed in this algorithm combines both the partitional and hierarchical approaches. It consists of a partitional clustering algorithm augmented by a splitting-and-merging step at each iteration. The split-and-merge clustering procedure is summarized as follows:

a) **Initialization**: Determine K_I vectors that are uniformly distributed throughout the data to serve as initial cluster centers (or mean vectors), where K_I is the number of initial clusters. A method to dynamically determine an optimal value of K_I for a given image has been devised.
b) **Clustering**: Assign each vector in the image to one of the clusters depending upon the selected *clustering criterion*. The selected criterion function is the square error criterion.
c) **Merging**: Check all possible groups of two clusters and combine them if they satisfy the chosen *merging criterion*. The chosen merging criterion is a threshold value of the squared Euclidean distance between cluster mean vectors, designated T_{MERGE}. Note, the algorithm is very insensitive to the choice of T_{MERGE}.

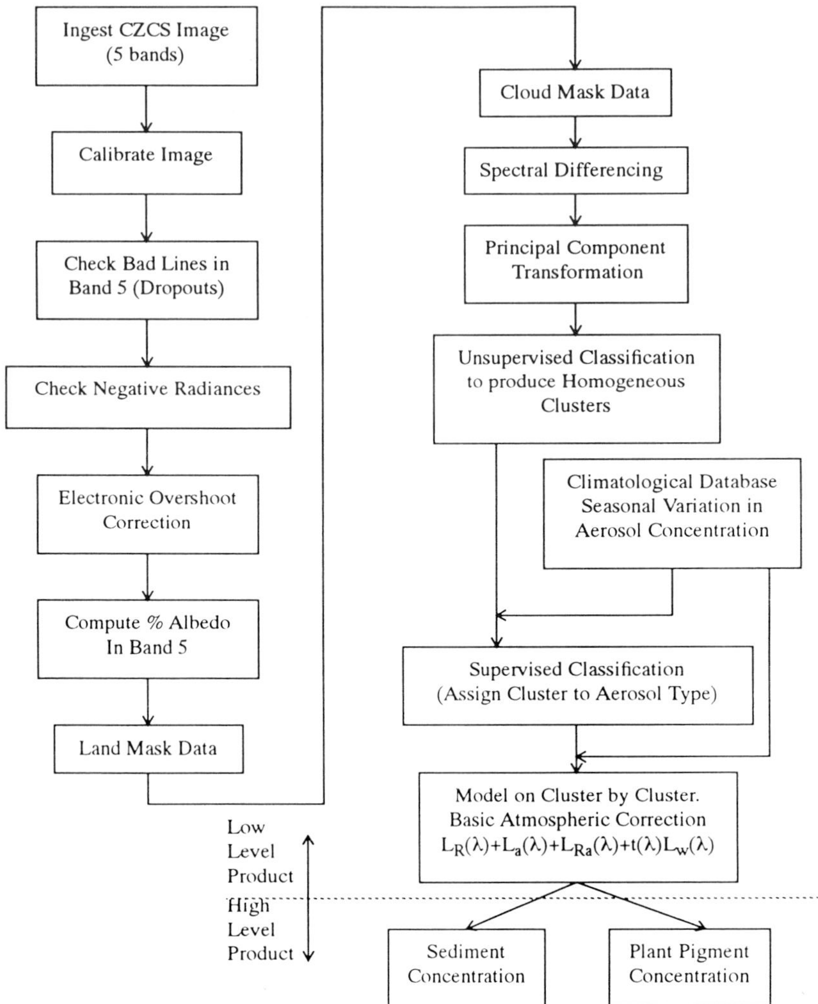

Figure 16. Proposed CZCS atmospheric correction procedure developed by Simpson. GOFS and GLOBEC concentrate on the high level product. This, however, requires accurate low level products.

d) **Splitting**: Check all clusters and split them in half if they satisfy the chosen *splitting criterion*. The chosen splitting criterion is a threshold value of the squared Euclidean distance between cluster maximum and minimum vectors, designated T_{SPLIT}. Note, the algorithm also is very insensitive to the choice of T_{SPLIT}.

e) **Recompute New Cluster Centers**: This is performed for the K_I' clusters that result from Steps b) - d), where K_I' is now a new number of clusters.

f) **Repeat Until Converged**: Beginning with Step b), continue the procedure until the *convergence criterion* is satisfied. The selected convergence criterion is the minimization of within-cluster scatter.

g) **Complete Cluster Assignment**: After the algorithm has converged, assign every vector to its appropriate cluster such that within-cluster scatter is minimized.

The proposed algorithm, unlike most other segmentation methods, combines a partitional with a hierarchical method of clustering. This approach has several advantages over the use of either method alone: (1) Pure hierarchical methods are not appropriate for complex data (Muerle and Allen, 1968; Fukada, 1980; Jain and Dubes, 1988); (2) Pure hierarchical methods are more appropriate for data that is to be partitioned on the basis of both local and global information, rather than global information only (Jain and Dubes, 1988); (3) Pure hierarchical methods impose a taxonomic structure on the data (Anderberg, 1973), which is not characteristic of CZCS imagery; (4) Pure hierarchical methods are *order* dependent - *i.e.*, the resulting segmentation will vary depending upon the order that the regions are split and merged (Cheevasuvit *et al.*, 1986); this often results in less than optimal segmentations of the data; (5) Pure partional algorithms often converge to local minima of the clustering criterion function (Pairman and Kittler, 1986; Jain and Dubes, 1988); (6) The combined approach is more efficient than pure merging or pure splitting methods of region detection (Pavlidis, 1977; Richards, 1986); and (7) The combined approach is less dependent on the initial segmentation, and therefore is more capable of recovering all of the natural clusters in the data (Seddon and Hunt, 1985; Jain and Dubes, 1988); this is because the number of clusters in the initial segmentation does not need to be the same as those that actually exist in the given data (Seddon and Hunt, 1985; Pairman and Kittler, 1986).

7. Calibration

Calibration of satellite-based ocean color sensors is a complex process and should be done in two separate stages: (1) calibration of the low-level products (*e.g.* atmospheric aerosols and the water-leaving radiances (or reflectances)); and (2) calibration of the high-level products (*e.g.* plant pigment concentration). The CZCS experience has shown clearly that the fundamental quantities of interest are the water-leaving radiances and the principal source of contamination is uncorrected atmospheric aerosol scattering. Future calibration procedures developed for next generation ocean color sensors (*e.g.* SeaWIFS, ADEOS/OCTS) should carefully and clearly incorporate knowledge gained from the CZCS experience into their calibration protocols. Therefore, such calibration procedures should provide for ground-truth activity in, at the minimum, the following two critical areas: (1) provide direct lidar measurements of atmospheric aerosol distributions, especially in coastal zones; and (2) provide high quality spectral measurements of upwelled and downwelled radiances and irradiance.

Biological measurements of phytoplankton and phaeophytin pigment concentration should not be used as a primary standard to evaluate the success of pigment extraction algorithms used with remotely-sensed ocean color data. Biological measurements have an important role, however, in phase II of the calibration process. Once the accuracy and precision of the retrieval of the water-leaving radiances (reflectances, if Eq. 35 is reformulated in terms of reflectances) from the satellite data have been established by physical-optical measurements, the biological calibration can be used to assess the accuracy and precision of the models used to convert water-leaving radiances (reflectances) to pigment concentration. For this purpose, however, it is equally important that reliable error estimates be obtained for the *in situ* observations. Moreover, a carefully followed, rigorous protocol for taking such measurements needs to be developed and enforced; otherwise, there will be no end to the potential grab bag of intercalibration problems.

8. Archive and Distribution of Data

The CZCS experience also tells us that there were long delays in making data generally available to the user community. The purpose of this comment is not to assign blame. Rather, it is to draw attention to the compelling need to make satellite data as readily available as possible to the community at large in near-real time.

At Scripps, we have developed a new tool to archive, browse, and distribute satellite data in near-real time which can be accessed by anyone directly connected to Scripps via Internet. This tool, the Scripps Satellite Archive and Browse for Localized Environments (SSABLE) system, currently handles Scripps AVHRR data, EDIS AVHRR data, and CZCS data; it could easily be expanded to handle future ocean color data sets such as those from SeaWIFS or OCTS. Therefore, a brief description of SSABLE is given below.

SSABLE is a new interactive tool for browsing the Scripps Satellite Oceanography Center's (SSOC) archive of satellite data in your laboratory or office. The archive contains data from the NOAA/AVHRR series of satellites (TIROS-N, NOAA6-12) dating from 1979 to the present as well as global EDIS AVHRR data and CZCS data off the west coast of North America. SSOC maintains a capture station and data are continually added to the SSABLE system within ten minutes of capture completion. Typically 4-5 new passes are captured each day.

SSABLE provides an easy to use point-and-click, mouse-driven user interface developed around the OSF/Motif widget set and the X Window protocol. All options are selected by either directly clicking the mouse on the desired pushbutton, or pulling down a sub-menu from a push-button. A "fill-in-the-blanks" window appears to guide the user in defining the operation via Internet. Data are available via SSABLE within 10 minutes of capture. All SSABLE option menus have "Cancel" push-button to abort the operation. Users can search for data by time and day, geographic coordinate, satellite, or if the user knows the structure of the Scripps archive, by explicit archive number. A typical SSABLE display screen (figure 17) shows satellite data, area of coverage, and some of SSABLE's point-and-click operations.

Once data have been selected which meet the user's needs, it can be ordered on a variety of media. A "scratch pad" is included inside SSABLE to temporarily save an image and re-browse later. In this manner, complex composite searches can be performed. By viewing the SSABLE scratch pad,

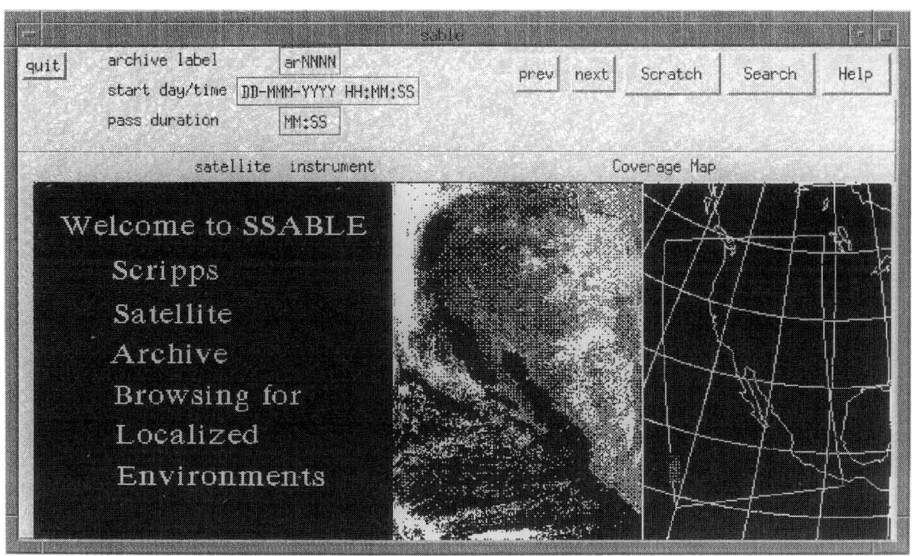

Figure 17. SSABLE X-Window scene showing an AVHRR browse product. SSABLE provides a user-friendly graphical interface for searching, ordering, and distributing on-line satellite data (*e.g.* AVHRR, CZCS) to the end-user.

more operations become defined. Each operates on either the entire scratch pad or only the selected images in the scratch pad. Images are selected by high-lighting them with the mouse. Among the scratch pad defined operations are mailing and ordering. The contents of the scratch pad can be mailed to a user if a final decision on ordering has not yet been reached. If certain images are desired, they can be ordered from the scratch pad.

All data available to browse are level_1 data with a spatial resolution of 1.1km at nadir.At present all data captured at SSOC since January 1, 1991 have been incorporated into SSABLE. Since 15 September 91 four to six AVHRR passes per day have been captured and automatically integrated into SSABLE. The historical SSOC archives also are being incorporated into SSABLE. At present all available 1979, 1980 and 1981, 1982, 1983, 1984, 1985 and 1986 data can be accessed through SSABLE. Within the next two years all archive data at Scripps will be incorporated into SSABLE.

Unlike many other digital browse systems, SSABLE does not require users to invest in expensive hardware and/or software. The only requirements to use SSABLE are: 1) have a bitmapped display (monochrome or greater); 2) be running the X Window system; and 3) be on a network directly reachable from the SSABLE system. Merely use *telnet* to determine whether a given workstation or XWindow terminal is directly connected to the Internet.

SSABLE is network and machine independent; it will run identically on any machine which meets the above requirements. SSABLE has been tested at Scripps Institution of Oceanography, University of British Columbia, University of Washington, Goddard Space Flight Center, NOAA Southwest Fisheries Center, Carnegie-Mellon University and Stanford University using HP workstations, SUN workstations, Silicon Graphics workstations, Macintosh PCs with X Window, DEC 3100 workstations, IBM 6000 workstations, and NCD X terminals.

Users may order images on the following media: 9-track tape (1600 bpi or 6250 bpi), digital audio tape, cartridge tape, or optical disk. All data are in the Archive Tape Format as described in NOAA Technical Memorandum NESS 107.

SSABLE is merely one example of how modern data archive, catalog, browse, and distribution systems might be designed to meet the near-real time needs of the global user community; a set of needs likely to expand in the upcoming decade of Global Change studies.

9. Conclusions and Recommendations

Computation of chlorophyll-like pigments from CZCS data is a complex process. Water-leaving radiance, a small portion of the total signal, controls the value of the pigment computed but fully 80 to 90% of the total radiance measured by the sensor is due to Rayleigh and aerosol scattering. In addition, prior to pigment retrieval a series of pre-processing steps must be performed to: (1) correct for instrumentation problems (*e.g.* drop outs, zero counts, electronic overshoot; and (2) eliminate clouds and land from the scene. Three of the major problems associated with the use of the traditional CZCS algorithm in the preparation of large CZCS data sets (*e.g.* NASA Goddard Global Ocean Color Data Set; West Coast Time Series CZCS data set) are: (1) lack of rigorous cloud-screening procedures; (2) use of an image-wide constant set of ε values to perform the atmospheric correction; and (3) neglect of the combined Rayleigh-aerosol scattering term $(L_{Ra}(\lambda))$. These three sources of error in the pigment retrieval process can lead to large errors in the level 3 plant pigment concentration. Therefore, care should be exercised when using these products to determine oceanic variability, especially if large time-space sequences of level 3 images are used.

Based on the collective community experience with CZCS data, the following specific recommendations can be made with regard to future ocean color sensors:

1. Maintain a continuous time series of sensor calibrations to evaluate issues such as sensor drift.
2. Perform a continuous set of *in situ* optical calibrations and *in situ* biological/chemical calibrations throughout the lifetime of the sensor. The optical calibrations should be used to calibrate measured brightness counts to low-level products (*e.g.* optical radiances and/or reflectances). The biological/chemical calibrations should be used to calibrate higher level derived products (*e.g.* plant pigment concentration, sediment concentration).
3. Perform rigorous electronic overshoot corrections for both cloud and land. The statistical database upon which the present equations for electronic

overshoot is based needs to be expanded in order to improve the statistical reliability of the coefficients used in the equations.
4. Perform rigorous land- and cloud-masking. Simple threshold models of cloud screening are inadequate. Rigorous cloud-masking also should use precision orbit geometric criteria to eliminate glint radiances from the scene.
5. Results indicate that the Clear Water Assumption generally will not be valid for an arbitrary scene. Its use should be avoided as a closure procedure for the radiative transfer equations.
6. Ideally, ε values used in atmospheric correction procedures should be done on an image by image, pixel by pixel basis. Computational constraints, however, may preclude this option, especially for either near-real time processing or the processing of large data sets. Alternatively, ε values could be computed on an image by image basis and for individual clusters within an image which have optically homogeneous properties as determined by a rigorous statistical analysis. This would represent a reasonable compromise between the competing efficiency and accuracy constraints on the pigment retrieval process.
7. level_3 products should be distributed with error estimates.
8. Data from second generation ocean color sensors need to be distributed to the community in near-real time. This requires the use of modern database management concepts and network protocols by groups funded to collect and archive future ocean color data.

References

Abbott, M. (1988) 'West Coast Time Series Release Notes', Version 0.5., College of Oceanography, Oregon State University, pp. 1-5.

Anderberg, M.R. (1973) 'Cluster Analysis for Applications', Academic Press, Inc., New York, NY.

Andre, J. M. and Morel, A. (1991) ' Atmospheric corrections and interpretation of marine radiances in CZCS imagery', revisited, Oceanologica Acta 14, 3-22.

Austin, R. W. and Petzold, T. J. (1981) 'The determination of the diffuse attenuation coefficient of sea water using the Coastal Zone Color Scanner', in J. F. R. Gower (ed.), Oceanography from Space, Plenum Press, New York, pp.239-256.

Blumel, K. and Tonn, W. (1986) 'Satellite-derived vegetation index as an indicator of biomass, primary production, and net CO_2 flux', Vegetation 80, pp. 71-89.

Bricaud, A. and Morel, A. (1987) 'Atmospheric corrections and interpretation of marine radiances in CZCS imagery: use of a reflectance model', Oceanologica Acta 7, 33-50.

Cheevasuvit, F., Maitre, H., and Vidal-Madjar, D. (1986) ' A robust method for picture segmentation based on a split and merge procedure', Computer Vision, Graphics, and Image Processing 34, 268-281.

Coakley, J. A., Jr., and Bretherton, F. P. (1982) 'Cloud cover from high resolution scanner data: Detecting and allowing for partially filled fields of view', Journal of Geophysical Research 87, 4917-4932.

Deschamps, P. Y., Herman, M., and Tanre, D. (1983) 'Modeling of the atmospheric effects and its application to remote sensing of ocean color', Applied Optics 22, 3751.

Eckstein, B.A. and Simpson, J.J. (1991a.) 'Aerosol and Rayleigh radiance contributions to Coastal Zone Color Scanner images', International Journal of Remote Sensing 12, 135-168.

Eckstein, B.A. and Simpson, J.J. (1991b.) 'Cloud Screening Coastal Zone Color Scanner images using channel 5', International Journal of Remote Sensing 12, 2359-2377.

Fargion, G. S. (1989) 'Physical and biological pattern during El Niño and non-El Niño episodes in the California current. Ph.D. dissertation', (unpublished), Scripps Institute of Oceanography, University of California, San Diego.

Feldman, G., Kuring, N., Esaias, C. Ng, W., McClain, C., Elrod, J., Maynard, N., Emdres, D., Evans, R., Brown, J., Walsh, S., Carle, M., and Podesta, G. (1989) 'Ocean Color: Availability of the Global Data Set', EOS Transactions AGU 70 (23), 634-635 and 640-641.

Fiedler, P. (1984) ' Satellite observations of the 1982-83 El Niño along the U.S. Pacific Coast', Science 224, 1251-1254.

Fiedler, P. (1986) 'CZCSCAL: a FORTRAN program for converting raw CZCS data to pigment concentrations', Scripps Satellite Oceanography Facility, 11 pp.

Fiedler, P. (1989) Personal communication. Southwest Fisheries Center, National Oceanographic and Atmospheric Administration, P.O. Box 271, La Jolla, CA, 92038-0271.

Fukada, Y. (1980) 'Spatial clustering procedures for region analysis', Pattern Recognition 12, 395-403.

Gallaudet, T. C. and Simpson, J. J. (1991) 'Automating cloud screening of AVHRR imagery using split-and-merge clustering', Remote Sensing of Environment 38, 77-121.

Gordon, H. R. and Castano, D. J. (1987) 'Coastal Zone Color Scanner atmospheric correction algorithm: multiple scattering effects', Applied Optics 26, 2111-2122.

Gordon, H. R. and Morel, A. Y. (1983) 'Remote assessment of ocean color for interpretation of satellite visible imagery', a review in Lecture Notes on Coastal and Estuarine Studies. SpringerVerlag, New York, 114 pp.

Gordon, H.R. and Clark, D.K. (1981) 'Clear water radiances for atmospheric correction of Coastal Zone Color Scanner imagery', Applied Optics 20, 4175-4180.

Gordon, H. R., Brown, J. W., and Evans, R. H. (1988a.) 'Exact Rayleigh scattering calculations for use with the Nimbus-7 coastal zone color scanner', Applied Optics 27 (5), 862-871.

Gordon, H. R., Brown, O.B., Evans, R. H., Brown, J. W., Smith, R. C., Baker, K. S., and Clark, D. K. (1988a.) 'A semianalytic radiance model of ocean color', Journal of Geophysical Research 93, 10909-10924.

Gordon, H.R., Clark, D.K., Brown, J.W., Brown, O.B., and Evans, R.H. (1983a,) 'Phytoplankton pigment concentrations in the Middle Atlantic Bight: comparison of ship determinations and CZCS estimates', Applied Optics 22 (1), 20-36.

Gordon, H. R., Brown, J. W.,Brown, O. B., Evans, R. H., and Clark, D. K. (1983b.) 'Nimbus 7 CZCS: reduction of its radiometric sensitivity with time', Applied Optics 22 (24), 3929-3931.

Gordon, H.R., Clark, D.K., Mueller, J. L., and Hovis, W.A. (1980) 'Phytoplankton pigments from Nimbus-7 Coastal Zone Color Scanner; comparison with surface measurements', Science 210, 63-66.

Guan, F., Pelaez, J., and Stewart, R. H. (1985) 'The atmospheric correction and measurement of chlorophyll concentration using the coastal zone color scanner', Limnology and Oceanography 30 (2), 273-285.

Haury, L.R., Simpson, J.J., Pelaez, J., Koblinsky, C.J., and Wiesenhahn, D. (1986) 'Biological consequences of a persistent eddy off Point Conception, California', Journal of Geophysical Research 91, 12937-12956.

Herring, W. S. (1985) 'Extended analysis of CZCS radiometric sensitivity decay', (preliminary report). Visibility Laboratory, Scripps Institute of Oceanography, University of California, San Diego, 22 pp.

Jain, A. K., and Dubes, R. C. (1988) 'Algorithms for Clustering Data', Prentice Hall, Englewood Cliffs, NJ.

Jerlov, N.G. (1976) 'Marine Optics', (2nd Edition), Elsevier Scientific Publishing Company, New York.

Jerlov, N.G., and Steemann Nielsen, E. (eds.) (1974) 'Optical Aspects of Oceanography', Academic Press, New York.

Khosraviani, G. and Cracknell, A.P. (1987) 'A two-look technique for studying atmospheric effects in optical scanner data from the ocean', International Journal of Remote Sensing 8, 291-308.

Morel, A., and Prieur, L. (1977) 'Analysis of Variations in Ocean Color', Limnology and Oceanography 22, 709-722.

Muerle, J.L. and Allen, D.C. (1968) 'Experimental evaluation of techniques for automatic segmentation of objects in a complex scene', in G. Cheng et. al. (eds.), Pictorial Pattern Recognition, Thompson, Washington, D.C. pp. 3-13.

Mueller, J. L. (1985) 'Nimbus-7 CZCS: confirmation of its radiometric sensitivity decay rate through 1982', Applied Optics 24 (7), 1043-1047.

Mueller, J. L. (1988) 'Nimbus-7 CZCS: electronic overshoot due to cloud reflectance', Applied Optics 27 (3), 438-440.

Mukai, S. (1990) 'Atmospheric correction of remote sensing images of the ocean based on multiple-scattering calculations', IEEE Trans. on Geoscience and Remote Sensing 28, 696-702.

Mukai, S. and Ueno, S. (1978) 'Apparent contrast of an atmosphere-ocean system with an oil polluted sea surface', Applied Mathematical Modelling 2, 254-260.

Mukai, S., Takemata, K., Kusaka, T., and Sasano, Y., (1990) 'Optical thickness of aerosol in Tsukaba area: Comparison of lidar measurements with Lowtran-6', Journal of Environmental Science Japan 3, 15-21.

Neumann, G. and Pierson, W. J., Jr. (1966) 'Principles of Physical Oceanography', Prentice-Hall, Inc., Englewood Cliffs, N.J.

NOAA polar orbiter data users guide. 1985. National Environmental Satellite, National Climatic Data Center, Satellite Data Service Division, U.S. Dept. of Commerce, Washington, D.C.

Pairman, D. and Kittler, J. (1986) 'Clustering algorithms for use with images of clouds', International Journal of Remote Sensing 7, 855-866.

Pavlidis, T., (1977) 'Structural Pattern Recognition', Springer-Verlag, Berlin.

Payne, R.E. (1972) 'Albedo of the sea surface', J. Atmos. Sci. 29, 959-970.

Pelaez, J. and McGowan, J. A. (1986) 'Phytoplankton pigment patterns in the California current as determined by satellite', Limnology and Oceanography 31 (5), 927-950.

Ramanthan, V. (1987) 'The role of earth radiation budget studies in climate and global circulation research', Journal of Geophysical Research 92, 4075-4095.

Richards, J.A., (1986) 'Remote Sensing Digital Image Analysis: An Introduction', Springer-Verlag, New York.

Robinson, I.S. (1985) 'Satellite Oceanography: an Introduction for Oceanographers and Remote Sensing Scientists', Ellis Horwood Limited, New York, 455 pp.

Seddon, A. M. and Hunt, G. E. (1985) 'Segmentation of clouds using cluster analysis', International Journal of Remote Sensing 6, 717-731.

Shannon, L.V. (1985) 'Description of the ocean colour and upwelling experiments', in L.V. Shannon (ed.), South African Ocean Colour and Upwelling Experiment, Sea Fisheries Research Institute (pub.), Cape Town, South Africa, pp. 1-12.

Simpson, J.J. (1985) 'Air-sea exchange of carbon dioxide and oxygen induced by phytoplankton: Methods and interpretation', in A. Zirino (ed.), Mapping Strategies in Chemical Oceanography, American Chemical Society, Washington, D.C., pp. 409-450.

Simpson, J.J. (1986) 'Processes affecting upper ocean chemical structure in an eastern boundary current', in J.D. Burton, P.G. Brewer, and R. Chesselet (eds.), Dynamic Processes in the Chemistry of the Upper Ocean, Plenum Publishing Corp., New York, pp. 53-77.

Simpson, J.J. (1992a.) 'Remote sensing in fisheries: a tool for better management in the utilization of a renewable resource', Canadian J. Fish. Aquatic Sci., in press.

Simpson, J. J., (1992b.) 'Image masking using polygon fill and morphological operations', Remote Sens. Environ., 40, 161-183.

Simpson, J.J. and Humphrey, C. (1990) 'An automated cloud screening algorithm for daytime AVHRR imagery', Journal of Geophysical Research 95, 13459-13481.

Simpson, J.J., Koblinsky, C.J., Pelàez, J., Haury, L.R., and Wiesenhahn, D. (1986) 'Temperature-plant pigment-optical relations in a recurrent offshore eddy near Point Conception, California', Journal of Geophysical Research 95, 12915-12936.

Simpson, J.J. and Paulson, C.A. (1979) 'Mid-ocean observations of atmospheric radiation', Quarterly Journal of the Royal Meteorological Society 105, 487-502.

Smith, R.C. and Wilson, W.H. (1981) 'Ship and satellite bio-optical research in the California Bight', in J.F.R. Gower (ed.), Oceanography from Space, Plenum Press, New York, pp. 281-294.

Sobolev, V.V. (1963) 'Treatise of Radiative Transfer', D. van Nostrand, New York, NY.

Stewart, Robert H. (1985) 'Methods of Satellite Oceanography', University of California Press, Los Angeles, California, 360 pp.

Strub, P.T., James, C., Thomas, A.C., and Abbott, M.A. (1990) 'Seasonal and non-seasonal variability of satellite-derived surface pigment concentration in the California Current', Journal of Geophysical Research 95, 11501-11530.

Sturm, B. (1981) 'The atmospheric correction of remotely sensed data and the quantitative determination of suspended matter in marine water surface layers', in Arthur P. Cracknell (ed.), Remote Sensing in Meteorology, Oceanography, and Hydrology, D. Reidel Publishing Co., Boston, pp. 163-197.

Sturm, B. (1983) 'Selected topics of coastal zone color scanner (CZCS) data evaluation', in Arthur P. Cracknell (ed.), Remote Sensing Applications in Marine Science and Technology, D. Reidel Publishing Co., Boston, pp.137-167.

Walters, N.M. (1985) 'Algorithms for the determination of near-surface chlorophyll and semi-quantitative total suspended solids in South African coastal waters from Nimbus-7 CZCS data', in L.V. Shannon (ed.), South African Colour and Upwelling Experiment, Sea Fisheries Research Institute, Cape Town, South Africa, pp. 175-182.

U. S. Navy, (1981) 'U. S. Navy Marine Climatic Atlas of the World, Volume IX, World-Wide Means and Standard Deviations', Naval Oceanographic Command Detachment, Asheville, N. C., May 1981.

Zege, E.P. and Prikhach, A.S. (1991) 'Iterative procedure for atmospheric correction', Rem. Sens. Environ. Submitted.

Zege, E.P., Prikhach, A.S., and Chaikovskaya, L.I. (1991) 'Modeling of visible radiation propagation in the ocean-atmosphere system: Concepts models and programs', Rem. Sens. Environ. Submitted.

Zion, P.M. (1983) 'Description of algorithms for processing coastal zone color scanner (CZCS) data', JPL Publications 83-98, 29 pp.

REVIEW OF MAJOR CZCS APPLICATIONS:
U.S. CASE STUDIES

C. R. McCLAIN
Oceans and Ice Branch
Goddard Space Flight Center
National Aeronautics and Space Administration
Greenbelt, MD 21114
USA

ABSTRACT. A chronological and geographic bibliography of publications on Nimbus-7 Coastal Zone Color Scanner (CZCS) applications by investigators working in the United States is presented. The literature survey concentrates on oceanographic research and does not include a large volume of work which focuses on atmospheric correction algorithms, bio-optical algorithms and sensor performance. The survey shows that the number of papers published in the refereed literature rapidly expanded in the late 1980's and that the data have been used to address a broad spectrum of scientific questions in a wide variety of oceanic regimes. A discussion of CZCS application case studies for the southeastern continental shelf is provided.

1. Introduction

The CZCS data set has revolutionized the manner in which the oceanographic community views the ocean and has inspired many to pursue interdisplinary research. Initially, most ocean scientists were quite skeptical, even critical, of the CZCS mission, but today, largely because of the Nimbus Experiment Team's (NET) successful development of atmospheric and bio-optical algorithms, most researchers recognize the value of satellite ocean colour observations. Indeed, the usefulness of these observations has been demonstrated in a wide variety of application studies and are now regarded to be essential for programs addressing issues involving bio-geochemical processes from meso-scales to global scales. The results of these studies have also demonstrated the coupling of biological processes to physical processes and, in a few cases, have suggested feedback mechanisms where biological processes influence physical processes. Because of the mission's success, a number of future missions are scheduled for launch before the end of the decade, *e.g.* Sea-viewing Wide Field-of-view Sensor (SeaWiFS), Ocean Colour and Temperature Scanner (OCTS), and Moderate-Resolution Imaging Spectrometer - Nadir (MODIS-N) (these are discussed in a companion paper by van der Piepen and Doeffer).

Work on CZCS data by individuals outside the NET was initially hampered by several circumstances. The CZCS mission was proposed as a proof-of-concept experiment and was never intended to be an operational mission.

167

V. Barale and P.M. Schlittenhardt (eds.),
Ocean Colour: Theory and Applications in a Decade of CZCS Experience, 167–188.
© 1993 *ECSC, EEC, EAEC, Brussels and Luxembourg. Printed in the Netherlands.*

Thus, the ground data system was not scoped for near-realtime generation of level-1, level-2 and level-3 data or for the distribution of data to the research community. Secondly, the development and validation of operational algorithms could not be completed until post-launch validation and verification cruises and data analysis were completed. Also, a serious time- and wavelength-dependent deterioration of the sensor's sensitivity was recognized to be occurring not long after launch and required the development of methodologies for quantifying its magnitude. Thirdly, the oceanographic community was not equipped with the computational resources (cpu's, mass storage, processing and analysis software) for handling data of this nature. However, by the mid-1980's, this situation had begun to reverse and the number of publications began to rapidly increase (table 1). Also, this literature survey does not include a large volume of work which focuses on atmospheric correction algorithms, bio-optical algorithms and sensor performance.

TABLE 1. Temporal distribution of publications on CZCS oceanographic applications authored by investigators working in the U. S. Publications dealing with atmospheric correction algorithms, bio-optical algorithms and sensor performance are not included.

YEAR	NUMBER	
1979-1981	0	
1982	1	
1983	0	
1984	6	
1985	6	
1986	11	
1987	6	
1988	13	
1989	9	
1990	12	
1991	8*	
TOTAL	71	*Incomplete survey.

As indicated in table 2 and figure 1, U.S.-based investigators have used CZCS data in a wide variety of regional studies. Initially, publications tended to focus on one or two images such as those shown in figures 2 and 3, but in 1984, the number of publications began to increase. This increase in publications was assisted by a number of developments. One particularly noteworthy activity was the West Coast Time Series processing effort initiated in the early 1980's at the Jet Propulson Laboratory (JPL) by Mark Abbott using software (called "DSP") developed at the University of Miami Rosenstiel School of Marine and Atmospheric Science (RSMAS). This effort included the processing of local-area-coverage (LAC) Advanced Very High Resolution Radiometer (AVHRR) imagery for sea surface temperature as well as the CZCS data from the U.S. west coast collected at the Scripps Institution

TABLE 2. Catagorized chronological refereed publication list of CZCS oceanographic applications authored by investigators working in the United States.

GENERAL OCEANOGRAPHY
North & Central America
  East Coast, Mid-Atlantic
        Gordon *et al.*, 1982
        Yentsch, 1984
        Yentsch & Phinney, 1985
        Brown *et al.*, 1985
        Evans *et al.*, 1985
        Smith *et al.*, 1987
        Walsh *et al.*, 1987
        Elrod, 1988
        Eslinger *et al.*, 1989
  East Coast, South Atlantic
        McClain et al, 1984, 1985
        McClain & Atkinson, 1985
        Yoder *et al.*, 1987
        McClain *et al.*, 1988
        McClain *et al.*, 1990a
        Ishizaka, 1990a, b, c
  Gulf of Mexico
        Muller-Karger *et al.*, 1991
  West Coast
        Fiedler, 1984
        Laurs *et al.*, 1984
        Abbott & Zion, 1985
        Thomson & Gower, 1985
        Barale & Fay, 1986
        Haury *et al.*, 1986
        Pelaez & McGowan, 1986
        Simpson *et al.*, 1986
        Abbott & Zion, 1987
        Pelaez, 1987
        Denman & Abbott, 1988
        Michaelsen *et al.*, 1988
        Roughgarden *et al.*, 1988
        Smith *et al.*, 1988
        Perry *et al.*, 1989
        Thomas & Strub, 1989
        Strub *et al.*, 1990
        Thomas & Strub, 1990
        Abbott & Barksdale, 1991
        Chelton & Schlax, 1991
  Bering Sea
        Maynard & Clark, 1987
        Muller-Karger *et al.*, 1990

  Great Lakes
    Mortimer, 1988
  Caribbean Sea
        Hallock *et al.*, 1988
        Muller-Karger *et al.*, 1989
North Atlantic Ocean
        Maynard, 1986
        McClain *et al.*, 1986
        Deuser *et al.*, 1988
        Muller-Karger *et al.*, 1988
        Wroblewski *et al.*, 1988
        Dugdale *et al.*, 1989
        Yentsch, 1989
        Deuser *et al.*, 1990
        McClain *et al.*, 1990b
        Thompson *et al.*, 1990
Mediterranean Sea
        Barale *et al.*, 1984
        Barale *et al.*, 1986
        Arnone & LaViolette, 1986
        Lohrenz *et al.*, 1988
        Arnone *et al.*, 1990
Black Sea
        Hay and Honjo, 1989
Arabian Sea
        Banse & McClain, 1986
        Brock *et al.*, 1991a
        Brock and McClain, 1991b
Southern Ocean
        McClain *et al.*, 1991c
Polar Oceans
        Sullivan *et al.*, 1988
        Walsh, 1989
        Comiso *et al.*, 1990
        Mitchell *et al.*, 1991a
        Mitchell *et al.*, 1991b
North & Equatorial Pacific
        Feldman *et al.*, 1984
        Mueller and Lang, 1989

GLOBAL CZCS PROCESSING
& GENERAL REVIEWS
        Esaias *et al.*, 1986
        Feldman *et al.*, 1989
        Abbott & Chelton, 1991
        McClain *et al.*, 1992

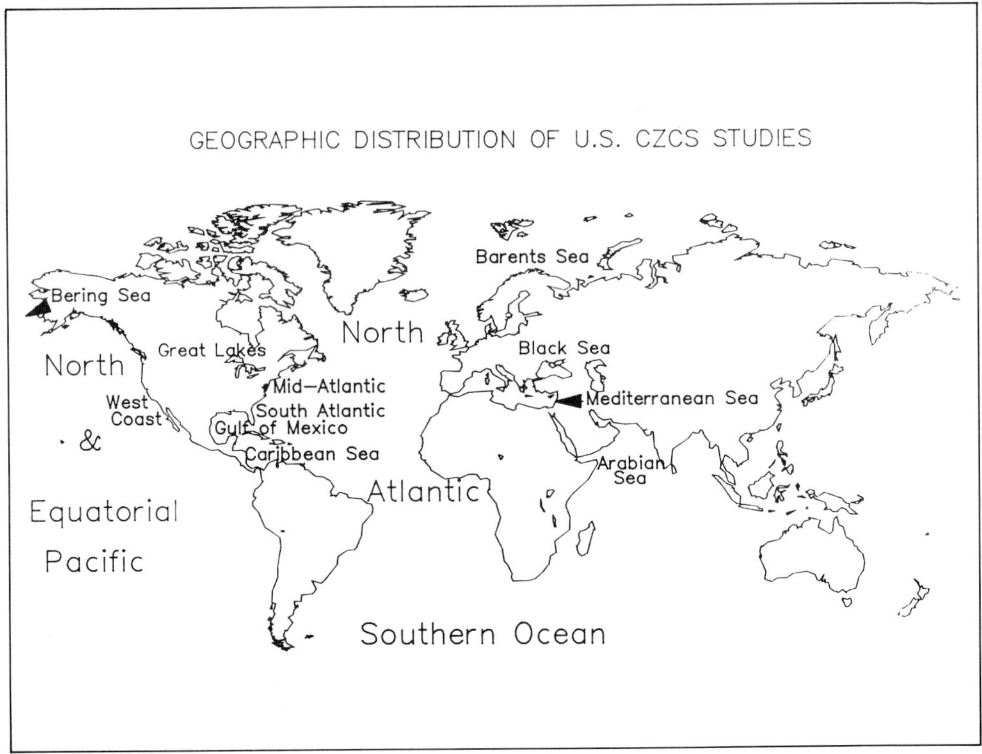

Figure 1. Global map showing regions where U.S.-based investigator CZCS studies were conducted.

of Oceanography in La Jolla, California. figure 4 shows sample sea surface temperature (SST) and pigment fields from the West Coast Time Series dataset. Access to the data was also improved by the development of image analysis systems such as SEAPAK (McClain *et al.*, 1991a) during the 1980's. Finally, the global CZCS processing project at NASA/Goddard Space Flight Center (Esaias *et al.*, 1986; Feldman *et al.*, 1989) has made all the CZCS data collected by NASA readily available.

Planning for that project was initiated in 1985 and the processing was completed in 1990. figure 5 presents the initial processing product, a monthly mean pigment data set for the North Atlantic which was used in preparation for the Joint Global Ocean Flux Study (JGOFS) field study in 1989.

Figure 6 shows a final product, a global mean pigment field (combined with vegetation index data by J. Tucker). The global CZCS processing project was also a collaboration with RSMAS. Thus, the data handling capabilities of the ocean colour community has evolved from the single image level of effort to the processing of the global dataset (over 66,000 scenes) in less than a decade. Recent reviews of CZCS related work have been written by Abbott and Chelton (1991) and McClain *et al.* (1992).

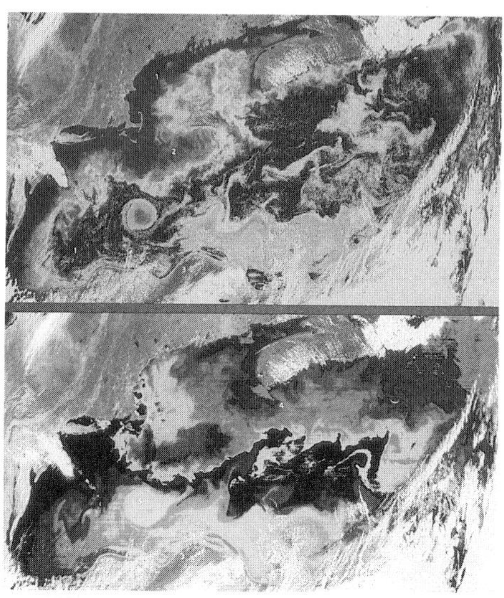

Figure 2. The classic CZCS pigment and sea surface temperature scenes from the U.S. east coast produced by the Nimbus Experiment Team. Colorphotograph on p. 350

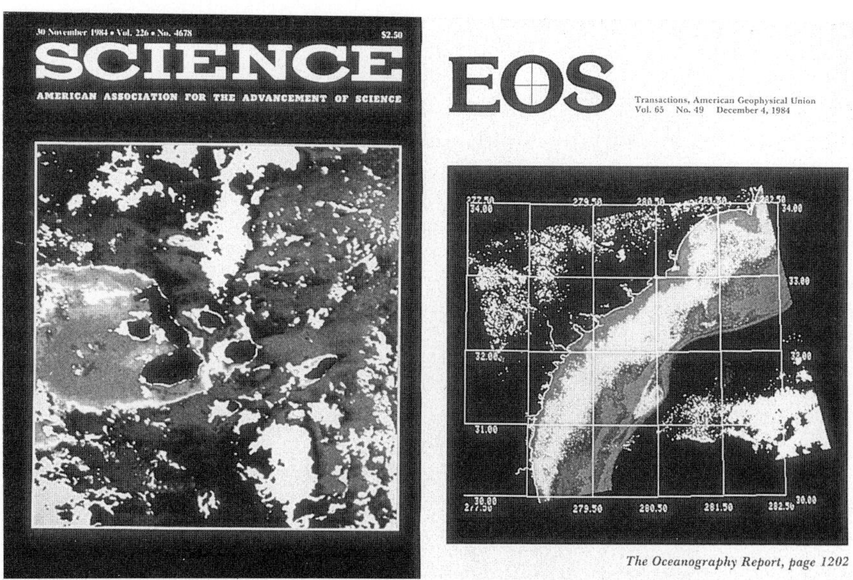

Figure 3. Covers of Science (Feldman *et al.*, 1984) and Eos (Blanton *et al.*, 1984; see also McClain *et al.*, 1984) which illustrate early investigations by non-NET researchers which focused on a single image or a small number of images. Colorphotograph on p. 351

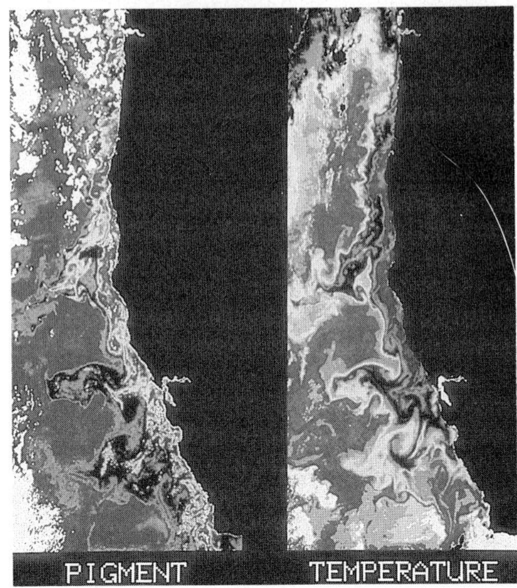

Figure 4. Temperature and pigment images from the West Coast Times Series project, the first regional CZCS processing project. Colorphotograph on p. 351

Figure 5. The first basin-scale monthly mean composite of CZCS pigment data which was the initial objective in the NASA/GSFC global processing project (Esaias *et al.*, 1986). Colorphotograph on p. 352

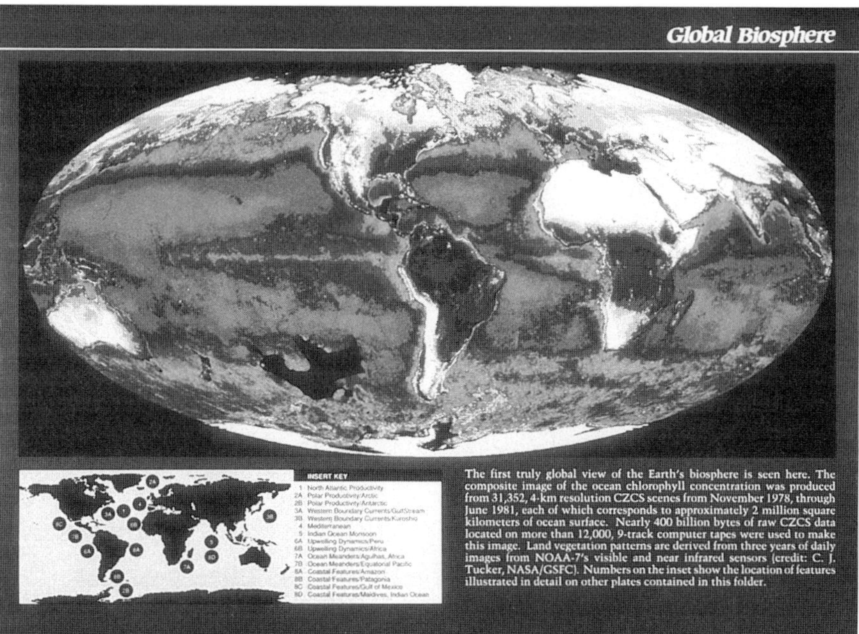

Figure 6. The mean pigment concentration derived from all the CZCS data as a result of the NASA/GSFC global processing project (Feldman *et al.* 1989). The data is combined with vegetation index data by J. Tucker (NASA/GSFC).

Colorphotograph on p. 352

2. Review of CZCS Applications in the South Atlantic Bight

As an example of the kind of CZCS applications typical of the US-based oceanographic community, a series of case studies concerning the South Atlantic Bight (SAB) shall be considered in some detail. The SAB (see figure 7) is the shelf region between Cape Canaveral, Florida (28°N) and Cape Hatteras, North Carolina (35.5°N). The SAB consists of four large embayments which are, from south to north, the Georgia Bight (Cape Canaveral to Cape Romain), Long Bay (Cape Romain to Cape Fear), Onslow Bay (Cape Fear to Cape Lookout) and Raleigh Bay (Cape Lookout to Cape Hatteras).

The influence of the Gulf Stream is particularly significant in these embayments with the exception of Long Bay where the continental shelf is wide and the Gulf Stream is disconnected from the shelf break due to a deflection of the current by a topographic feature called the Charleston Bump (Brooks and Bane, 1978; Chao and Janowitz, 1979). Along the coast, fresh water discharge for the entire SAB is largely confined to rivers in South Carolina and Georgia. The papers in a special issue of the Journal of Geophysical Research, Oceans (Volume 88:C8, 1983) and in Atkinson *et al.* (1985) provide a multi-disciplinary review of the biological, physical and chemical oceanography of the SAB.

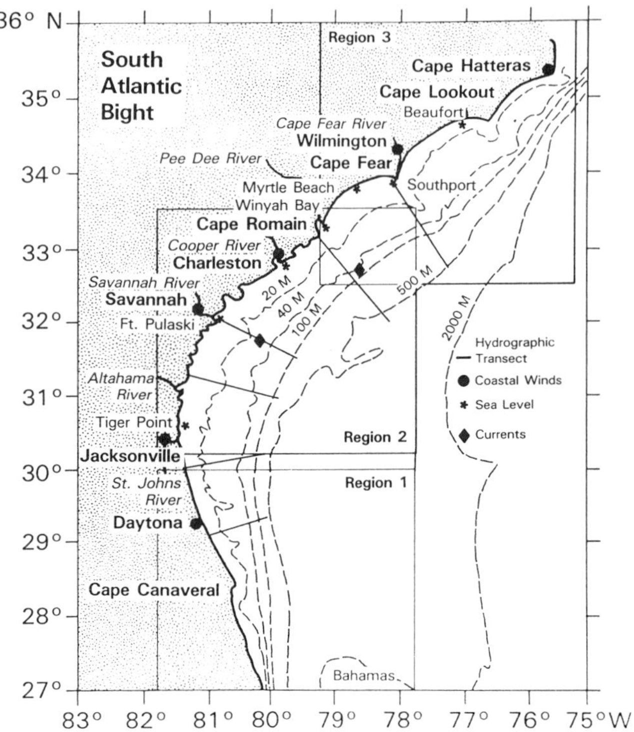

Figure 7. Map of the South Atlantic Bight (McClain *et al.*, 1988)

Upwelling in the South Atlantic Bight occurs as a result of (1) the shelf break flow encountering divergent isobaths on the lea side of capes (Blanton *et al.*, 1981), (2) Gulf Stream meander and eddy processes (Yoder *et al.*, 1981; Lee and Atkinson, 1983; McClain *et al.*, 1984, 1985), (3) wind-driven coastal upwelling (McClain *et al.*, 1988) and (4) the deflection of the Gulf Stream by bathymetric features (McClain and Atkinson, 1985). The Eos cover in figure 2 shows the pigment distributions of the Georgia Bight coast and of a Gulf Stream cold-core eddy (Blanton *et al.*, 1984). North of the Georgia Bight, phytolankton blooms in the winter are observed as the result of convective mixing of the water column in response to periodic cold air outbreaks which are characterized by high offshore winds, clear skies and large fluxes of heat to the atmosphere. These wintertime events were studied extensively during the Genesis of Atlantic Lows Experiment (GALE). The results of oceanographic studies conducted during GALE are presented in a special issue of the Journal of Geophysical Research, Oceans (Volume 94:C8, 1989). Studies on the influence of cold-air outbreaks on biological processes in the SAB which incorporate CZCS data are underway, but have not yet been published.

Prior to the late 1970's, it was generally believed that biological processes in the SAB were confined to the estuaries and nearshore regimes (Turner *et al.*, 1979; Turner, 1981). The nearshore region is separated from the continental shelf waters by a density front (Blanton, 1981). Thus, it was believed that the primary nutrient source (*i.e.* land) was isolated from the offshore regimes, at least in the Georgia Bight. Previous studies in Onslow Bay to the north had documented large nutrient-rich intrusions of subsurface Gulf Stream water onto the shelf (Blanton, 1971), but their contribution to biological productivity had not been evaluated and were thought to be too episodic to have a significant impact. It is now known that these events play a major role in the continental shelf's biological productivity (Lee *et al.*, 1991).

By the late 1970's, field programs supported by the Department of Energy and the Minerals Management Service were designed to quantify the frequency and magnitude of the intrusions and the physical, chemical and biological processes associated with them. A preliminary field survey was conducted during April-May 1979 off the Florida-Georgia coast (the CZCS was launched in October 1978). During that cruise, an extensive phytoplankton bloom was observed along the Gulf Stream front (figure 8). The bloom extended from Cape Canaveral to the Charleston Bump (Georgia) and was several hundred kilometers long. While the feature's greatest width was only about 20 kilometers, concentrations exceeded 7 mg/m^3 in some locations (Yoder *et al.*, 1981; McClain *et al.*, 1984, 1985). Concentrations in

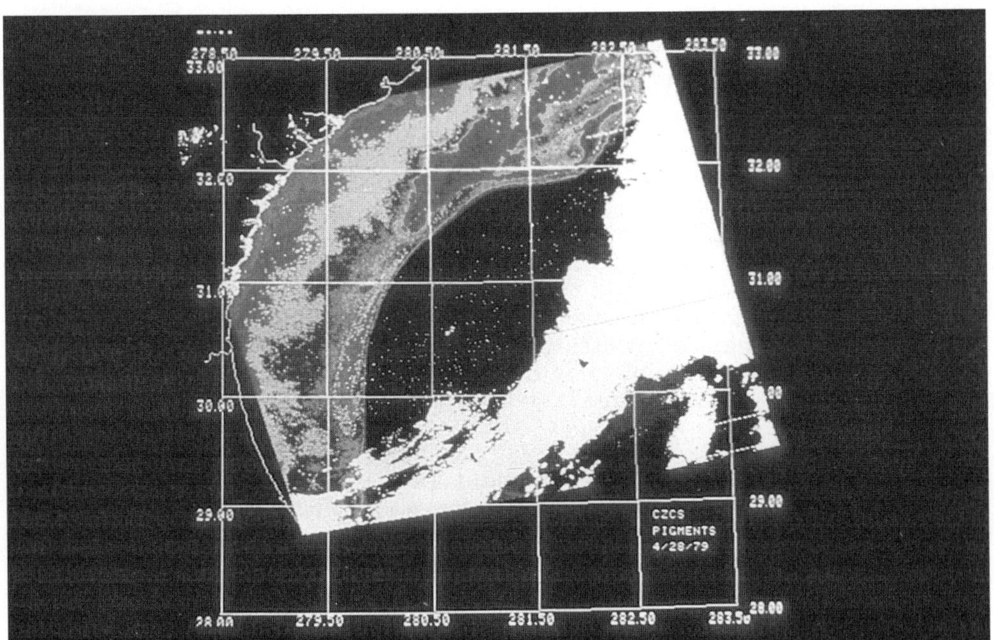

Figure 8. CZCS pigment image showing a Gulf Stream frontal upwelling event on April 28, 1979 (McClain *et al.*, 1984, 1985). Color bar for the image is shown in figure 12. Colorphotograph on p. 353

the adjacent Gulf Stream were ≈ 0.1 mg/m^3 and were a few tenths of a mg/m^3 over the continental shelf. This data provided an early test of the CZCS algorithms as detailed pigment data from transects across the feature could be matched with CZCS imagery for the same day. As indicated in figure 9, the agreement was excellent. The surface bloom was the result of upwelling associated with a cold-core eddy (figure 3, right panel, is another example). These frontal wave features propagate northward and are commonly observed along the Gulf Stream front in the SAB.

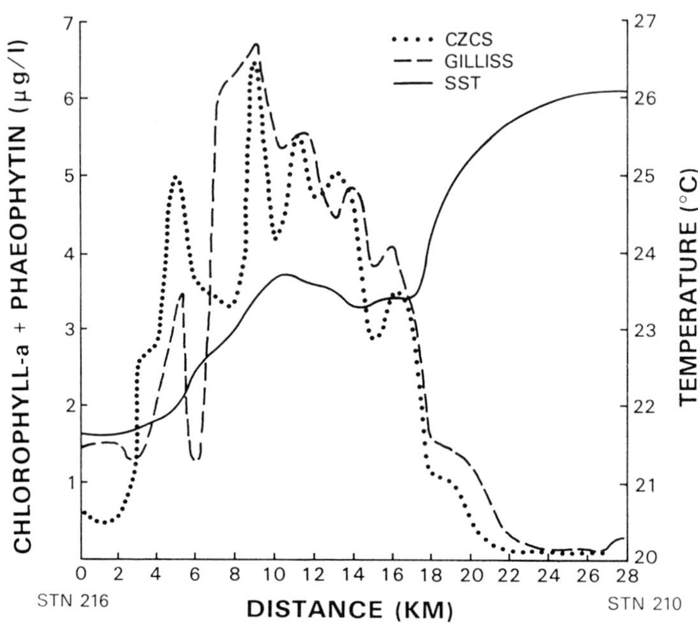

Figure 9. Horizontal profiles of in situ surface pigment concentration (RV Gilliss), CZCS pigment concentration and SST from a transect across the Gulf Stream frontal upwelling shown in figure 8 (McClain *et al.*, 1984, 1985).

During the following spring 1980, a comprehensive field study was conducted, the Georgia Bight Experiment-I (GABEX-I, Blanton *et al.*, 1984). The experiment incorporated two survey vessels and a moored array of current meters and temperature probes. During the experiment, a number of frontal eddies and upwelling events were observed (McClain *et al.*, 1984, 1985; McClain *et al.*, 1990a; Ishizaka 1990a, b, c). These studies include detailed analyses of a sequence of CZCS scenes collected during a one week period in April, 1980 (figure 10). The Ishizaka papers are discussed in greater detail in a companion article in this volume by Ishizaka and Hofmann. McClain *et al.* used pigment concentration differences between consecutive images, spatial derivatives and current meter data to estimate the variability in space and time of processes which control pigment distributions (figure 11).

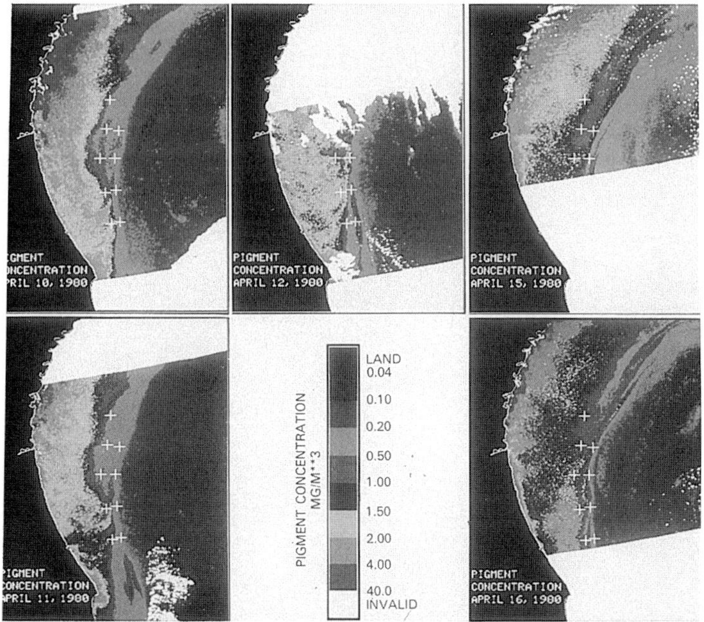

Figure 10. Sequence of CZCS pigment images during the period of April 10 - 16, 1980 (McClain *et al.*, 1990a). The current meter moorings are indicated by + signs and are numbered from left to right, bottom to top in the sequence 3, 4, 5, 6, 9, 10, 11, 12 and 15 (referenced in figure 11). Colorphotograph on p. 353

The biological-vertical (BV) component is the summation of net phytoplanton growth and vertical fluxes which could not be explicitly quantified in the analyses. The result shows that all processes are important at the shelf break and that the magnitudes are quite variable.

Another unique feature of the SAB is the Charleston Gyre. The deflection of the Gulf Stream by the Charleston Bump results in a quasi-permanent cyclonic eddy over the outer continental shelf and slope offshore of Cape Romain. Singer *et al.* (1983) provide a historical perspective on observations in the vicinity of Cape Romain and the Charleston Gyre. The most vivid illustrations of the gyre are CZCS and SeaSat Synthetic Aperture Radar (SAR) images in McClain and Atkinson (1985, figure 12). The enhanced pigment concentrations in the eddy are the result of upwelling produced by domed isopycnals in its core. The satellite data was obtain during a hydrographic survey of the gyre (the transect lines are shown in figure 12. Pigment data from the survey and the CZCS compared very well.

Yoder *et al.* (1987) and McClain *et al.* (1988) used a 13-month time series (October 1978 to November 1979) of imagery covering the continental shelf from Cape Canaveral, Florida to Cape Hatteras, North Carolina to examine spatial and temporal variability of pigment concentrations. They separated the SAB coverage into the three regions shown in figure 7. The data set included 143 full- resolution scenes extracted from 71 orbits. Each scene was

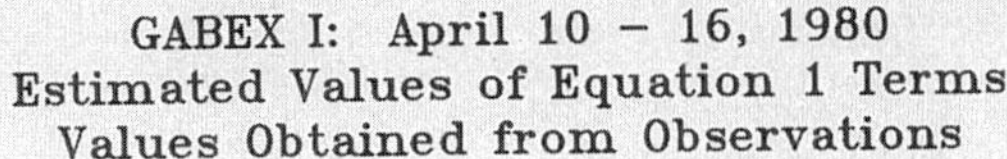

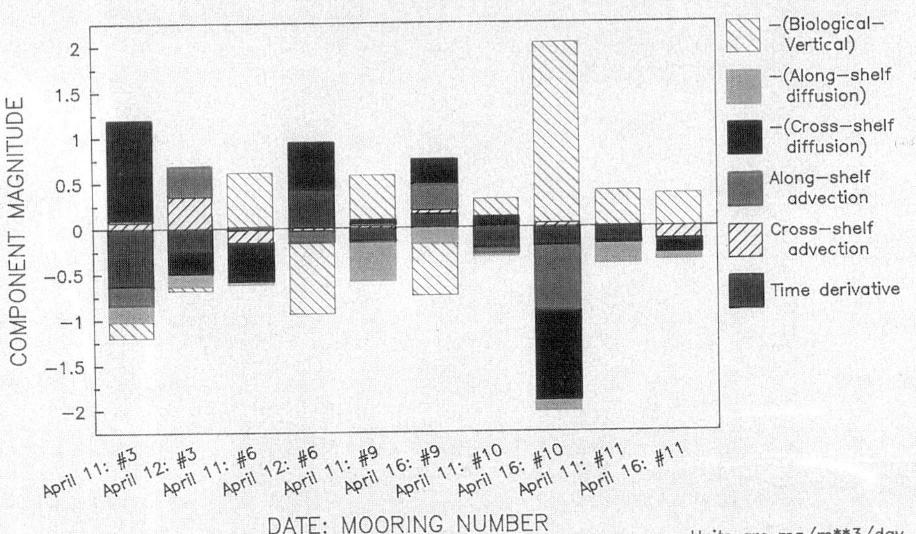

Figure 11. Estimation of the magnitude of various terms in the pigment balance equation at different moorings and days (McClain *et al.*, 1990a). The equation is: time derivative $(\partial C/\partial t)$ + cross-shelf advection $(u\partial C/\partial X)$ + along-shelf advection $(v\partial C/\partial Y)$ - cross-shelf diffusion $(K\partial 2C/\partial 2X)$ - along-shelf diffusion $(K\partial 2C/\partial 2Y)$ - biological- vertical (BV) processes = 0. All terms except BV are computed from the data and BV is the residual. Colorphotograph on p. 354

processed to level-2 using Ångström exponents determined for each orbit using the interactive procedure described in Barale *et al.* (1986). The mean Ångström exponents, as defined in Gordon *et al.* (1983), were -0.38, -0.41 and -0.34 for 443, 520 and 550 nm, respectively, indicating a significant influence from continental haze. The Ångström exponents used in the CZCS global processing (0.12, 0, 0 for 443, 420 and 550 nm, respectively) are typical of marine haze. Yoder *et al.* used scale analyses (variograms) to estimate the variation of dominant length scales with season and location (nearshore, mid-shelf, shelf break) for the three regions in figure 7. Ishizaka (1990a, b, c) used the results of the scale analyses in the objective analysis of the GABEX current meter data. McClain *et al.* correlated temporal variability of the CZCS pigment concentrations with river discharge, coastal sea level, coastal wind stress, currents (from fixed moorings) and stratification (from 5 general surveys of the Georgia Bight, standard transects shown in figure 7). One analysis performed by McClain *et al.* was the computation of six-month seasonal mean concentrations for each region. Differences in these means

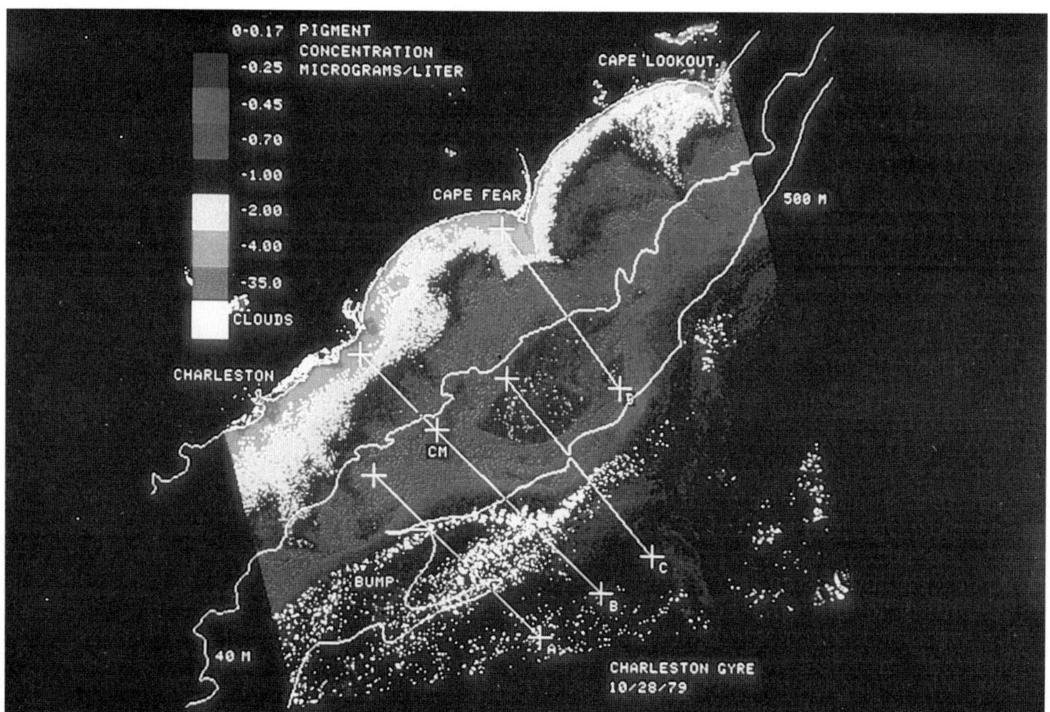

Figure 12. Pigment concentration image of the Charleston Gyre (McClain and Atkinson, 1985) showing hydrographic transects and bathymetry contours.

Colorphotograph on p. 354

helped identify when and where various processes were active. figure 13 is an image showing the difference image for Region 2 (winter minus summer). The positive values imply higher concentrations in winter and indicate that much of the late winter river discharge exited the shelf off Cape Romain. The negative values are the result of summertime wind- driven coastal upwelling which enhanced phytoplankton concentrations.

In conclusion, the SAB provides a wealth of physical and biological processes that can be observed using the CZCS. Because of the extensive field programs that were being conducted there during the CZCS mission and the relatively dense CZCS coverage, a variety of studies have been successfully completed and others are underway. To date, most of the CZCS data from the SAB has not been thoroughly analyzed. However, researchers at Old Dominion University and at the University of Rhode Island have recently received large quantities of full resolution data from the CZCS archive at NASA/GSFC and are processing the data using the SEAPAK and Miami DSP systems.

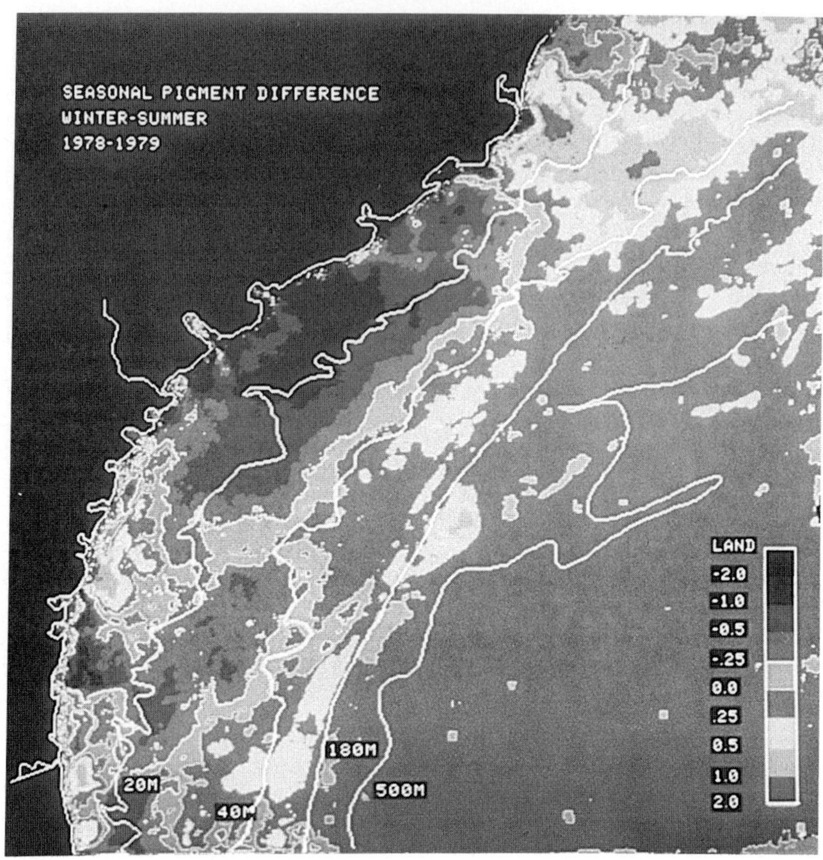

Figure 13. Seasonal pigment difference image (winter minus summer) for Region 2 (McClain *et al.*, 1988). Seasons were defined as winter (November - April) and summer (May - October). Colorphotograph on p. 355

References

Abbott, M. R. and Zion, P. M. (1985) 'Satellite observations of phytoplankton variability during an upwelling event', Continental Shelf Research 4, 661-680.

Abbott, M. R. and Zion, P. M. (1987) 'Spatial and temporal variability of phytoplankton pigment off northern California during Coastal Ocean Dynamics Experiment 1', Journal of Geophysical Research 92(C2), 1745-1755.

Abbott, M. R. and Chelton, D. B. (1991) 'Advances in passive remote sensing of the ocean, U. S. National Report to International Union of Geodesy and Geophysics 1987-1990', Contributions in Oceanography, EOS Transactions AGU, 571-589.

Abbott, M. R. and Barksdale, B. (1991) 'Phytoplankton pigment patterns and wind forcing off central California', Journal of Geophysical Research 96 (C8), 14,649-14,667.

Arnone, R. A. and LaViolette, P. E. (1986) ' Satellite definition of the bio-optical and thermal variation of coastal eddies associated with the African Current', Journal of Geophysical Research 91 (C2), 2351-2364.

Arnone, R. A., Wiesenburg, D. A., and Saunders, K. D. (1990) 'The origin and characteristics of the Algerian Current', Journal of Geophysical Research 95 (C2), 1587-1598.

Atkinson, L. P., Menzel, D. W., and Bush, K. A. (1985) 'Oceanography of the Southeastern U. S. Continental Shelf', AGU, Washington, D. C., pp. 156.

Banse, K. and McClain, C. R. (1986) 'Satellite-observed winter blooms of phytoplankton in the Arabian Sea', Marine Ecology Progress Series 34 (3), 201-211.

Barale, V., Malanotte Rizzoli, P., and Hendershott, M. C. (1984) 'Remote sensing the surface dynamics of the Adriatic Sea', Deep-Sea Research 3 (12), 1433-1459.

Barale, V. and Wittenburg Fay, R. (1986) 'Variability of the ocean surface colour field in central California near-coastal waters as observed in seasonal analysis of CZCS imagery', Journal of Marine Research 44, 291-316.

Barale, V., McClain, C. R., and Malanotte-Rizzoli, P. (1986) 'Space and time variability of the surface colour field in the northern Adriatic Sea', Journal of Geophysical Research 91, 12957-12974.

Blanton, J. O. (1971) 'Exchange of Gulf Stream water with North Carolina shelf water in Onslow Bay during stratified conditions', Deep-Sea Research 18, 167-178.

Blanton, J. O. (1981) 'Ocean currents along a nearshore frontal zone on the continental shelf of the southeastern United States', Journal of Physical Oceanography 11, 1627-1637.

Blanton, J. O., Atkinson, L. P., Pietrafesa, L. J., and Lee, T. N. (1981) 'The intrusion of Gulf Stream Water across the continental shelf due to topographically-induced upwelling', Deep-Sea Research 28A, 393-405.

Blanton, J. O., and others (1984) 'A multidisciplinary oceanography program on the southeastern U.S. continental shelf', EOS 65, 1202-1203.

Brock, J. C., McClain, C. R., Luther, M. E., and Hay, W. W. (1991) 'The phytoplankton bloom in the northwest Arabian Sea during the southwest monsoon of 1979', Journal of Geophysical Research 96 (C11), 20623-20642.

Brock, J. C. and McClain, C. R. (1992) 'Interannual variability in phytoplankton blooms observed in the northwestern Arabian Sea during the southwest monsoon', Journal of Geophysical Research 97 (C1), 733-750.

Brooks, D. A. and Bane, J. M., Jr., (1978) 'Gulf Stream deflection by a bottom feature off Charleston, South Carolina', Science 201, 1225-1226.

Brown, O. B., Evans, R. H., Brown, J. W., Gordon, H. R., Smith, R. C., and Baker, K. S. (1985) 'Phytoplankton blooming off the U. S. East Coast: a satellite description', Science 229, 163-167.

Chao, S.-Y. and Janowitz, G. S. (1979) 'The effect of a localized topographic irregularity on the flow of a boundary current along the continental margin', Journal of Physical Oceanography 9, 900-910.

Chelton, D. B. and Schlax, M. G. (1991) 'Estimation of time averages from irregularly spaced observations: with application to Coastal Zone Color Scanner estimates of chlorophyll concentrations', Journal of Geophysical Research 96 (C8), 14669-14692.

Comiso, J. C., Maynard, N. G., Smith, W. O., Sullivan, Jr. and C. W. (1990) 'Satellite ocean colour studies of Antarctic ice edges in summer and autumn', Journal of Geophysical Research 95 (C6), 9481-9496.

Denman, K. L. and Abbott, M. R. (1988) 'Time evolution of surface chlorophyll patterns from cross-spectrum analysis of satellite colour images', Journal of Geophysical Research 93 (C6), 6789-6798.

Deuser, W. G., Muller-Karger, F. E., and Hemleben, C. (1988) 'Temporal variations of particle fluxes in the deep subtropical and tropical North Atlantic: Eulerian versus Lagrangian effects', Journal of Geophysical Research 93 (C6), 6857-6862.

Deuser, W. G., Muller-Karger, F. E., Evans, R. H., Brown, O. B., Esaias, W. E., and Feldman, G. C. (1990) 'Surface-ocean colour and deep-sea carbon flux: how close a connection?', Deep-Sea Research 37 (8), 1331-1343.

Dugdale, R. C., Morel, A., Bricaud, A., and Wilkerson, F. P. (1989) 'Modeling new production in upwelling centers: a case study of modeling new production from remotely sensed temperature and colour', Journal of Geophysical Research 94 (C12), 18119-18132.

Elrod, J. A. (1988) 'CZCS view of an oceanic acid waste dump', Remote Sensing of Environment 25, 245-254.

Esaias, W., Feldman, G., McClain, C. R., and Elrod, J. (1986) 'Satellite observations of oceanic primary productivity', EOS 67 (44), 835-837.

Eslinger, D. L. and Iverson, R. L. (1986) 'Wind effects on Coastal Zone Color Scanner chlorophyll patterns in the U. S. Mid-Atlantic Bight during spring 1979', Journal of Geophysical Research 91 (C11), 12,985-12,992.

Eslinger, D. L., O'Brien, J. J., and Iverson, R. L., (1989) 'Empirical orthogonal function analysis of cloud-contaminated Coastal Zone Color Scanner images of northeastern North American coastal waters', Journal of Geophysical Research 94 (C8), 10884-10890.

Feldman, G., Clark, D., and Halpern, D. (1984) 'Satellite colour observations of the phytoplankton distribution in the eastern equatorial Pacific during the 1982-1983 El Niño', Science 226 (4678), 1069-1071.

Feldman, G. (1986) 'Variability of the productive habitat in the eastern equatorial Pacific', EOS 67, 106-108.

Feldman, G. (1989) 'Ocean Color: Availability of the global data set', EOS 70 (23), 634.

Fiedler, P. C. (1984) 'Satellite observations of the 1982-1983 El Niño along the U. S. Pacific coast', Science 224, 1251-1254.

Gordon, H. R., Clark, D. K., Brown, J. W., Brown, O. B., and Evans, R. H. (1982) 'Satellite measurements of phytoplankton pigment concentration in the surface waters of a warm core Gulf Stream ring', Journal of Marine Research 40 (2), 491-502.

Gordon, H. R., Clark, D. K., Brown, J. W., Brown, O. B., Evans, R. H., and Broenkow, W. W. (1983) 'Phytoplankton pigment concentrations in the Middle Atlantic Bight: Comparison of ship determinations and CZCS estimates', Applied Optics 22 (1), 20-36.

Hallock, P., Hine, A. C., Vargo, G. A., Elrod, J. A., and Jaap, W. C. (1988) 'Platforms of the Nicaraguan Rise: examples of the sensitivity of carbonate sedimentation to excess trophic resources', Geology 16, 1104-1107.

Haury, L. R., Simpson, J. J., Pelaez, J., Koblinsky, C., and Wiesenhahn, D. (1986) 'Biological consequences of a recurrent eddy off Point Conception, California', Journal of Geophysical Research 91 (C11), 12937-12956.

Hay, B. and Honjo, S. (1989) 'Particle deposition in the present and holocene Black Sea', Oceanography, 26-31.

Ishizaka, J. (1990a) 'Coupling of Coastal Zone Color Scanner data to physical-biological model of the southeastern U. S. continental shelf ecosystem, 1. CZCS data description and Lagrangian particle tracing experiments', Journal of Geophysical Research 95 (C11), 20167-10181.

Ishizaka, J. (1990b) 'Coupling of Coastal Zone Color Scanner data to physical-biological model of the southeastern U. S. continental shelf ecosystem, 2. an Eulerian model', Journal of Geophysical Research 95 (C11), 20183-10199.

Ishizaka, J. (1990c) 'Coupling of Coastal Zone Color Scanner data to physical-biological model of the southeastern U. S. continental shelf ecosystem, 3.

nutrient and phytoplankton fluxes and CZCS data assimilation', Journal of Geophysical Research 95 (C11), 20201-10212.

Laurs, R. M., Fiedler , P. C., and Montgomery, D. R. (1984) 'Albacore tuna catch distributions relative to environmental features observed from satellites', Deep-Sea Research 31 (9), 1085-1099.

Lee, T. N. and Atkinson, L. P. (1983) 'Low-frequency current and temperature variability from Gulf Stream frontal eddies and atmospheric forcing along the southeastern U. S. outer continental shelf', Journal of Geophysical Research 88, 4541-4567.

Lee, T. N., Yoder, J. A., and Atkinson, L. P. (1991) 'Gulf Stream frontal eddy influence on productivity of the southeastern U. S. continental shelf', Journal of Geophysical Research 96 (C12), 22191-22205.

Lohrenz, S. E., Arnone, R. A., Wiesenburg, D. A., and DePalma, I. P. (1988) 'Satellite detection of transient enhanced primary production in the western Mediterranean Sea', Nature 335 (6187), 245-247.

Maynard, N. G. (1986) 'Coastal Zone Color Scanner imagery in the marginal ice zone', Marine Technology Society Journal 20 (2),14-27.

Maynard, N. G. and Clark, D. K . (1987) 'Satellite colour observations of spring blooming in Bering Sea shelf waters during the ice edge retreat in 1980', Journal of Geophysical Research 92 (C7), 7127-7139.

McClain, C. R., Pietrafesa, L. J., and Yoder, J. A. (1984) 'Observations of Gulf Stream-induced and wind-driven upwelling in the Georgia Bight using ocean colour and infrared imagery', Journal of Geophysical Research 89, 3705- 3723.

McClain, C. R., Pietrafesa, L. J., and Yoder, J. A. (1985) correction to 'Observations of Gulf Stream-induced and wind-driven upwelling in the Georgia Bight using ocean colour and infrared imagery', Journal of Geophysical Research 90, 12015-12018.

McClain, C. R. and Atkinson, L. P. (1985) 'A note on the Charleston Gyre', Journal of Geophysical Research 90, 11857-11861

McClain, C., Chao, S.-Y., Atkinson, L., Blanton, J., and de Castillejo, F. (1986) 'Wind-driven upwelling in the vicinity of Cape Finisterre', Spain, Journal of Geophysical Research 91 (C7), 8470-8486.

McClain, C. R., Yoder, J. A., Atkinson, L. P., Blanton, J. O., Lee, T. N., Singer, J. J., and Muller-Karger, F. (1988) 'Variability of Surface Pigment Concentrations in the South Atlantic Bight', Journal of Geophysical Research 93 (C9), 10675-10697.

McClain, C. R., Ishizaka, J., and Hofmann, E. (1990a) 'Estimation of phyto-plankton pigment changes on the Southeastern U. S. continental shelf from a sequence of CZCS images and a coupled physical- biological model', Journal of Geophysical Research 95 (C11), 20213-20235.

McClain, C. R., Esaias, W. E., Feldman, G. C., Elrod, J., Endres, D., Firestone, J., Darzi, M., Evans, R., and Brown, J. (1990b) 'Physical and biological processes in the North Atlantic during the First Global GARP Experiment', Journal of Geophysical Research, 95 (C10), 18027-18048.

McClain, C. R., Darzi, M., Firestone, J., Yeh, E.-Y., Fu, G., and Endres, D., (1991a) 'SEAPAK Users Guide, Version 2.0', NASA Tech. Memo 100728, Vol. I-System Description, 158 pp., Vol. II-Descriptions of Programs, 586 pp., Natl. Aero. and Space Admin., Goddard Space Flight Center, Greenbelt, MD.

McClain, C. R., Koblinsky, C. J., Firestone, J., Darzi, M., Yeh, E.-N., and Beckley, B. D. (1992) 'Examining several Southern Ocean data sets', EOS Transactions AGU 72 (33), 345.

McClain, C. R., Feldman, G., and Esaias, W. (1992) 'Oceanic biological productivity', in R. Gurney, J. Foster and C. Parkinson (eds.), Global Change Atlas, Cambridge University Press, in press.

Michaelsen, J., Zhang, X., and Smith, R. C. (1988) 'Variability of pigment biomass in the California Current system as determined by satellite imagery, 2. temporal variability', Journal of Geophysical Research 93 (D9), 10883- 10896.

Mitchell, B. G., Brody, E. A., Holm-Hansen, O., McClain, C., and Bishop, J. (1991) 'Light limitation of phytoplankton biomass and macronutrient utilization in the Southern Ocean', Limnology and Oceanography., 36 (8), 1662-1677.

Mitchell, B. G., Brody, E. A., Yeh, E.-N., McClain, C., Comiso, J., and Maynard, N. G. (1992) 'Meridional zonation of the Barents Sea ecosystem inferred from satellite remote sensing and in situ bio-optical observations', Polar Research, 10 (1), 147-167.

Mortimer, C. H. (1988) 'Discoveries and testable hypotheses arising from Coastal Zone Color Scanner imagery of southern Lake Michigan', Limnology and Oceanography 33 (2), 203-226.

Mueller, J. L. and Lang, R. E. (1989) 'Bio-optical provinces of the northeast Pacific Ocean: a provisional analysis', Limnology and Oceanography 34 (8), 1572-1586.

Muller-Karger, F., McClain, C. R., and Richardson, P. (1988) 'The dispersal of the Amazon water', Nature 333, 56-59.

Muller-Karger, F. E., McClain, C. R., Fisher, T. R., Esaias, W. E., and Varela, R. (1989) 'Pigment distribution in the Caribbean Sea: Observations from space', Progress in Oceanography 23, 23-64.

Muller-Karger, F. E., McClain, C. R., Sambrotto, R. N., and Ray, G. C. (1990) 'A comparison of ship and CZCS-mapped distributions of phytoplankton in the Southeastern Bering Sea', Journal of Geophysical Research 95 (C7), 11483-11499.

Muller-Karger, F. E., Walsh, J. J., Evans, R. H., and Meyers, M. B. (1991) 'On the seasonal phytoplankton concentration and sea surface temperature cycles of the Gulf of Mexico as determined by satellites', Journal of Geophysical Research 96 (C7), 12645-12665.

'Oceanography of the Southeast U. S. Continental Shelf and Adjacent Gulf Stream', reprinted from the Journal of Geophysical Research 88, 4539-4738, 1983.

Pelaez, J. and McGowan, J. A. (1986) 'Phytoplankton pigment patterns in the California Current as determined by satellite', Limnology and Oceanography 31 (5), 927-950.

Pelaez, J. (1987) 'Satellite images of a "red tide" episode off Southern California, Oceanologica Acta 10 (4), 403-410.

Perry, M. J., Bolger, J. P., and English, D. C. (1989) 'Primary production in Washington coastal waters', in M. R. Landry and B. M. Hickey (eds.), Coastal Oceanography of Washington and Oregon, Elsevier, Amsterdam, pp. 117-138.

Roughgarden, J., Gaines, S., and Possingham, H. (1988) 'Recruitment dynamics in complex life cycles', Science 241, 1460-1466.

Simpson, J. J., Koblinsky, C. J., Pelaez, J., Haury, L. R., and Weisenhahn, D. (1986) 'Temperature-plant pigment-optical relations in a recurrent offshore mesoscale eddy near Point Conception, California', Journal of Geophysical Research 91 (C11), 12919-12936.

Singer, J. J., Atkinson, L. P., Blanton, J. O., and Yoder, J. A. (1983) 'Cape Romain and the Charleston Bump: Historical and recent hydrographic observations', Journal of Geophysical Research 88, 4685-4697.

Smith, R. C., Brown, O. B., Hoge, F. E., Smith, K. S., Evans, R. H., Swift, R. N., and Esaias, W. E. (1987) 'Multiplatform sampling (ship, aircraft, and satellite) of a Gulf Stream warm core ring', Applied Optics 26 (11), 2068-2081.

Smith, R. C., Zhang, X., and Michaelsen, J. (1988) 'Variability of pigment biomass in the California Current system as determined by satellite imagery, 1. spatial variability', Journal of Geophysical Research 93 (D9), 10863-10882.

Strub, P. T., James, C., Thomas, A. C., and Abbott, M. R. (1990) ' Seasonal and nonseasonal variability of satellite-derived surface pigment concentration in the California Current', Journal of Geophysical Research 95 (C7), 11501-11530.

Sullivan, C. W., McClain, C. R., Comiso, J. C., and Wood, W. O. (1988) 'Phytoplankton standing crops within an Antarctic ice edge assessed by satellite remote sensing', Journal of Geophysical Research 93 (C10), 12487-12498.

Thomas, A. C. and Strub, P. T. (1989) 'Interannual variability in phytoplankton pigment distribution during the spring transition along the West Coast of North America', Journal of Geophysical Research 94 (C12), 18095-18117.

Thomas, A. C. and Strub, P. T. (1990) ' Seasonal and interannual variability of pigment concentrations across a California Current frontal zone, Journal of Geophysical Research 95 (C8), 13023-13042.

Thompson, A. M., Esaias, W. E., and Iverson, R. L. (1990) 'Two approaches to determining the sea-to-air flux of dimethyl sulfide: satellite ocean colour and a photochemical model with atmospheric measurements', Journal of Geophysical Research 95 (D12), 20551-20558.

Thomson, R. E. and Gower, J. F. R. (1985) 'A wind-induced mesoscale eddy over the Vancouver Island continental slope', Journal of Geophysical Research 90 (C5), 8981-8993.

Turner, R. E., Woo, S. W., and Jitts, H. R. (1979) 'Estuarine influences on a continental shelf plankton community', Science 206, 218-220.

Turner, R. E. (1981) 'Plankton productivity and the distribution of fishes on the southeastern U. S. continental shelf', Science 214, 351-354.

Walsh, J. J., Dieterle, D. A., and Esaias, W. E. (1987) 'Satellite detection of phytoplankton export from the mid-Atlantic Bight during the 1979 spring bloom', Deep-Sea Research 34 (5/6), 675-703.

Walsh, J. J. (1989) 'Arctic carbon sinks: present and future', Global Biogeochemical Cycles 3 (4), 393-411.

Wroblewski, J. S., Sarmiento, J. L., and Flierl, G. R. (1988) 'An ocean basin scale model of plankton dynamics in the North Atlantic 1. solutions for the climatological oceanographic conditions in May', Global Biogeochemical Cycles 2 (3), 199-218.

Yentsch, C. S. (1984) 'Satellite representation of features of ocean circulation indicated by CZCS colorimetry', in J.C.J. Nihoul (ed.), Remote Sensing of Shelf Sea Hydrodynamics, Elsevier Science Publishers B. V., Amsterdam, pp. 337-354.

Yentsch, C. S. and Phinney, D. A. (1985) 'Rotary motion and convection as a means of regulating primary production in warm core rings', Journal of Geophysical Research 90 (C2), 3237-3248.

188

Yentsch, C. S. (1989) 'An overview of mesoscales distribution of ocean colour in the North Atlantic', Advances in Space Research 9 (7), 435-442.

Yoder, J. A., Atkinson, L. P., Lee, T. N., Kim, H. H., and McClain, C. R. (1981) 'Role of Gulf Stream frontal eddies in forming phytoplankton patches on the outer southeastern shelf', Limnology and Oceanography 26, 1103-1110.

Yoder, J. A., McClain, C. R., Blanton, J. O., and Oey, L.-Y. (1987) 'Spatial scales in CZCS-chlorophyll imagery of the southeastern U. S. continental shelf', Limnology and Oceanography 32, 929-941.

OCEAN COLOUR AND CZCS APPLICATIONS
IN AND AROUND EUROPE

V. BARALE
Institute for Remote Sensing Applications
Joint Research Centre, Commission of the European Communities
21020 Ispra (VA), Italy

R. DOERFFER
Institute of Physics
GKSS Forschungszentrum Geesthacht
Postfach 1160
2054 Geesthacht, Germany

ABSTRACT. Ocean optics have an important place in the European oceanographic tradition. This has led to substantial contributions to the Coastal Zone Color Scanner (CZCS) experiment, with the development of algorithms and models to handle data calibration, atmospheric correction, pigment value retrieval. Special requirements have been imposed on such developments by the peculiarity of the European basins. The Seas of Europe present an heterogeneous mixture of water constituents, including different planktonic populations, dissolved organic compounds, and a considerable suspended load. As a consequence, their optical properties range from those of Case 1 waters, where planktonic agents provide the dominant pigments, as in the open Mediterranean Sea or in the North Atlantic Ocean, to the most typical Case 2 waters of the North Sea or of many near-coastal areas, which in general are not optically dominated by plankton alone, but also by other suspended particulate and dissolved organic compounds, partly because of the impact of human activities, resources exploitation, and environmental pollution. The complex ensamble of environmental conditions to be approached has determined the development of specific bio-optical, reflectance, and light/photosynthesis models, as well as the application of methods such as factor analysis and inverse modeling techniques for the evaluation of CZCS data. The major applications of CZCS in and around Europe range from early descriptive oceanography work, looking at pigment distribution as related to circulation features, to the quantification of pigment concentration patterns, plankton dynamics, as well as coastal runoff and river plumes. Further, the potential for application of ocean colour data in primary production assessments has been explored in some detail, and systematic analyses based on CZCS data are now being conducted, particularly in the Mediterranean Sea. These involve the assessment of primary production from remote estimates of biomass and from a suite of auxiliary data on plankton distribution, properties and physiological state. The major European CZCS test sites have been the North Sea and the Baltic Sea, both basins with special water constituents characteristics; the Mediterranean Sea, and the Adriatic Sea in particular, providing small scale models for many oceanic processes; the North Atlantic Ocean and the upwelling region off North West Africa. The complete CZCS image archive on these sites of European interest is being used to generate the historical data base, know-how and scientific tools needed to exploit bio-optical information, in support of current research activities and in prepation of future ocean colour missions.

V. Barale and P.M. Schlittenhardt (eds.),
Ocean Colour: Theory and Applications in a Decade of CZCS Experience, 189–211.

1. Introduction

European waters comprise a large variety of optical properties, which reflect very different physical and bio-geo-chemical situations. Typical conditions range from the clear, oligotrophic waters in the North Atlantic Ocean, or the open Mediterranean Sea, to the turbid waters of estuaries and tidal flats in the North Sea, or to the unique phenomena of the Baltic Sea - where on occasion massive blooms of blue-green algae float on the water surface. As a consequence of this variety of environments, optical remote sensing has become an essential tool to investigate and monitor the complex mosaics of different water types of the European Seas, and their high variability in space and time. On the other hand, simple methods of data analysis, which have been developed and successfully used for a number of oceanic regions, do not apply in many European cases: rather, these require special techniques for the analysis of spectral radiances observed over the sea surface.

Considerable experience has been gained in this field by virtue of the Coastal Zone Color Scanner (CZCS) experiment in the 1980's (Hovis *et al.*, 1980; Feldman *et al.*, 1989). Substantial contributions to this experiment originated from studies carried out in Europe, where both theoretical and field activities have been conducted, during and after the CZCS mission. Naturally, European scientists did not work in isolation: on the contrary, the CZCS experimental results, shared at the international level through the NIMBUS-7 Experimental Team (NET), represent a classical example of success in science reached through international cooperation. However, specific environmental issues have received particular attention in Europe, such as that of coastal waters with optical properties dominated by suspended sediments and/or dissolved organic substances, a typical condition recurring in most of the enclosed basins of the European Seas (ESA Ocean Colour Working Group, 1987).

Different methods of analysis have been developed for different areas and different applications, ranging from the use of simple contrast-enhanced raw data, which can provide a qualitative impression of water masses distribution, to navigated quantitative maps of water constituents, time series of chlorophyll-like pigment concentration, and maps of primary production. The most sophisticated ocean colour analyses, based on radiative transfer calculations, require assessments of the specific optical properties of water constituents. Thus, the development of *in situ* techniques and laboratory measurements have been, and still are, a major part of European research programmes concerning visible remote sensing of the sea.

The following sections will concentrate on such issues, referring in particular to the scientific literature generated within the European community. On one hand, algorithms have been developed, by several European groups, so that measurements operated by the CZCS could be optimized for their particular tasks. On the other, measurement campaigns (using both *in situ* and remote sensing techniques) were organized to develop, and later feed, algorithms and numerical models with realistic data. The main case studies reported in the international scientific literature concern various European CZCS test sites, such as the North Sea and the Baltic Sea, the Mediterranean Sea and its many sub-basins, the North Atlantic Ocean and the upwelling region off North West Africa.

2. Optical Oceanographic Research

Ocean optics have traditionally received considerable attention in European oceanographic research (see *e.g.* Van Der Piepen *et al.*, 1991, and references therein). Famous precursors of modern optical oceanographers include P.A. Secchi, who conducted his first experiments on the disappearence of a circular disk in the Tyrrhenian Sea in 1865, as well as F.A. Forel, who developed a color scale for the optical classification of Swiss lakes in 1890. In more recent times, *i.e.* the 1930's, airborne remote sensing of surface colour was already used to classify rivers, lakes and coastal zones, by visual inspection and comparison with colour cards, from low flying aircrafts (Wasmund 1930). Early operational applications of that period are reported by fishermen of the Shetland Islands, who financed aerial surveys of the North Sea to search for perspectives fishing grounds. The use of airborne electro-optical sensors for remote sensing of ocean colour started in Europe at the end of the 1960's. About at the same time, classical works were published by Jerlov (1968), Jerlov and Steemann Nielsen (1974) and later again by Jerlov (1976), with the formulation of the general principles of optical oceanography and with the establishment of robust scientific basis for remote sensing of ocean colour.

More airborne experiments were carried out in the North Sea, in 1976, during the International Fladenground Experiment (FLEX 76), in order to map the temporal and spatial development of the spring plankton bloom, over a period of one month, using spectrometers mounted on low-altitude aircrafts as a supplement to ship surveys (Amann and Doerffer, 1983). As of 1977, a large European CZCS pre-launch experiment called EURASEP - combining surveys from ship and aircraft, and using the airborne Ocean Colour Scanner (OCS) - took place in the southern North Sea and the northern Adriatic Sea (Sorensen, 1978a, 1978b, 1979a, 1979b). The main step forward in ocean colour studies, obviously, came in October 1978 with the launch of the CZCS, which opened the possibility of mapping the distribution of water constituents over entire basins. Subsequently, much effort has been devoted to the assessment of quantitative correlations between optical properties of surface waters and various biological, geochemical and physical parameters, indispensable for a correct use of CZCS data (see *e.g.* Hojerslev, 1981; Sturm, 1981; Bricaud *et al.*, 1983; Bricaud and Morel, 1986; Bricaud *et al.*, 1987; Dirk and Spitzer, 1987; Siegel, 1987; Morel, 1988; Sathyendranath *et al.*, 1989).

In its modern interpretation, the term ocean colour is used to indicate the visible spectrum of upwelling radiance as observed at the sea surface. This radiance is related to the presence, nature and abundance of substances dissolved and suspended in the surface layer of the sea, the so-called water constituents, by the processes of absorption and scattering. Based on the spectral properties of the upward-directed radiance, or irradiance, different classification systems for marine waters have been proposed (Jerlov, 1976; Morel, 1980; Prieur and Sathyendranath, 1981). For remote sensing purposes, it has become common practice to classifly different water types according to the particular nature of the water constituents shaping their optical properties. In particular, two major water types, referred to as Case 1 and Case 2 waters (Morel and Prieur, 1977), can be identified (and are now widely used in the scientific literature concerning marine optics).

In Case 1 waters, the optical properties are dominated by biological constituents (correlated in such a way that they may be represented by a single factor only). These include the photosynthetic pigments of phytoplankton, both from living algal cells and from associated debris originated by natural decay or zooplankton grazing, as well as other dissolved organic matter liberated by the algae and their debris. This water type is usually found in the pelagic zone, but it occurs also in coastal regions with arid climate, considerable depth and limited coastal runoff. The range of Case 1 water spans from blue, oligotrophic waters (with pigment concentration below 0.1 mgm-3), to biologically active waters (pigments concentration around 1 mgm-3), as well as to green, eutrophic waters (pigment concentration as high as 10 mgm-3) like those found in upwelling areas.

In Case 2 waters, instead, the same constituents found in Case 1 water are accompanied by the presence of terrigenous particles from coastal runoff, supended sediments of variable nature, dissolved organic matter (mostly degrading organic remains from land drainage: the so-called yellow substance) and other particulate or dissolved substances originated from anthropogenic influx. Analogous deviations from the characteristics of Case 1 water may also be originated by exceptional plankton blooms, which discolour the water in abnormal ways. The Case 2 water type may occur in coastal zones with high coastal and fluvial runoff, or near mud flats, estuaries and river deltas, as well as in shallow offshore zones. The total area covered by Case 2 waters is just a small fraction of the global marine surface. However, their profound impact on marine biology and particularly coastal ecology makes this an important topic of research.

Most open sea regions, in the European Seas, show the usual Case 1 waters. However, as mentioned above, many European basins (the Baltic Sea in particular, but other areas in the North Sea and some enclosed basins of the Mediterranean Sea as well) do present Case 2 waters, particularly in the near-coastal zone, as a regular environmental feature.

3. Calibration, Atmospheric Correction and Pigment Algorithms

Due to the technological problems of assessing marine optical properties, to the fact that remote sensing of such properties must be performed through the Earth's atmosphere, and to the complexity of the environmental conditions to be considered, the analysis of ocean colour data requires the development of specific algorithms and models. Three main kinds of algorithms must be used to estimate marine environmental parameters from the total signal measured from an optical sensor, like the CZCS, operating in the visible part of the spectrum (see Gordon and Morel, 1983, and references therein). First, the sensor recorded digital counts are converted into apparent radiance values by means of a calibration algorithm - which, in the CZCS case, also had to account for the progressive sensor sensitivity degradation. Second, the total apparent radiance values, which are contaminated by a substantial atmospheric contribution, must be corrected to derive water upwelling radiances (or else sub-surface reflectances, which depend only on the nature and concentration of water constituents in the marine surface layer) by means of an atmospheric correction algorithm. And, third, these estimates are used to calculate pigment concentration in surface waters (*i.e.* the concentration of chlorophyll-like pigments of phytoplankton) or, in alternative, other parameters such as total suspended matter concentration or diffuse attenuation coefficient, at specific wavelengths, by means of a pigment algorithm.

The processing of CZCS data, to extract information on marine environmental parameters from the radiances measured by the sensor over the sea, depends in a crucial way on the solution of problems induced by incertitudes in the sensor calibration and sensitivity degradation on water pixels. The CZCS calibration algorithm requires that, for each water pixel selected in a scene, the raw value of each spectral channel (*i.e.* the digital counts) be converted into a total radiance value. However, the sensitivity of the sensor has been known to be decreasing with time, particularly in the latter part of its mission. Such degradation in the sensor response occurred in different measure in all spectral channels, in an unknown, presumably non-linear manner. Attempts at evaluating the sensitivity decay have been made by comparing actual measured radiances with their expected values over clear water open ocean areas - where the contribution of water constituents to the local optical properties should be minimun, if not absent. This problem has been recognized and approached by a number of European groups, with a cosiderable degree of success, as reported by Singh and Cracknell (1982), Viollier (1982), Sturm (1983), Singh *et al.* (1985), Sturm (1986).

Once calibrated, the total radiance seen by an orbiting sensor, looking at the sea surface in the visible spectral range, is the sum of various contributions of both atmospheric and marine origin. The atmospheric component of this total radiance, due to scattering by air molecules (Rayleigh scattering) and by suspended particles (aerosol scattering), is by far the largest contribution, accounting for up to 90% of the signal reaching the remote sensor. The marine component, constituted by the water leaving radiance, further attenuated while traveling through the atmosphere from the sea surface to the sensor, is therefore only a minor part of the total signal recorded. Yet, it is the one which carries the necessary information for inferring the water constituents concentration in the surface layer of the ocean, and therefore must be evaluated with great care. Notable examples of atmospheric algorithms, and models, developed over the last decade in Europe include those reported by Tanre *et al.* (1979), Sturm (1980), Viollier *et al.* (1980), Bricaud and Morel (1987), Deschamps and Viollier (1987), Andre' and Morel (1991).

It should be noted that 'classical' atmospheric algorithms developed for Case 1 waters (as summarized, *e.g.*, in Gordon and Morel, 1983) find limited application in many European Seas, due to the complex optical properties of their waters, as already mentioned before. A peculiar problem of atmospheric correction algorithms is the determination of the path radiance, which includes light scattered into the field of view of the sensor both by air molecules and by aerosols. The first component of the total signal - due to Rayleigh scattering - is usually calculated from air pressure and from the angles of observation and of sun illumination. The derivation of the second component is based on the radiance of CZCS channel 4 (center wavelength around 670 nm). Within the corresponding spectral range, for an atmosphere of oceanic type and for Case 1 waters, the water leaving radiance can be neglected, due to the strong absorption by pure water. Thus, all of the radiance measured in this channel can be assumed to be path radiance, including that part of skylight which is specularly reflected by the sea surface. By knowing the spectral distribution of the light scattered by aerosol (in most algorithms the low governing such distribution is expressed by wavelength ratios elevated to the so-called Angstrom exponent), the amount of path radiance in the other CZCS channels can be assessed by extrapolation. Such a procedure, obviously, does not work in the case of turbid waters (as those of many European near-coastal sites), for which the radiance of CZCS channel 4 includes the contribution of light backscattered by suspended matter. To solve this problem, Sturm (1980) developed an iterative procedure to determine path radiance and water constituents concentration simultaneously. However, the procedure requires knowledge of the Angstrom exponent as well as of the spectral distribution of light backscattered by suspended matter. For areas with high concentration of suspended matter, yellow substance and planktonic pigments, all three water constituents have to be taken into account, when performing the atmospheric correction. The problem can be approached by means of inverse modelling of the radiance transfer, as described below.

The CZCS pigment retrieval algorithms are used for the direct transformation from water leaving radiance - or subsurface reflectance - ratios to chlorophyll-like pigment concentration (*i.e.* including both chlorophyll and its associated degradation products, which cannot be disciminated by the CZCS), or other parameters such as total supended matter. This is a topic on which the work developed in Europe has been most specialized to the peculiar environmental characteristics of the European Seas (Bricaud *et al.*, 1981; Sturm, 1981; Viollier and Sturm, 1984; Sathyendranath *et al.*, 1989).

For Case 1 waters, where chlorophyll-like pigments are the single dominant optical variable to be determined, algorithms based on colour ratios have been proven to be robust, reliable and rather accurate. Such algorithms are based on empirically-derived relationships between pigment concentration (measured in water samples) and the ratio of water leaving radiance - or subsurface reflectance - obtained from CZCS channels 1, 2 and 3 (center wavelength around 443, 520 and 550 nm). This technique requires sets of simultaneous ship measurements and satellite data, for the comparison and regression of corresponding samples and pixel values. In areas with strong currents, as *e.g.* in many coastal zones of the North sea, it is very difficult - even impossible, at times - to collect such a simultaneous data set, since significant changes in the abundance and distribution of water constituents occur on time scales shorter than one hour. The situation becomes even more complex when Case 2 waters are present, or dominate within a certain area, since simple weighted colour ratios often cannot be used in the determination of pigment concentration. Alternative procedures, such as factor analysis (Doerffer, 1981), have been used to separate the contribution of the particular substances shaping the radiance spectra of Case 2 waters. One of the most promising methods is the inverse modelling techique. The method calls for the use of a radiative transfer model to describe the process of remote sensing, *i.e.* the propagation of light downwards through the atmosphere, into the water, and back upwards to the sensor. By varying the water constituent concentration and other parameters in the model, with the aid of an optimization procedure, a condition is approached for which the deviation between measured and computed spectrum is minimum (see Figure 1). The application of this method requires to determine the specific optical constants of the constituents as used by the model, such as the absorption and

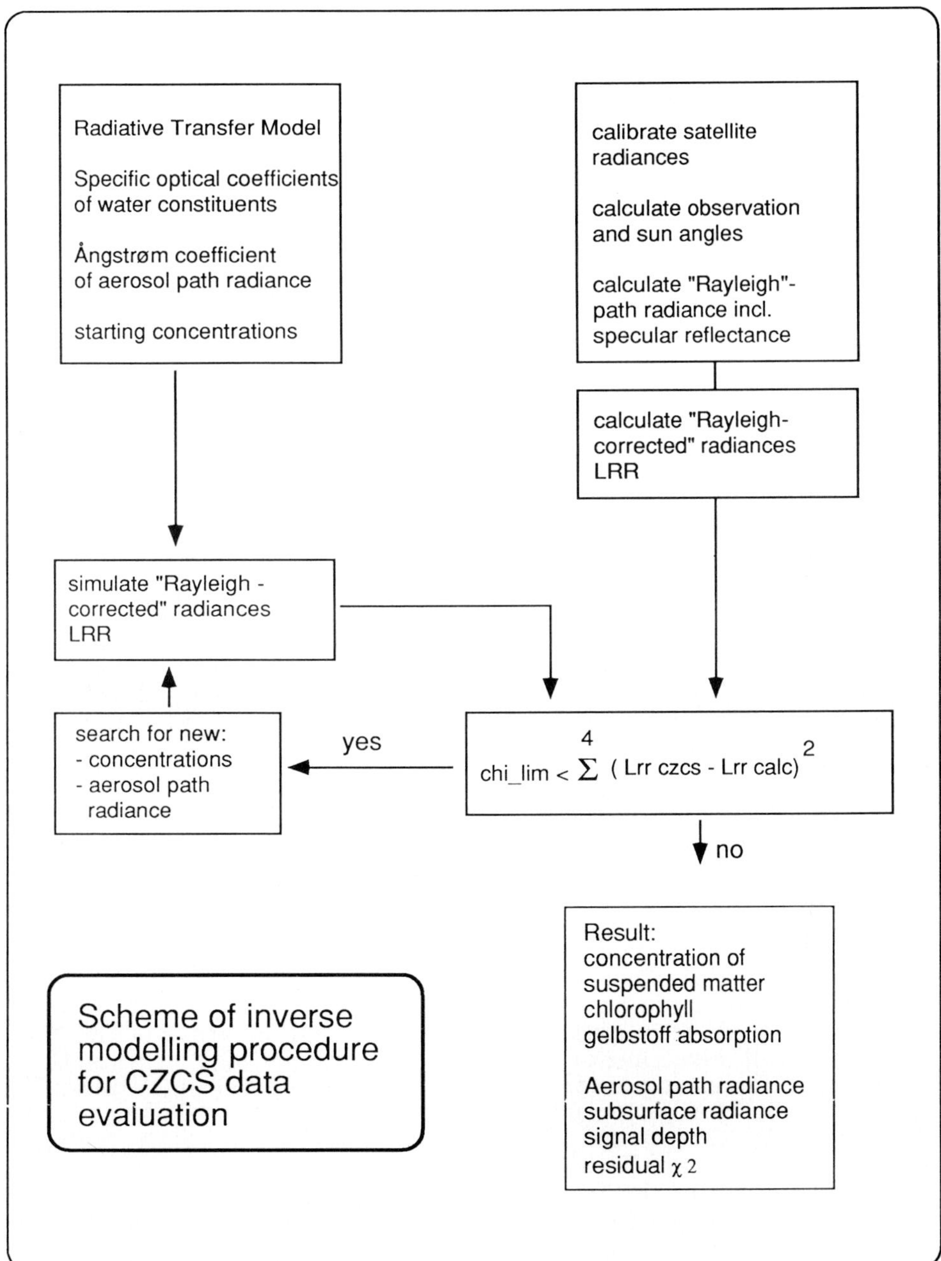

$$\text{chi\_lim} < \sum^{4} (Lrr\ czcs - Lrr\ calc)^{2}$$

Figure 1. Schematics illustrating the inverse modelling procedure for CZCS data.

backscattering coefficient of suspended matter, yellow substance and planktonic pigments. Since such constants depend on the type of constituent (*e.g.* of phytoplankton) different values may have to be used for different places and seasons. Early applications of the inverse modelling technique to the analysis of CZCS data, making use of a radiative transfer model based on a matrix-operator method, are described by Fischer and Doerffer (1987). Further developments of this procedure, using a two-flow model and the simplex optimization method, have been obtained calibrating the model by regression with data from the matrix-operator method. This, then, can be used to calculate subsurface CZCS radiances, by assuming a large variety of different concentrations of suspended matter, yellow substance and chlorophyll-like pigments (Doerffer, 1990).

4. Pigment Distribution, River Plumes and Coastal Upwelling

The remote assessment of optical properties of the sea surface finds several applications in the fields of marine biology and ecology at large, water pollution and sediment transport, water circulation and dynamical processes, as well as energy transfer, carbon cycling and climatology in general (see, *e.g.*, Barale, 1991, and references therein). Applications in European case studies range from early descriptive oceanography studies, just looking at pigment distribution as related to circulation features, to the quantification of pigment concentration patterns, as well as to that of patterns related to coastal runoff and river plumes. Much work has been done in various European test sites of CZCS, such as the North Sea and the Baltic Sea, both marine regions with special water constituents characteristics, the Mediterranean Sea, and the Adriatic Sea in particular, the North Atlantic Ocean and the North West Africa upwelling region (figure 2).

The use of CZCS data for the northern European basins (*i.e.* primarily the North Sea and the Baltic Sea, but also the English Channel and the Irish Sea) has been hampered somewhat by some inherent difficulties in their analysis. First, usable data are sparse for these northern regions, mainly due to cloud cover, and limited to those periods of the year when sun elevation is suitable to have a meaningful signal from the water surface. Second, the particular kinds of waters (mainly Case 2 waters) present in these regions require proper processing techniques and render sometimes unreliable the kind of information derived from such processing. The shallow areas of the North Sea, *e.g.*, are well mixed most of the year, due to strong tidal currents. Due to the high -and increasing- nutrient supply, both of fluvial and atmospheric origin, phytoplankton can occur in very dense patches even during summer, at time forming exceptional blooms such as red tides, brown waters of *Nocticula*, foam covers of *Phaeocystic*, and white waters of *Coccolithophorides*. On the other hand, in the transition zone between the North Sea and the Baltic Sea, *i.e.* the Skagerrak, a dense haline stratification supports plankton growth very early in the year, while in other well mixed areas growth is still limited by light. As noted above, efforts have been made to solve the problems posed by the consistant presence of Case 2 waters or by exceptional phenomena (see figure 3), and the literature does report several examples of successful interpretation of CZCS images of the northern European Seas, as related, *e.g.*, to plankton blooms, environmental impact of river discharges,

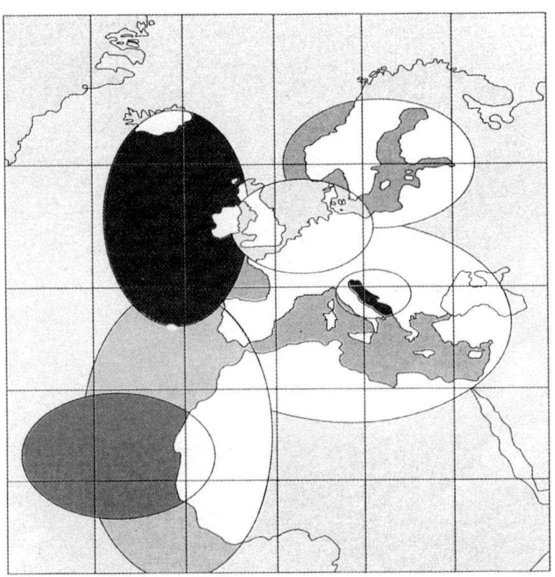

Figure 2. Main European basins adopted as CZCS test sites. Colorphotograph on p. 355

distribution and transport of suspended matter, shallow water dynamics (Holligan *et al.*, 1983a; Singh *et al.*, 1983; Horstmann *et al.*, 1986; Mitchelson *et al.*, 1986; Boxall and Robinson, 1987; Horstmann, 1988; Aarup *et al.*, 1990; Doerffer, 1990).

In the region of the Mediterranean Sea, a number of CZCS applications have been devoted to exploring and understanding processes related, one way or the other, to coastal and river runoff. The basin, in fact, presents relatively clear, oligotrophic waters in the pelagic region, where the signal due to planktonic pigments is reduced to a bare minimum (see below for applications involving the assessment of primary production over entire Mediterranean sub-basins). On the other hand, it does present also several points of particular interest for the study of enclosed basins and coastal phenomena. Major rivers such as the Nile, Danube, Rhone, Po, Ebro, Guadalquivir, for example, have all distinct plumes interacting with the marine dynamical and bio-geo-chemical environment. Within the range of the plumes -as with coastal runoff in general- it is often impossible for the CZCS to distinguish the signature of biogenic pigments from that of the total load of dissolved and suspended materials present in the water. However, the observation of such features provides important clues on frontal dynamics and potential correlations with nutrient enrichment or pollution (see *e.g.* Caraux and Austin, 1983; Barale *et al.*, 1987; Morel *et al.*, 1990). In particular, the extensive studies conducted on the Adriatic Sea (figure 4), and on the fate of the Po river outflow, provide examples of the results obtainable in this field from the combination of remote sensing, *in situ* campaigns, and theoretical modelling (Barale *et al.*, 1984; Alberotanza *et al.*, 1985; Barale *et al.*, 1986; Clement *et al.*, 1987; Sturm, 1987; Bergamasco and Barale, 1988). Similar

198

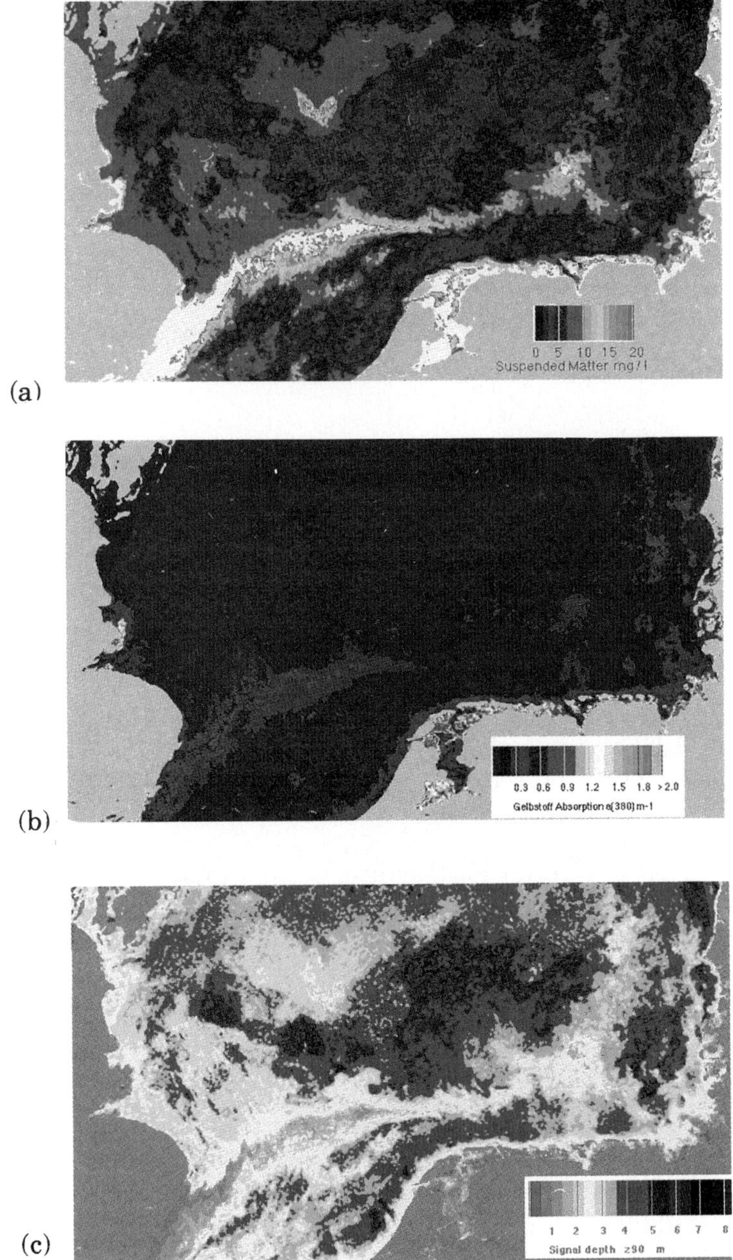

Figure 3. Concentration of suspended matter (a) and yellow substance (b), and depth Z_{90} (c) from which 90 % of the measured radiance at 550 nm stems from; southern North Sea; calculated with the inverse modelling procedure from a CZCS scene (September 6, 1979). Colorphotograph on p. 356

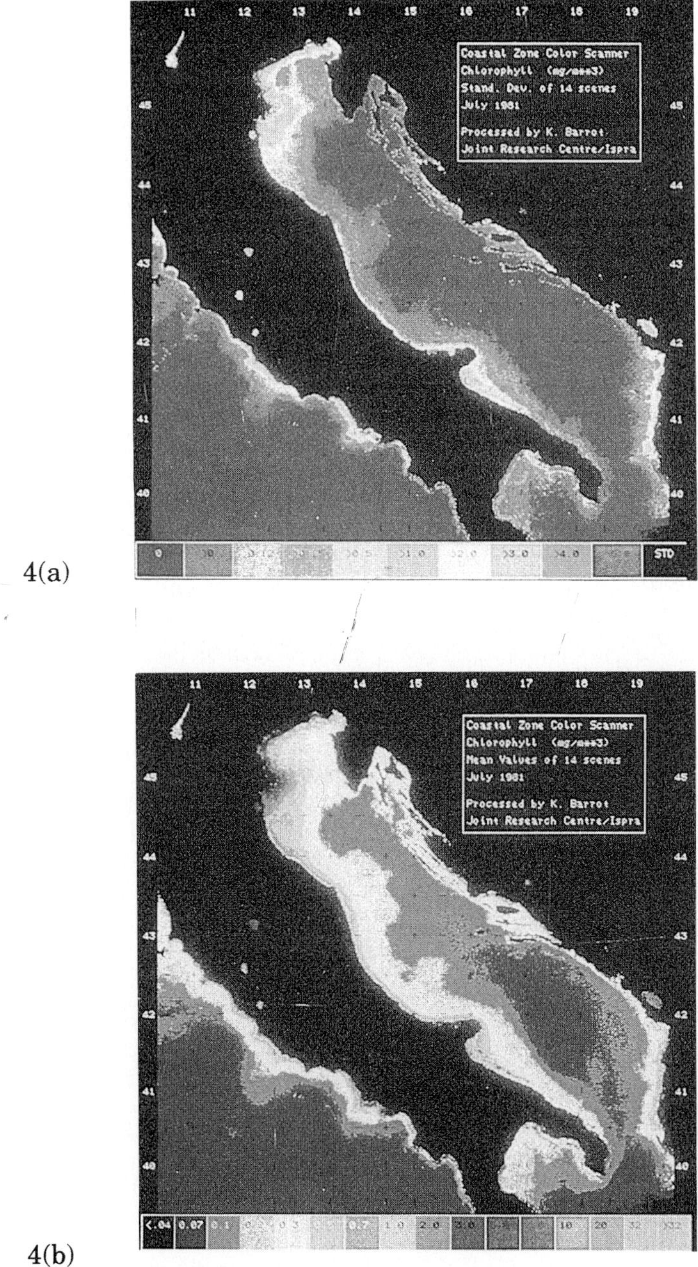

4(a)

4(b)

Figure 4. Adriatic Sea, surface pigments mean (a) and standard deviation (b) images for July 1981. Colorphotograph on p. 357

applications appear in the literature concerning the observation of coastal currents and their interaction with the basin interior, as in the case *e.g.* of the Algerian current and the eddy field of the western Mediterrean basin (Arnone and La Violette, 1986; Arnone *et al.*, 1990).

Finally, the North Atlantic Ocean (as *e.g.* in Groom and Holligan, 1987) and, in particular, the North West African upwelling region (see Van Camp *et al.*, 1991, and references therein) must be reported as sites of European attention for the application of CZCS data. Of particular interest is the temporal occurrence and spatial distribution of *Coccolithophorides* massive blooms in the North Atlantic -as well as in areas of the North Sea- because of their potential role in the carbon cycle (Holligan *et al.*, 1983b; Groom and Holligan, 1987). Time series of CZCS images have been used by a number of researchers for studies of coastal upwelling off North West Africa, which is known for being a region of high biological productivity and high fishing potential (Bricaud *et al.*, 1987; Dupouy and Demarcq, 1987). The physical process of coastal upwelling is relatively well known, while the relationship between upwelling and high trophic levels still needs in depth investigations. For this reason, CZCS data have been -and still are- used in the evaluation of biological conditions of importance for fisheries, by analysing the relationship between the standing stock of phytoplankton, as seen in the images, and other features influencing fish populations (figure 5). The latter may comprise parameters such as location of spawing grounds, migration patterns, catch statistics, as well as, sea surface temperature, wind field anomalies, circulation patterns, and the like (Van Camp, 1987).

5. Ocean Colour and Primary Production

The assessment of primary production from estimates of biomass distribution, such as those offered -even though with serious limitations- by CZCS, is the latest application of ocean colour data developed in the past decade. Unfortunately, it demands knowledge of a considerable amount of auxiliary information. Assuming that it is required to evaluate the production per unit area of sea surface per day (*i.e.* the daily water column production), the supplementary information needed are estimates of the irradiance at the sea surface, resolved with respect to wavelength, and corrected for cloud cover; at least two parameters of the photosynthesis light curve; and at least three parameters of the curve describing the shape of the biomass profile as a function of depth. Normally, it is assumed that the supplementary information are characteristic of a particular site. These data can also be combined with equations describing light penetration in the sea and utilization of submarine light in photosynthesis, which may be integrated over depth and over the day for a particular site in a 'local algorithm' (Platt and Sathyendranath, 1988).

Naturally not all difficulties have been overcome, for this application of ocean colour data. There is a need, *e.g.*, to test local algorithms at a number of particular sites for which CZCS data exist and for which the relevant supplementary data are available. Further problems arise in the extrapolation of the results from local algorithm to larger horizontal scales. This is the real puzzle of how to group single, punctual biomass assessments with respect to the assignement of the supplementary parameters. To solve it,

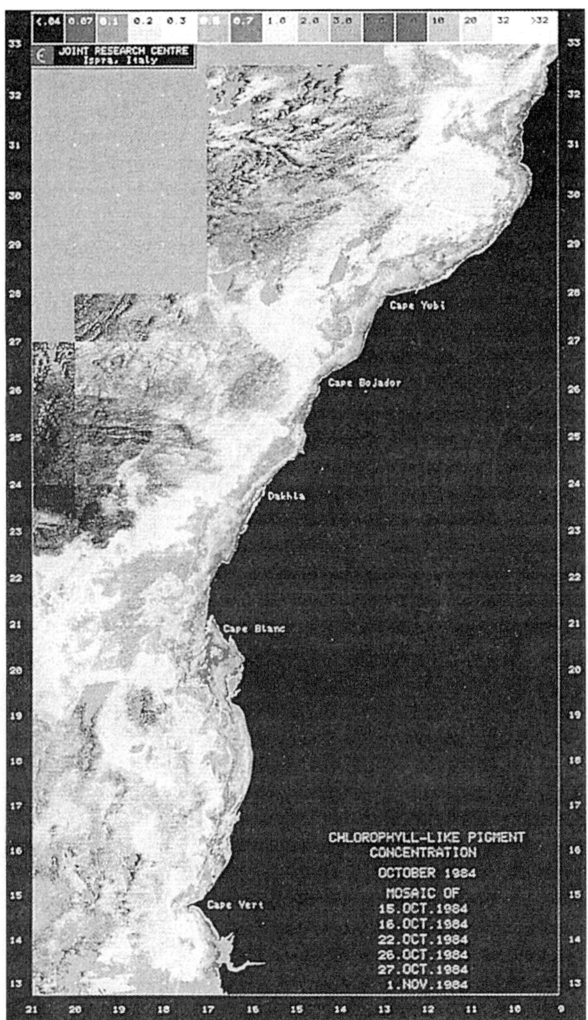

Figure 5. Upwelling area off North West Africa. Colorphotograph on p. 358

knowledge is required of how these vary in space and with seasons - in essence knowledge about dimensions, characteristics and dynamics of biogeographical provinces.

Early developments in Europe on this subject include empirical approches to the correlation of pigments, biomass profiles and production (Morel and Berthon, 1989), and the development of spectral light / photosynthesis models (as, *e.g.* in Morel, 1991, with reference to previous work on a bio-optical model by Morel, 1988, and an atmospheric transmission model derived from Tanre' *et al.*, 1979). Applications have also been reported in the literature for the

Mediterranean Sea, in particular for what its western basin is concerned (Lohrenz *et al.*, 1988; Morel and André, 1991). At present, the problem of using an analogous approach to the evaluation of primary production in the North Sea and Baltic sea, dominated by the presence of Case 2 waters, has not been solved. It must be noted that, in contrast to Case 1 water areas, most of the coastal zones, particularly in the North Sea, show only little stratification, so that the surface concentration of phytoplankton pigments provides in fact an estimate of the mean water column concentration. For these regions, however, it would be necessary to separate the light absorption of planktonic pigments from that of other substances. This requires the distinction of all water constituents of optical significance.

Although the inverse modelling scheme mentioned above does provide the necessary data, its potential for assessing primary production has not been tested, up to now.

Progress in estimating primary production by optical remote sensing can also be expected from the determination of natural, *i.e.* sunlight-induced, fluorescence of chlorophyll. This signal has been measured using airborne spectrometers to determine chlorophyll-like pigment concentration (Figure 6 and Figure 7) in the northern North Sea (Doerffer, 1981; Amann and Doerffer, 1983), as well as in the German Bight and the Baltic sea (GKSS, 1986). The methodology adopted involves use of the so-called fluorescence line height algorithm, which calculates the fluorescence energy at 685 nm above a baseline constructed from two channels, around 630 nm and 710 nm. By incorporating the fluorescence signal into the inverse modelling scheme, originally developed for CZCS data, it is possible to compute also the fluorescence efficency, *i.e.* the ratio of the emitted fluorescence energy to the

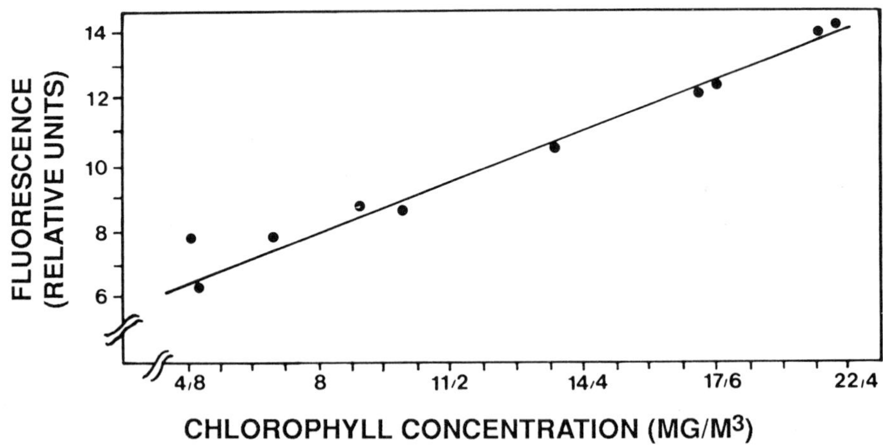

Figure 6. Correlation between the chlorophyll concentration sampled at 2 m depth and the fluorescence line height as measured with an airborne spectrometer from 600 m altitude along a 100 km track in the Fladenground, North sea, during FLEX 76.

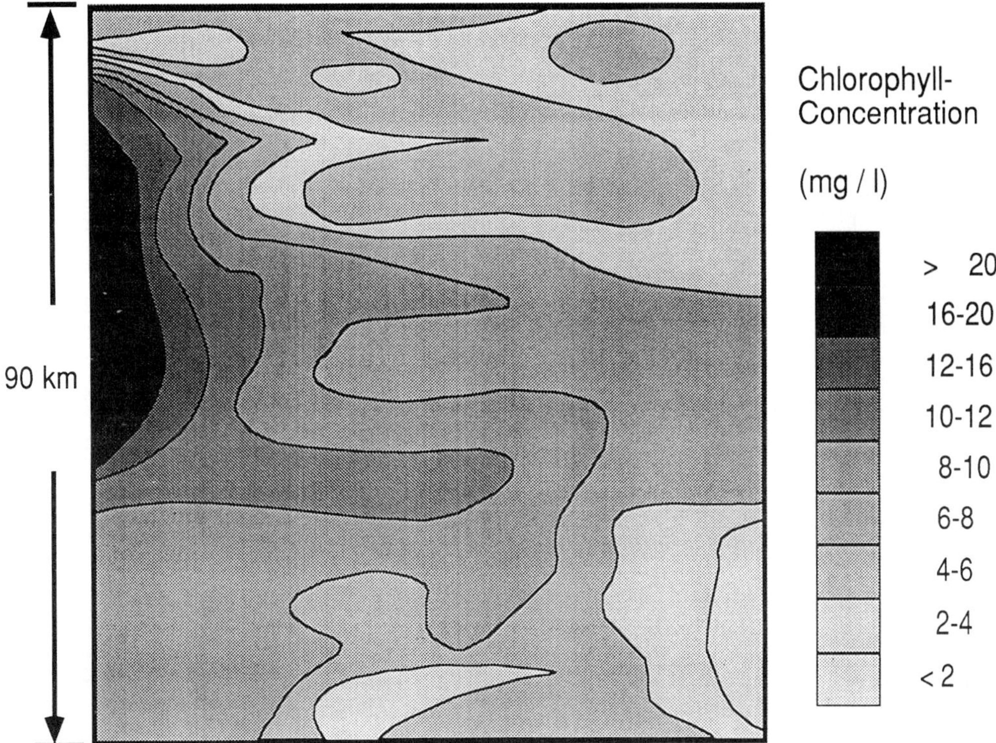

Figure 7. The chlorophyll distribution in the Fladenground as derived from remote measurements of the fluorescence line height.

energy absorbed by phytoplankton pigments. Such quantity is related to the quantum efficency of primary production (Topliss and Platt, 1986; Doerffer, 1992). Looking beyond the CZCS era, it is expected that both airborne and orbital imaging spectrometers will enable to use this method for improved remote assessments of primary production.

6. Conclusion

Optical remote sensing of the Seas around Europe, and of its enclosed basins in particular, requires special algorithms and measurement strategies. In the past, most of the European research groups involved in the CZCS experiment have had to concentrate on the special requirements of their test sites. As a consequence, although the analysis of Case 1 waters, where simple colour ratio techniques could be applied, progressed more rapidly, quantitative maps of constituents for Case 2 waters appeared rather late in the scientific literature. A number of peculiar scientific problems presented by the European basins can now be approached - from an historical point of view -

with the help of synoptic data sets, such as those generated by the CZCS experiment. It is now clear that the real wealth of information, in the case of oceanographic remote sensing, is in the long-term, large-scale monitoring of interacting bio-geo-chemical and physical processes of the sea. In this context, time series of ocean colour assessments play a crucial role for the solution of the marine environmental puzzle, and shall undergo further developments.

At present the complete CZCS image archive covering the sites of European interest, which were mentioned above, is being exploited in the framework of the Ocean Colour European Archive Network (OCEAN) Project. This effort is centered on the development - and utilization - of methodologies and algorithms to extract geophysical parameters from marine optical data, as well as the organization - and exploitation - of reliable, documented, accessible archives of CZCS-derived ocean colour data. The OCEAN Project integrates three main long-term objectives, devoted respectively to generate an European data base of bio-optical information on the marine environment using the historical CZCS data, to promote the use of ocean colour data in an Application Demonstration Programme devoted to marine regions of European interest, and to develop a 'network' of scientific groups and facilities capable of supporting in Europe current research activities and future ocean colour missions.

The main requirement for future ocean colour sensors and evaluation procedures is that of reducing the number of unknowns when retrieving the biomass, and eventually productivity, from radiance spectra. In particular, the factors which predominate in limiting the accuracy of primary productivity estimates with present satellite data (*i.e.* CZCS data, essentially) are uncertainities in: (*i*) the determination of the aerosol contribution to the signal seen from orbit; (*ii*) the separation of the contribution by the other different sea water constituents beside phytoplankton chlorophyll, such as other pigments of planktonic origin, yellow substance, suspended matter in general; (*iii*) the retrieval of information about the vertical distribution of the phytoplankton under observation, its specific absorption, and its physiological state, which determines the relation between irradiance and productivity per unit of chlorophyll concentration.

Imaging spectrometers, having a high number of narrow-band spectral channels in the visible and near infrared spectral range, will provide improvements with respect to such problems (see *e.g.* Van der Piepen *et al.*, 1991, for an overview of future technological perspectives, and Doerffer, 1992, for an overview of optics, radiative transfer modelling and evaluation techniques for imaging spectrometer data). A general consensus is also emerging, in the European scientific community, calling for increased investements in three main research components, *i.e.* the solution of residual problems for data pre-processing and archival, for high-level data processing and scientific analysis, and for coordination of research activities conducted in various European Institutions.

Acknowledgements

This paper is the result of research sponsored by the International Space Year 1992 (ISY '92) Activity "Productivity of the Global Ocean". The authors wish to express their gratitude to B. Sturm, L. Van Camp, L. Nykjaer and P.

Schlittenhardt for generously providing the CZCS images of the Adriatic Sea and of the North West Africa upwelling region.

References

Aarup, T., Groom, S., and Holligan, P.M. (1990) 'The processing and interpretation of North Sea CZCS imagery', Netherlands Journal of Sea Research 25 (1/2), 3-9.

Alberotanza, L., Barale, V., and Bergamasco, A. (1985) 'Nimbus-7 CZCS images as inputs in circulation modelling of the North Adriatic Sea', Il Nuovo Cimento 8C (6), 621-630.

Amann, V. and Doerffer, R. (1983), 'Aerial survey of the temporal and spatial distribution of phytoplankton during FLEX 76', in J. Sündermann and W. Lenz (eds.), North Sea Dynamics, Springer-Verlag, Berlin and Heidelberg, pp. 517-529.

André, J.M. and Morel, A. (1991), 'Atmospheric corrections and interpretation of marine radiances in CZCS imagery: revisited', Oceanologica Acta 14, 3-22.

Arnone, R.A. and La Violette, P.E. (1986) 'Satellite definition of the bio-optical and thermal variation of coastal eddies associated with the African Current', Jounal of Geophysical Research 91 (C2), 2351-2364.

Arnone, R.A., Wiesenburg, D.A., and Saunders, K.D. (1990) 'The origin and characteristics of the Algerian Current', Jounal of Geophysical Research 95 (C2), 1587-1598.

Barale, V. (1991) 'Sea surface colour in the field of biological oceanography', International Journal of Remote Sensing 12 (4), 781-793.

Barale, V., Alberotanza, L., and Bergamasco, A. (1987) 'Coastal runoff patterns in the Northern Adriatic Sea from CZCS imagery', in R. Reuter and R.H. Gillot (eds.), Remote Sensing of Pollution of the Sea, CEC DG XII JRC Ispra and Bibliotheks- und Informationssystem der Universitat Oldenburg, S.P.I. 87.46, Ispra - Oldenburg, pp. 162-172.

Barale, V., Malanotte Rizzoli, P., and Hendershott M.C. (1984) 'Remotely sensing the surface dynamics of the Adriatic Sea', Deep Sea Research 31, 1433-1459.

Barale, V., McClain, C.R., and Malanotte Rizzoli, P. (1986) 'Space and time variability of the surface color field in the northern Adriatic Sea', Jounal of Geophysical Research 91 (C11), 12957-12974.

Bergamasco, A. and Barale, V. (1988) 'Comparison between coastal runoff pattern from CZCS imagery and from a general circulation model', in A. Marani (ed.), Advances in Environmental modelling, Elsevier, Amsterdam, pp. 395-404.

Boxall, S.R. and Robinson, I.S. (1987) 'Shallow sea dynamics from CZCS imagery', Advances in Space Research 7 (2), 37-46.

Bricaud, A., Bedhomme, A.L., and Morel, A. (1987), 'Optical properties of diverse phytoplankton species: experimental results and theoretical interpretation', Journal of Plankton Research 10, 851-873.

Bricaud, A. and Morel, A. (1986), 'Light attenuation and scattering by phytoplankton cells: a theoretical modeling', Applied Optics 25, 571-580.

Bricaud, A. and Morel, A. (1987), 'Atmospheric corrections and interpretation of marine radiances in CZCS imagery: use of a reflectance model', Oceanologica Acta 7, 33-50.

Bricaud, A., Morel, A., and Prieur, L. (1981), 'Absorption by dissolved organic matter of the sea (yellow substance) in the UV and visible domains', Limnology and Oceanography 26, 43-53.

Bricaud, A., Morel, A., and Prieur, L. (1983), 'Optical efficiency factors of some phytoplankters', Limnology and Oceanography 28 (5), 816-832.

Bricaud, A., Morel, A., and André, J.M. (1987), 'Spatial/temporal variability of algal biomass and potential productivity in the Mauritanian upwelling zone, as estimated from CZCS data', Advances in Space Research 7 (2), 53-62.

Caraux, D. and Austin, R.W. (1983) 'Delineation of seasonal changes of chlorophyll frontal boundaries in mediterranean coastal waters with Nimbus-7 Coastal Zone Color Scanner data', Remote Sensing of Environment 13 (3), 239-249.

Clement, F., Franco, P., Nykjaer, L., and Schlittenhardt, P. (1987) 'Northern Adriatic Sea and River Po: oceanographic *in situ* measurements, remote sensing and modelling as a contribution to the understanding of dispersion phenomena', in R. Reuter and R.H. Gillot (eds.), Remote Sensing of Pollution of the Sea, CEC DG XII JRC Ispra and Bibliotheks- und Informationssystem der Universitat Oldenburg, S.P.I. 87.46, Ispra - Oldenburg, pp. 98-111.

Deschamps, P.Y. and Viollier, M. (1987) 'Algorithms for ocean colour from space and application to CZCS data', Advances in Space Research 7 (2), 11-19.

Dirk, R.W., and Spitzer, D. (1987) 'On the radiative transfer in the sea, including fluorescence and stratification effects', Limnology and Oceanography 32 (4), 942-953.

Doerffer, R. (1981) 'Factor analysis in ocean colour interpretation', in J.F.R. Gower (ed.), Oceanography from space, Marine Science Series, Vol. 13, Plenum Press, New York and London, pp. 339-345.

Doerffer, R. (1990) 'How to derive concentrations of chlorophyll, suspended matter and gelbstoff from multispectral radiances of Case 2 waters', ICES

Statutory Meeting, Copenhagen (DK), Theme Session P, paper C.M. 1990 E:19.

Doerffer, R. (1992) 'Imaging spectroscopy for detection of chlorophyll and suspended matter', in F. Toselli and J. Bodechtel (eds.), Imaging Spectroscopy: Fundamentals and Prospective Applications, EUROcourses Remote Sensing Vol. 2, Kluver Academic Publishers, Dordrecht - Boston - London, pp. 213-257.

Dupouy, C. and Demarcq, H. (1987), 'CZCS as an aid for understanding modalities of the phytoplankton productivity during upwelling off Senegal', Advances in Space Research 7 (2), 63-71.

ESA Ocean Colour Working Group (1987) 'Ocean Colour. Proceedings of the ESA Ocean Colour Workshop, Villefranche sur Mer (France), 5-6 November 1986', Report ESA SP-1083 (ISSN 0379-6566), pp. 118.

Feldman, G., Kuring, N., Ng, C., Esaias, W., McClain, C., Elrod, J., Maynard, N., Endres, D., Evans, R., Brown, J., Walsh, S., Carle, M., and Podesta, G. (1989) 'Ocean color: availability of the global data set', Eos Transactions, AGU, 70 (23), 634-635 and 640-641.

Fischer, J. and Doerffer, R. (1987) 'An inverse technique for remote detection of suspended matter, phytoplankton and yellow substance from CZCS measurements', Advances in Space Research 7 (2), 21-26.

GKSS (1986) 'The use of chlorophyll fluorescence measurements from space for separating constituents of sea water', ESA Contract No. RFO 3-5059/84/NL/MD, Vol. I (summary report) and Vol. II (appendices), GKSS Forschungszentrum, Geesthacht.

Gordon, H.R. and Morel, A. (1983) 'Remote assessment of ocean color for interpretation of satellite visible imagery: a review', in M. Bowman (ed.), Lecture Notes on Coastal and Estuarine Studies, Vol. 4, Springer-Verlag, New York - Berlin - Heidelberg - Tokio, pp. 1-114.

Groom, S.B. and Holligan, P.M. (1987) 'Remote sensing of coccolithophore blooms', Advances in Space Research 7 (2), 73-78.

Hojerslev, N.K. (1981) 'Assessment of some suggested algorithms on sea colour and surface chlorophyll', in J.F.R. Gower (ed.), Oceanography from Space, Marine Science Series, Vol. 13, Plenum Press, New York and London, pp. 347-353.

Holligan, P.M., Viollier, M., Dupouy, C., and Aiken, J. (1983a) 'Satellite studies on the distribution of chlorophyll and dinoflagellate blooms in the western English Channel', Continental Shelf Research 2, 81-96.

Holligan, P.M., Viollier, M., Harbour, D.S., Camus, P., and Champagne-Philippe, M. (1983b) 'Satellite and ship studies of coccolithophore production along a continental shelf edge', Nature 304, 339-342.

Horstmann, U. (1988) 'Satellite remote sensing for estimating coastal offshore transports', in B.O. Jansson (ed.), Coastal-Offshore Ecosystem Interactions, Lecture Notes on Coastal and Estuarine Studies, Vol. 22, SpringerVerlag, Berlin - Heidelberg, pp. 50-66.

Horstmann, U., Van der Piepen, H., and Barrot, K.W. (1986) 'The influence of river water on the southeastern Baltic Sea as observed by Nimbus 7 CZCS imagery', AMBIO A Journal of the Human Environment 15 (5), 286-289.

Hovis, W.A., Clark, D.K., Anderson, F., Austin, R.W., Wilson, W.H., Baker, E.T., Ball, D., Gordon, H.R., Mueller, J.L., El-Sayed, S.Z., Sturm, B., Wrigley, R.C., and Yentsch, C.S. (1980) 'Nimbus-7 Coastal Zone Color Scanner system description and initial imagery', Science 210, 60-63.

Jerlov, N.G. (1968) 'Optical Oceanography', Elsevier Oceanography Series 5, Elsevier Scientific Publishing Company, Amsterdam, Oxford and New York.

Jerlov, N.G. (1976) 'Marine optics', second edition, Elsevier Scientific Publishing Company, Amsterdam, Oxford and New York.

Jerlov, N.G. and Steemann-Nielsen, E., eds. (1976) 'Optical aspects of oceanography', Academic Press, London and New York.

Lohrenz, S.E., Arnone, R.A., Wiesenburg, and De Palma, I.P. (1988) 'Satellite detection of transient enhanced primary production in the western Mediterranean Sea', Nature 335 (6187), 245-247.

Mitchelson, E.G., Jacob, N.J., and Simpson, J. (1986) 'Ocean colour algorithms from the Case 2 waters of the Irish Sea in comparison to algorithms from Case 1 waters', Continental Shelf Research 5, 403-415.

Morel, A. (1980) 'In-water and remote measurements of ocean colour', Boundary-Layer Meteorology 18, 177201.

Morel, A. (1988) 'Optical modeling of the upper ocean in relation to its biogenous matter content (Case I waters)', Journal of Geophysical Research 93, 10749-10768.

Morel, A. (1991) 'Light and marine photosynthesis: a spectral model with geochemical and climatological implications', Progress in Oceanography 26, 301-344.

Morel, A. and André, J.M. (1991) 'Pigment distribution and primary production in the Western Mediterranean as derived and modelled from space (CZCS) observations', Journal of Geophysical Research 96 (C7), 12685-12698.

Morel, A. and Berthon, J.F. (1989) 'Surface pigments, algal biomass profiles and potential production of the euphotic layer: relationships re-investigated in view of remote sensing applications', Limnology and Oceanography 34, 1545-1562.

Morel, A., Bricaud, A., André, J.M., and Pelaez Hudlet, J. (1991) 'Spatial / temporal evolution of the Rhone plume as seen by CZCS imagery - Consequences upon the primary production in the Gulf of Lions', in J.M. Martin and H. Barth (eds.), EROS 2000 (European River Ocean System) Workshop, Water Pollution Research Reports 20, ISBN 2-87263-066-X, Commission of the European Communities, Brussels, pp. 45-62.

Morel, A. and Prieur, L. (1977) 'Analysis of variations in ocean colour', Limnology and Oceanography 22 (4), 709-722.

Platt, T. and Sathyendranath, S. (1988) 'Oceanic primary production: estimation by remote sensing at local and regional scales', Science 241, 1613-1629.

Prieur, L. and Sathyendranath, S. (1981) 'An optical classification of coastal and oceanic waters based on the specific spectral absorption curves of phytoplankton pigments, dissolved organic matter and other particulate materials', Limnology and Oceanography 26, 671-689.

Sathyendranath, S., Prieur, L., and Morel, A. (1987) 'An evaluation of the problems of chlorophyll retrieval from ocean colour, for Case 2 waters', Advances in Space Research 7 (2), 27-30.

Sathyendranath, S., Prieur, L., and Morel, A. (1989) 'A three-component model of ocean colour and its application to remote sensing of phytoplankton pigments in coastal waters', International Journal of Remote Sensing 10, 1373-1394.

Siegel, H. (1987) 'On the relationship between spectral reflectance and inherent optical properties of oceanic waters', Beitrage zur Meereskunde 56, 73-80.

Singh, S.M., and Cracknell, A.P. (1982) 'Coastal Zone Color Scanner: failure of active calibration', Journal of Physical E. Science Instrum. 15, 1003-1007.

Singh, S.M., Cracknell, A.P., and Charlton, J.A.. (1983) 'Comparison between CZCS data from July 10, 1979, and simultaneous *in situ* measurements for southeastern Scottish waters', International Journal of Remote Sensing 4, 755-784.

Singh, S.M., Cracknell, A.P., and Spitzer, D. (1985) 'Evaluation of sensitivity decay of Coastal Zone Color Scanner detectors by comparison with *in situ* near-surface radiance measurements', International Journal of Remote Sensing 6, 749-758.

Sorensen, B.M., ed. (1978a) 'Recommendation from the International Workshop on Remote Sensing Sea Tuth Data', JRC CEC Publication Service, Ispra (I).

Sorensen, B.M., ed. (1978b) 'Recommendation from the International

Workshop on Atmospheric Corrections of satellite Observation of Sea Water Colour', JRC CEC Publication Service, Ispra (I).

Sorensen, B.M., ed. (1979a) 'The North Sea Ocean Color Scanner Experiment 1977, Final Report', JRC CEC Publication Service, Ispra (I).

Sorensen, B.M., ed. (1979b) 'Workshop on the EURASEP Ocean Color Scanner Experiments 1977, Proceedings', JRC CEC Publication Service, Ispra (I).

Sturm, B. (1980) 'The atmospheric correction of remotely sensed data and the qualitative determination of suspended matter in marine water surface layers', in A.P. Cracknell (ed.), Remote Sensing in Meteorology, Oceanography and Hydrology, Ellis Horwood Ltd., Chichester, pp. 163-197.

Sturm, B. (1981) 'Ocean colour remote sensing and quantitative retrieval of surface chlorophyll in coastal waters using CZCS data', in J.F.R. Gower (ed.), Oceanography from Space, Marine Science Series, Vol. 13, Plenum Press, New York - London, pp. 267-279.

Sturm, B. (1983) 'Selected topics of Coastal Zone Color Scanner (CZCS) data evaluation', in A.P. Cracknell (ed.), Remote Sensing Applications in Marine Science and Technology, Reidel Plublishing Company, Dordrecht , pp. 137-167.

Sturm, B. (1986) 'Correction of the sensor degradation of the Coastal Zone Color Scanner on Nimbus-7', in Europe from Space, ESA-EARSeL Proceedings, Lingby (DK), June 1986, ESA SP-258.

Sturm, B. (1987) 'Application of CZCS data to productivity and water quality studies in the Northern Adriatic Sea', Advances in Space Research 7 (2), 47-51.

Tanré, D., Herman, M., Deschamps, P.Y., and De Leffe, A. (1979) 'Atmospheric modelling for space measurements of ground reflectances, including bidirectional properties', Applied Optics 18, 3587-3594.

Topliss, B., and Platt, T. (1986) 'Passive fluorescence and photosynthesis in the ocean: implications for remote sensing', Deep-sea Research 33 (7), 849-864.

Van Camp, L. (1987) 'Marine geographical information systems as a tool for the integrated processing of remote sensing data, *in situ* data and maps', in R. Reuter and R.H. Gillot (eds.), Remote Sensing of Pollution of the Sea, CEC DG XII JRC Ispra and Bibliotheks- und Informationssystem der Universitat Oldenburg, S.P.I. 87.46, Ispra - Oldenburg, pp. 130-139.

Van Camp, L., Nykjaer, L., Mittelstaedt, E., and Schlittenhardt, P. (1991) 'Upwelling and boundary circulation off Northwest Africa as depicted by infrared and visible satellite observations', Progress in Oceanography 26, 357-402.

Van Der Piepen, H., Amann, V. and Doerffer, R. (1991) 'Remote sensing of substances in water', GeoJournal 24, 27-48.

Viollier, M. (1982) 'Radiance calibration of the Coastal Zone Color Scanner: a proposed adjustment', Applied Optics 21, 1142-1145.

Viollier, M. and Sturm, B. (1984) 'CZCS data analysis in turbid costal water' Journal of Geophysical Research 89 (D4), 4977-4985.

Viollier, M., Tanré, D., and Deschamps, P.Y. (1980) 'An algorithm for remote sensing of water color from space', Boundary-Layer Meteorology 18, 247-267.

Wasmund, E. (1930) 'Flugbeobachtungen über Mittel- und Ost-Europäischen Gewässern', Geographische Zeitschrift, 36, 528-611.

SPECIAL FEATURES AND APPLICATIONS OF CZCS DATA IN ASIAN WATERS

H. FUKUSHIMA
School of High-Technology for Human Welfare
Tokai University
317 Nishino, Numazu
410-03 Japan

J. ISHIZAKA
National Institute for Resources and Environment
16-3 Onogawa, Tsukuba, Ibaraki
305 Japan

ABSTRACT. The paper summarizes recent work regarding ocean color remote sensing in Japan. It first explains atmospheric features in this area, one being moderately high aerosol concentration and another being the effect of Asian Dust (or KOSA - Yellow Sand Dust). A CZCS image affected by a KOSA event in May, 1981 will be shown. An attempt for a new atmospheric correction scheme tuned for that unique aerosol will also be presented. The paper then reviews an application of CZCS data to a study of local upwelling, based on a ship cruise that has matching CZCS data, enabling a comparison between ship-observed and satellite-derived pigment concentration. Although the initial result showed a factor 5 discrepancy in pigment concentration, the final satellite-derived pigment shows nearly perfect match with the ship-measured continuous pigment data, after adjustment using "sea-truth" data obtained at a ship station simultaneous to the satellite overpass. As another instance of CZCS data application, work observing "high-reflectance" waters in north-eastern sea of Japan is described. Because its magnitude and spectral dependency of reflectances sampled from 89 CZCS scenes agree well with those reported elsewhere, the high reflectance is considered due to detached coccoliths of *Emiliania huxleyi*, whose blooms seem to occur typically in early summer in this area. Finally, the paper demonstrate the necessity for algorithms adapted to waters with high concentration of suspended solids (SS), showing a CZCS image of the Yellow Sea with heavy SS loading.

1. Introduction

Japan is located in the far eastern region of the Asian continent, at the western boundary of the north Pacific Ocean. It is composed by series of islands (including the 4 large ones Hokkaido, Honshu, Shikoku, and Kyushu) separated from the continent by the Sea of Okhotsk, the Sea of Japan, and the East China Sea. Therefore, due to the island nature of their country, Japanese people depend on the sea, and Japan is known for its most intense fisheries. For many aspects of their life since ancient times, people have noticed the differences of ocean color, as differences of water mass characteristics. One

V. Barale and P.M. Schlittenhardt (eds.),
Ocean Colour: Theory and Applications in a Decade of CZCS Experience, 213–236.
© 1993 *ECSC, EEC, EAEC, Brussels and Luxembourg. Printed in the Netherlands.*

214

example is the word "Kuroshio" which literally means black water. Red tide also came from translation of Japanese "akashio" in which "aka" means red and "shio" means seawater and tide.

In terms of oceanographic features, the marginal seas of Japan show much variability (figure 1). The Kuroshio, known as one of the western boundary currents, flows between the shelfbreak of the East China Sea and the Ryukyu Islands, through the Tokara Strait south of Kyushu Island. Passing along the southern coast of Japan, the Kuroshio shifts its position frequently, and its often categorized by three different meander types. After the Kuroshio leaves the Japanese coast it is called Kuroshio Extension. The area off Sanriku, which is to the north-east of Honshu, is known as a complex but productive area, where warm core rings separate from Kuroshio and where the nutrient-rich Oyashio Current flows from northeast, while the relatively warmer Tsugaru Current comes in through the Tsugaru Strait. Part of the Kuroshio

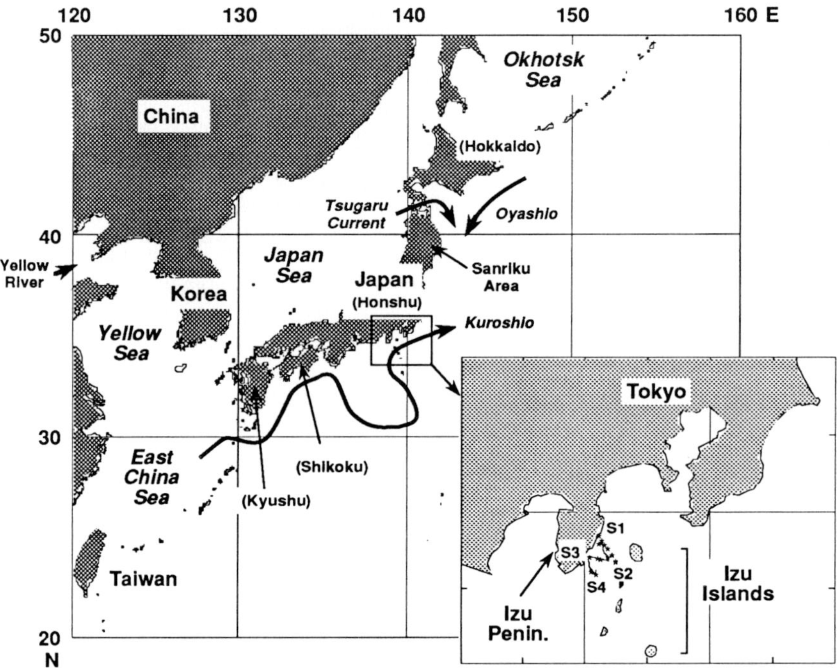

Figure 1 Geography of Japan and its surrounding seas. Kuroshio is a western boundary current which passes by the southern coast of Japan. Off the north-eastern coast, in the area off Sanriku, warm core rings are often formed by the Kuroshio stream. Oyashio, a nutrient-rich cold water current, comes down from the east of Hokkaido. In the Yellow Sea, the water is laden with high concentration of suspended soilds (SS). The inset map shows the area of ship-satellite observation mentioned in section 3. The line connecting S1 through S4 shows the ship track whereas the asterisks show ship stations. After Ishizaka *et al.* (In press).

flows north and west of Kyushu Island in the East China Sea. The East China Sea is a wide and shallow sea adjacent to the Yellow Sea where the Yellow River discharges much suspended materials. The Sea of Japan, finally, is a unique deep basin, separated from other seas by shallow straits. It has been considered to be of low productivity but this has been questioned recently. In comparison to other regions with analogous geographical characteristics, *e.g.* with the east coast of north America, the complicated bottom topography of the Japanese marginal seas causes a wide variety of oceanographic conditions.

There have been some excellent works regarding ocean optics in Japan. But not many satellite ocean color studies have been conducted so far. Sasaki *et al.* (1983) first attempted to obtain chlorophyll concentrations of the Yellow Sea by CZCS data. Ogishima *et al.* (1986) and Hiramatsu *et al.* (1987) pointed out problems in applying the standard atmospheric correction methods around Japan. Recently, there has been much more interest in satellite ocean color data because of the interest to study global biogeochemical cycles which is related to the sinking of CO_2 as well as to the study of fishing ground, and the human impact on the oceanic environment. Ishizaka and Harashima (1991) found about 4000 scenes of CZCS data of the northwest Pacific in the National Aeronautics and Space Administration (NASA) CZCS Video Browse System (Feldman *et al.*, 1989), and pointed out that many of the images seem useful to study the oceanography around Japan. Furthermore, the National Space Development Agency of Japan (NASDA) is to launch the Ocean Color and Temperature Scanner on the Advanced Earth Observing System (ADEOS) satellite in early 1995.

Here we summarize recent studies related to ocean color regarding Japan. The next section will describe characteristic (or special) features of Japanese aerosols, particularly about Chinese sand dust particles and their effect on CZCS imagery. Results from the development of a new atmospheric correction scheme will also be introduced. The third section will review a study of ocean color data in a local upwelling region. This is presently the only study comparing CZCS data with ship-observed pigment measurements as far as Japanese waters are concerned. In the forth section, a study on coccolithophore blooms in the area off Sanriku with the use of CZCS data will be reviewed. The fifth section will introduce suspended solid loading in the Chinese coastal area shown in CZCS imagery. Last section is the conclusion.

2. Asian Dust Aerosol and its Correction Algorithm

2.1 CHARACTERISTICS OF JAPANESE AEROSOLS

The Japanese aerosols can be characterized by moderately high average concentrations with occasional influence of Asian Dust (or KOSA). Fukushima *et al.*(1989) studied aerosol radiances in about 400 CZCS scenes that cover the Japan area. The data set was processed using the standard NASA CZCS atmospheric correction algorithm (Gordon *et al.*, 1983 ; Gordon *et al.*, 1988) to produce $L_A(670)$, the radiance due to aerosol scatterance in the CZCS channel 4. All the color-scaled $L_A(670)$ pictures were inspected and categorized into 3 groups, in terms of mean $L_A(670)$: (1) mean $L_A < 0.7$ μW/cm^2 nm sr (clear image), (2) mean L_A between 0.7 and 1.0 μW/cm^2 nm sr (hazy), and (3) mean $L_A > 1.0$ µW/cm² nm sr (very hazy). The selection was done empirically, and probably has some ± 0.1 µW/cm² nm sr uncertainty. Figure 2 shows the number of CZCS scenes obtained in each season during the study period (fall '78 to summer '82). In the figure, the seasons are abbreviated as follows; "S" for both spring (March-May) and summer (June-August), "F" for fall (September-November) and "W" for winter (December-February). Japanese spring-summer CZCS scenes have higher L_A in comparison to those of the eastern U.S. coastal area where typical $L_A(670)$ values of 0.5-1.0 µW/cm² nm sr are reported (Gordon and Castaño, 1987).

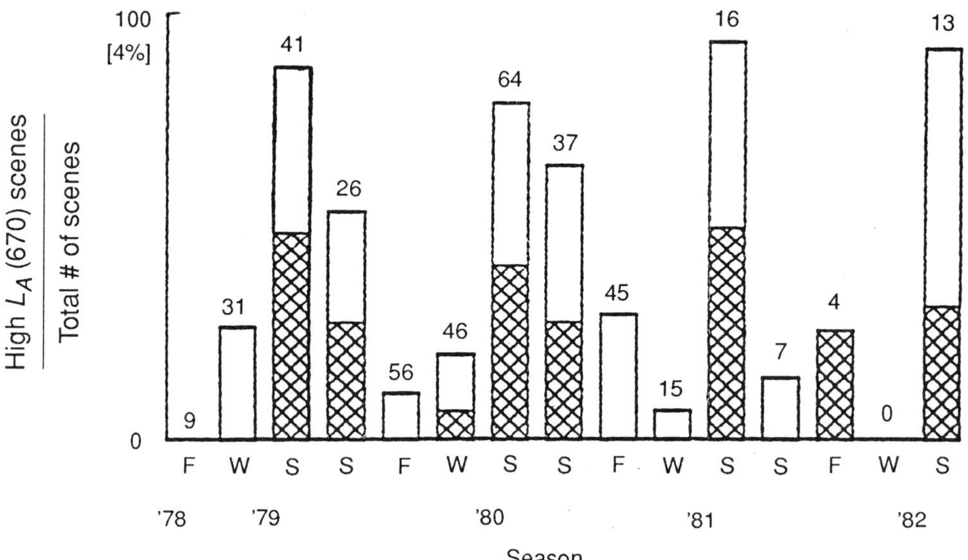

Figure 2 Seasonal change in the ratio of high aerosol scenes in the Japanese CZCS data set. The total height of each bar represents those scenes where $L_A(670)$ averaged more than 0.7µW/cm² nm sr, whereas shaded bars correspond to those with $L_A(670)$ higher than 1.5µW/cm² nm sr. The numbers shown indicate the total number of scenes analyzed for each season.

This high concentration of aerosol implies that the atmospheric correction for these scenes is vulnerable to the error in evaluating the aerosol parameters, namely ε values, as shown below. Under Gordon's scheme (Gordon *et al.*, 1983), the target water-leaving radiance at wavelength λ $(L_W(\lambda))$ is described as

$$L_W(\lambda) = \frac{1}{t(\lambda)}\left\{ L_T(\lambda) - L_M(\lambda) - \varepsilon(\lambda, 670) L_A(670) \frac{F_0'(\lambda)}{F_0'(670)} \right\} \qquad (1)$$

where L_T is the total radiance observed by satellite, L_M is the radiance due to gas molecular scattering, t is the transmittance between the sea-surface and the satellite and F'_0 is the extraterrestrial solar irradiance with absorptions by ozone and by aerosol taken into consideration. Suppose the true value for ε (443,670) was 1.0 and $L_A(670)$ was 1.5 μW/cm^2 nm sr. If we had a $+10\%$ error in estimating ε (443, 670), we would have about 0.2 less for $L_W(443)$, assuming $t(443)$ is 0.7. This will result in an erroneously high estimate of pigment concentration as shown in figure 3 where the standard in-water ("C13") algorithm and constant normalized $L_W(550)$ of 0.3 μW/cm^2 nm sr are assumed.

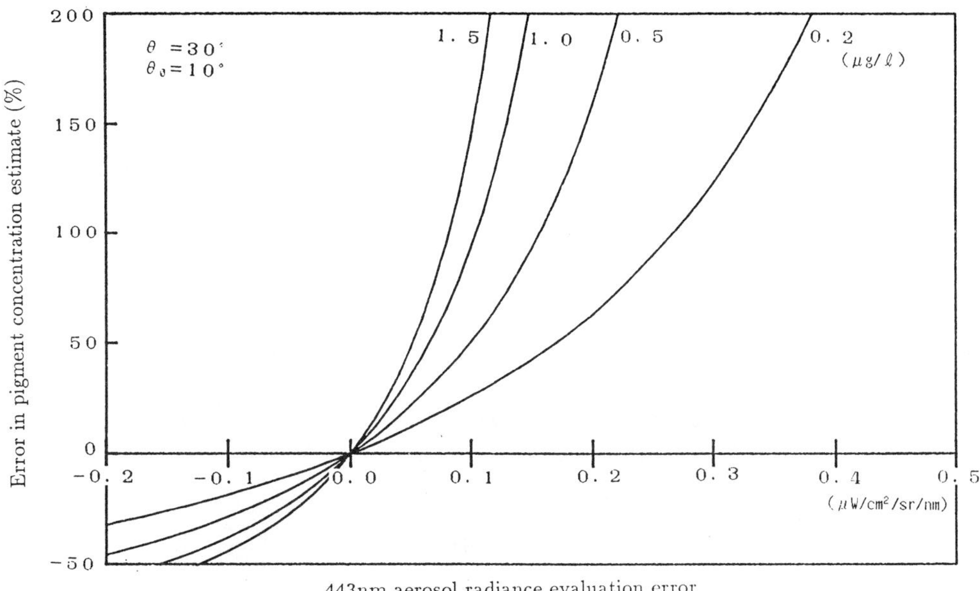

Figure 3 Model pigment concentration estimation error due to the error in evaluating aerosol radiance at 443 nm for satellite (θ) and sun (θ_0) zenith angles of 30° and 10°, respectively. The curves show the percentage estimated error for the "true" pigment concentration of 0.2-1.5 μg/l. The standard C13 algorithm which uses the atmospherically corrected channel 1 and 3 with constant $nL_W(550)$ is assumed. Note that the higher the true concentration, the larger the error that will result for higher estimates of 443 nm aerosol radiance ($L_A(443)$).

2.2 INFLUENCE OF THE ASIAN DUST

In spring, yellow sand dust ("KOSA") aerosols are occasionally observed in Japan, associated with yellow colored sky. The sand dust particles which originates from the Chinese desert area are lifted to the air by "sand storm" events which occur quite commonly in that region. Recent studies have revealed that a significant portion of those Asian dusts is transported at altitude of 1500-3000 m in the air, to fall down in the North Pacific Ocean (Iwasaka et al, 1988, Uematsu *et al.*, 1983 and Merrill *et al.*, 1988). It is reported by Betzer *et al.*(1988) that even giant particles of Chinese origin with diameter of about 100 µm are observed as far as the Hawaian islands. This transport of mineral aerosol is considered to be important in terms of the global geochemical cycle, not only as a source of sedimentary material but as a source of nutrients for phytoplankton.

Heavy sand dust events influence the CZCS atmospheric correction. Figure 4(a) shows a CZCS pigment concentration image of May 26, 1981, composed by the NASA standard atmospheric correction with ε values of 0.95, 1.0, 1.0 for channels 1, 2 and 3, respectively. Obviously, the image shows an erroneously high pigment concentration area to the east of the Korean Peninsula. The normalized L_W at 443 nm (figure 4(b)) records very low value (almost zero or below 0) in the area. Trying other sets of ε value does not improve the situation substantially. Judging from the spatial scale and pattern of the phenomenon, this is suspected to be caused by some air mass. Fukushima *et al.* (in preparation) have compared the image with meteorological records. On May 23 (3 days prior to the CZCS observation), dust storms were observed in the desert area of China (figure 5). They have conducted isentropic trajectory analysis using 700 mb and 800 mb weather charts. The dotted lines in the figure indicate estimated location of "the yellow sand dust" air mass front. Its position on May 26, the day the satellite observation was made, is in good agreement with the anomalous pattern in the image.

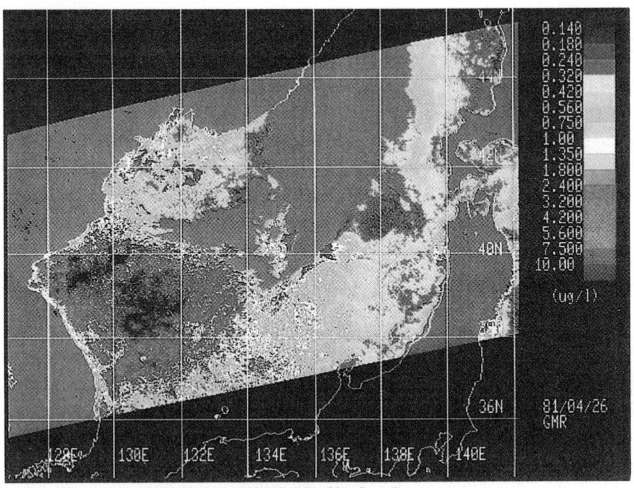

Figure 4 (a) Anomalous CZCS pigment image derived from satellite data of May 26, 1981, using the standard Gordon-Clark atmospheric correction. Land and cloud areas are masked in gray. Note the erroneously high pigment concentration area to the east of the Korean peninsula. This is considered to be caused by the Yellow Sand Dust (KOSA) event shown in figure 5.

Colorphotograph on p. 358

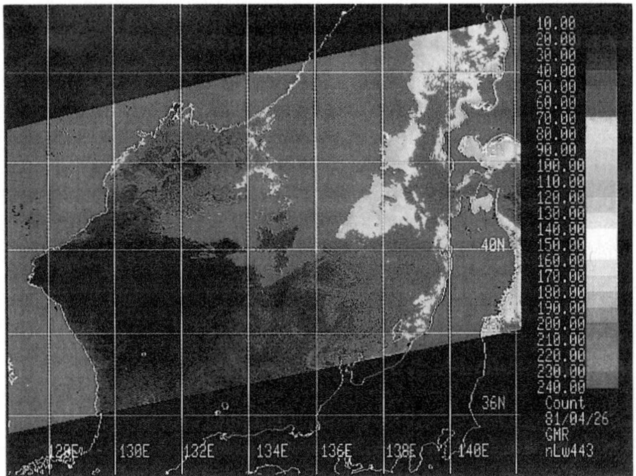

Figure 4 (b) Atmospherically corrected channel 1 data, or normalized water-leaving radiance nL_W image for the channel, corresponding to figure 4(a). Note that it results in erroneously low nL_W values near the Korean Peninsula.   Colorphotograph on p. 359

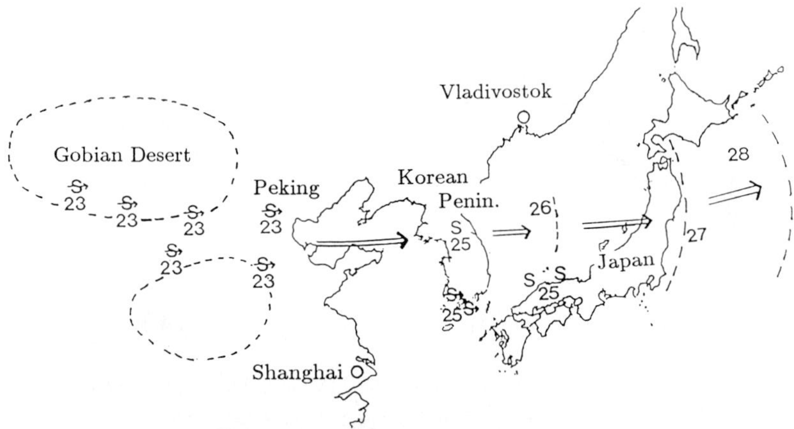

Figure 5 Estimated trajectory of Asian Dust-laden air mass at the end of May, 1981. "S" indicates locations where a "sand dust storm" was observed whereas "S" represents "dust", or presumably "KOSA (Yellow Sand Dust)". On the 23rd, sand dust storms were reported by local weather stations in continental China. By an isentropic trajectory analysis with 700mb and 800mb weather charts, the locations of the sand dust front (dashed line) were estimated for several days following the 23rd. On the 26th, when the CZCS observations were made, the air mass front is in the Japan Sea, well corresponding to the feature seen in figure 4.

The yellow sand dust contamination of this kind in the satellite imagery should be removed carefully since it will result in a bias in estimation of pigment concentration, and hence, productivity in this region. Although heavy KOSA events are occasional, recent studies revealed that there are weak but persistent "background KOSA" events (Iwasaka *et al.*, 1988) whose effect on ocean color observation is not yet known.

2.3 NEW ATMOSPHERIC CORRECTION SCHEME FOR ASIAN DUST AEROSOLS

Fukushima *et al.* (in preparation) attempted to modify the Gordon's algorithm so that
(1) it treats absorption by aerosol particles, introducing aerosol scattering albedo; and
(2) it calculates ε values for each pixel based on the assumption that normalized L_W at 550 nm is constant (0.3 µW/cm^2 nm sr).

In the NASA standard atmospheric correction algorithm (Gordon *et al.*, 1983), the transmittance of aerosol is approximated as unity, considering that aerosol particles have strong forward scatterance and hence negligibly small absorption. For the Asian Dust, however, the particles can be considered to be significantly absorptive when the particles are assumed to be spherical (Nakajima *et al.*, 1989). Therefore, Fukushima *et al.* introduced the following aerosol transmittance between the ocean surface and the satellite

$$t_A(\lambda) = exp\left\{-\left(1 - \omega_A(\lambda)\,\eta(\lambda)\right)\tau_A/cos\theta\right\} \tag{2}$$

where ω_A is the aerosol single scattering albedo, η is forward scattering probability, τ_A is aerosol optical thickness and θ is satellite zenith angle. The transmittance between the sun and the sea surface was also modified similarly. Further, $\eta(\lambda)$ is assumed to be unity. For ω_A, which is dependent on aerosol type, it is desirable to determine its value for each pixel automatically during the atmospheric correction, but such an algorithm is still to be studied.

The radiative transfer model is just the same as that of the standard algorithm

$$L_T(\lambda) = L_M(\lambda) + L_A(\lambda) + t_M(\lambda)\,t_A(\lambda)\,t_{OZ}(\lambda)\,L_W(\lambda) \tag{3}$$

In the equation above, t_M and t_{OZ} are transmittances from the sea surface to the satellite, accounting for molecular scattering and ozone absorption, respectively.

Because the introduction of t_A requires a new set of variables τ_A for each channel, Fukushima *et al.* relate L_A to τ_A, in order to keep the number of independent variables constant. This is done by using the relation

$$L_A(\lambda) = \frac{\omega_A(\lambda)\,\tau_A(\lambda)\,F_0(\lambda)\,P_A(\lambda,\Psi)}{4\pi cos\theta} \tag{4}$$

where P_A is the aerosol scattering phase function and ψ is the scattering angle. P_A is modeled by the two term Henyey- Greenstein function with parameters $\gamma = 0.938$, $g_1 = 0.82$ and $g_2 = -0.55$. The τ_A's for all the 4 CZCS visible channels are now taken as independent variables in place of L_A.

To solve equation (3) for each pixel, Fukushima *et al.* use an iterative procedure since the equation is nonlinear in τ_A. Figure 6 shows a flow diagram of the pixel-wise iterative atmospheric correction scheme which assumes constant normalized water-leaving radiance at 550 nm ($nL_W(550) = 0.3$ µW/cm^2 nm sr). The idea of constant $nL_W(550)$ was introduced by Gordon *et al.*, (1988b), allowing pixel-wise determination of ε value.

A resulting pigment image obtained by applying the new algorithm is shown in figure 7 together with an $L_W(443)$ image. The original CZCS data is the same as one in figure 4. The values for ω_A at 443, 520, 550 and 670 nm were set to 0.7, 0.8, 0.9 and 1.0, respectively. All the values were selected by trial and error so that the resulting water-leaving radiance varies in an expected range in target area. As shown in the figure, the new images represent a clear improvement on the old erroneous ones, although there still seems to remain a residual atmospheric pattern in the yellow sand dust contaminated area, urging further study.

Since the Japanese CZCS data set is rather sparse, with very little matched sea-truth data, the effort of developing an automated yellow sand dust algorithm is not straight forward. This kind of study, however, is of vital necessity in preparing for the next generation of ocean color satellite observations. Carder *et al.* (1991) have also studied a similar problem for Saharan dust aerosols.

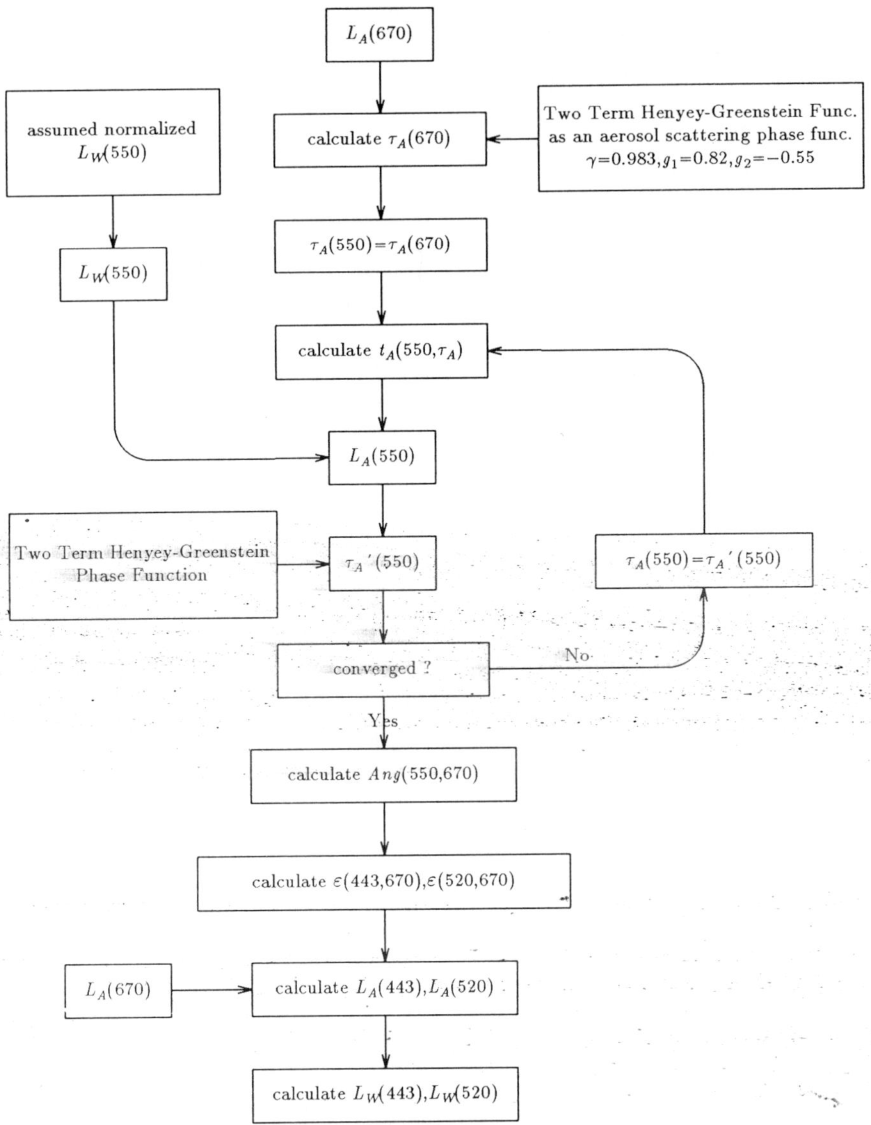

Figure 6 Iterative CZCS atmospheric correction scheme with $\tau_A(\lambda)$ (aerosol optical thickness at wanelength λ) being independent variables, based on a model with non-unity (absorptive) aerosol single scattering albedo for each channel(eqs. (2)-(4)). The two term Henyey-Greenstein function is used as a model for aerosol scattering phase function to relate $\tau_A(\lambda)$ and $L_A(\lambda)$ (aerosol scattering radiance at λ). Ang(550, 670) is the Angstrom exponent derived from $\tau_A(550)$ and $\tau_A(670)$. Constant normalized water-leaving radiance at 550 nm ($nL_W(550)=0.3\mu W/cm^2\,nm\,sr$) is assumed.

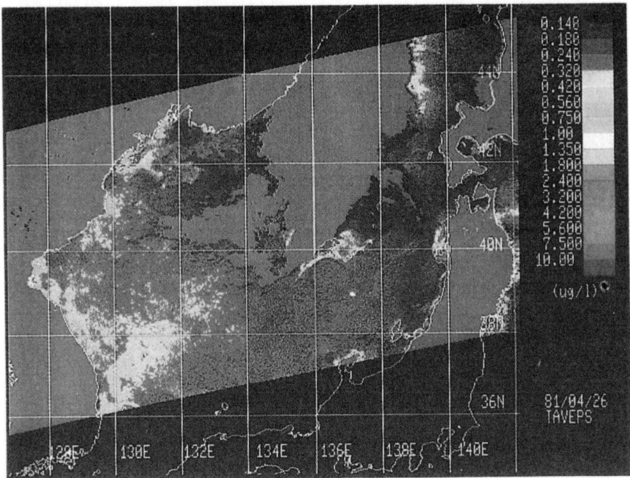

Figure 7 (a) Pigment concentration image derived by the new atmospheric correction scheme for the same CZCS data as in figure 4. The aerosol single scattering albedo (ω_A) is assumed to be 0.7, 0.8, 0.9 and 1.0 for 443, 520, 550 and 670 nm channels, respectively. Colorphotograph on p. 359

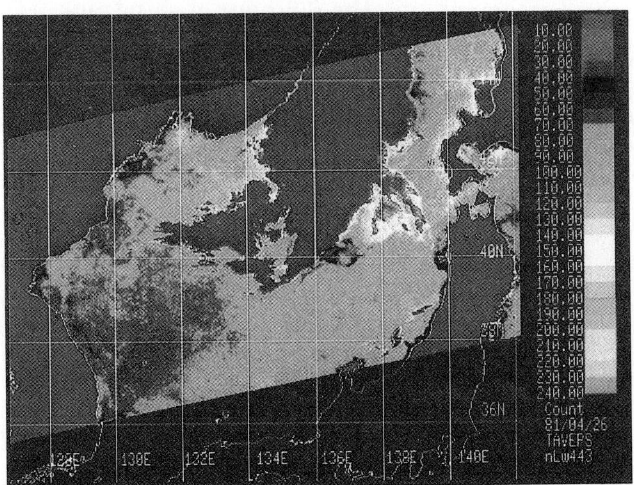

Figure 7 (b) Atmospherically corrected channel 1 ($nL_W(443)$) corresponding to the figure 7 (a) image under the new correction scheme. Note that,in comparison to figure 4 (b), $nL_W(443)$ now ranges in normal or more likely values. Colorphotograph on p. 359

3. Observation of Regional Upwelling around Izu Peninsula

Presently there is only one CZCS scene which was directly compared to pigment observation by ship (Ishizaka *et al.*, in press). The scene was taken on May 23, 1982, and the concurrent cruise was conducted for the observation of regional upwelling around Izu Peninsula. A detailed description of the observations can be found in Atkinson *et al.* (1987) and Takahashi *et al.* (1986).

Izu Peninsula is located (figure 1) about 100 km southeast of Tokyo. It was known that regional upwelling frequently occurs around the Izu Peninsula and the Izu Islands as an interaction between the Kuroshio, topography, and wind, and that phytoplankton responds to the upwelling (Takahashi *et al.*, 1981; Takahashi and Kishi, 1984).

Ishizaka *et al.* (in press) processed the CZCS data with ε values known for typical oceanic and land type aerosols. They also applied values estimated by the clear water algorithm (Gordon *et al.*, 1983). The resulted image showed most of the expected oceanographic features, and the pigment concentrations were correlated with ship-observed pigment concentrations; however, the magnitude of the satellite-derived concentrations was about a factor 5 smaller than measured values (Table 1). Therefore, alternative ε values were estimated so that the CZCS-derived pigment concentration was equivalent to the ship-observed pigment concentration at the location of the satellite overpass. Explanations of the method is in Ishizaka *et al.* (in press), and the resultant ε values are shown in Table 1. The estimated pigment concentrations show good agreement with the ship observed concentrations (Table 1). Figure 8 shows a comparison between the CZCS-derived pigment concentrations and the ship measured continuous pigment concentrations along the track shown in figure 1. Both pigment concentrations show surprisingly good agreement over the 6.5-hour of ship track.

TABLE 1. ε values for different atmospheric corrections used for the May 26 CZCS data, together with resultant correlation coefficients (r) and root mean square errors (RMSE) against ship observation (Ishizaka *et al.*, In press). Pigment concentrations were obtained by L2MULT command, whereas ε values for Clear Water were calculated with CLRWAM command, both of PC-SEAPAK system (McClain *et al.* 1990).

	Epsilon			r	RMSE
	443	520	550	(n＝9)	
Oceanic	0.95	1.00	1.00	0.696	0.767
Land	1.32	1.22	1.17	0.622	0.763
Clear Water	1.07	1.04	1.03	0.721	0.737
Adjustment	1.43	1.17	1.13	0.827	0.290

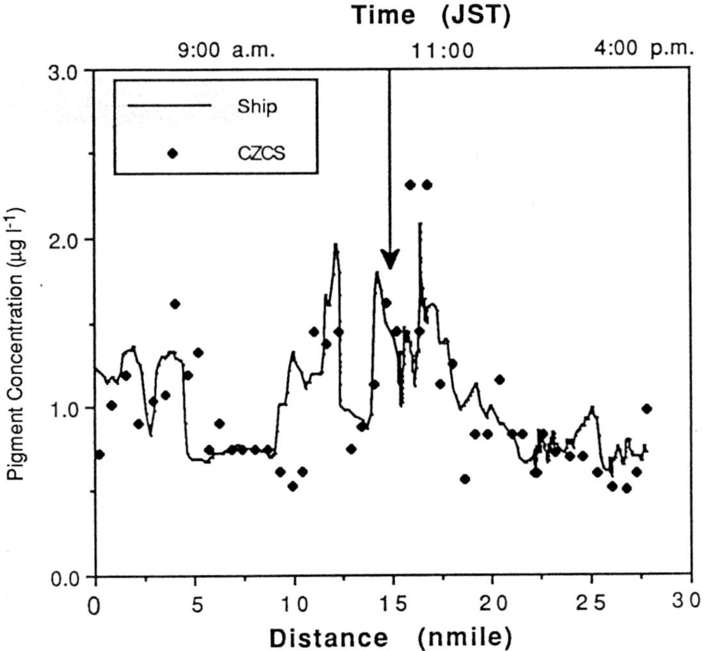

Figure 8 Continuously measured pigment concentrations along the ship track and CZCS-derived pigment concentrations at the corresponding locations. Pigment concentrations from CZCS were adjusted so that the ship-observed and CZCS-derived pigment concentrations at 11:00 a.m. (indicated by arrow) are equivalent. After Ishizaka *et al.* (In press).

One of the reasons speculated for the underestimation by the standard processing methods is the in-water algorithm; however, Ishizaka *et al.* (in press) found that the upward radiance data collected on the same cruise (Sugihara *et al.* 1985) fit well the result reported by Gordon *et al.* (1983), and argue that the in-water algorithm is not the reason for the underestimation. A second possible reason is sensor calibration. But since the calibration they have applied was the same one that NASA (updated one) was using (Evans, personal communication ; McClain, 1990) and cause no similar anomaly (*i.e.*, too low pigment estimates) for other CZCS data, this cannot be the cause. The most attributable interpretation was that local aerosol had different characteristics.

Ishizaka *et al.* (in press) showed that once the appropriate atmospheric correction parameters were chosen, the CZCS data exhibited a reasonable amount of variability in pigment concentration (figure 9). The CZCS image and Sea Surface Temperature (SST) image clearly showed low temperature and high pigment concentration in the regional upwelling area along Izu Peninsula where Takahashi *et al.* (1986) described the local phytoplankton growth in the low temperature upwelled water. Several cold upwelled water eddies with the diameter of 5-10 km were also identified west of the Izu

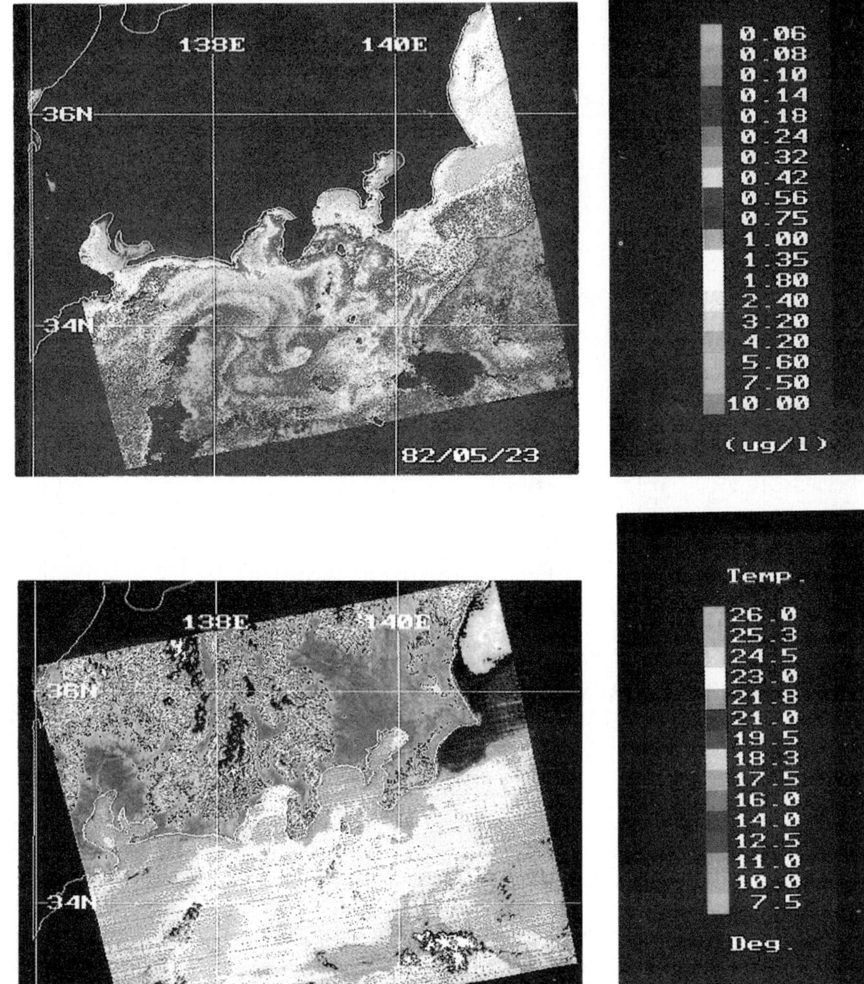

Fig. 9a

Fig.. 9b

Figure 9 Pigment concentration (a) and temperature (b) distributions around Izu Peninsula on May 23, 1982, obtained by CZCS (Ishizaka *et al.*, in press).

Colorphotograph on p. 360

Islands. However, in contrast to the regional upwelling along the peninsula, these cold eddies do not necessarily correspond to higher pigment waters. Ishizaka *et al.* (in press) speculate that the difference may be caused by the possible difference of interaction between physical regime and biological response.

Ishizaka *et al.* (in press) also showed the variability in pigment and SST at larger scale (figure 10). Kuroshio meanders were clearly shown in both SST and pigment images, separating from the south of Japanese coast, looping with about a 400 km diameter and then returning back to the coast, northeast along the Izu Peninsula. The Kuroshio water could be identified as a warm stream looping to the south of Japan. In general, pigment concentrations were lower in the Kuroshio waters. However, the area around the Izu Peninsula and the Islands showed relatively higher pigment concentrations than any other region of the Kuroshio. Furthermore, inside the Kuroshio pigment concentrations appear higher than in south of Kuroshio, clearly distinguishable from the coastal waters. From these features, Ishizaka *et al.* (in press) suggested that the Kuroshio enhances the productivity around Japan and possibly exports organic materials to the open ocean.

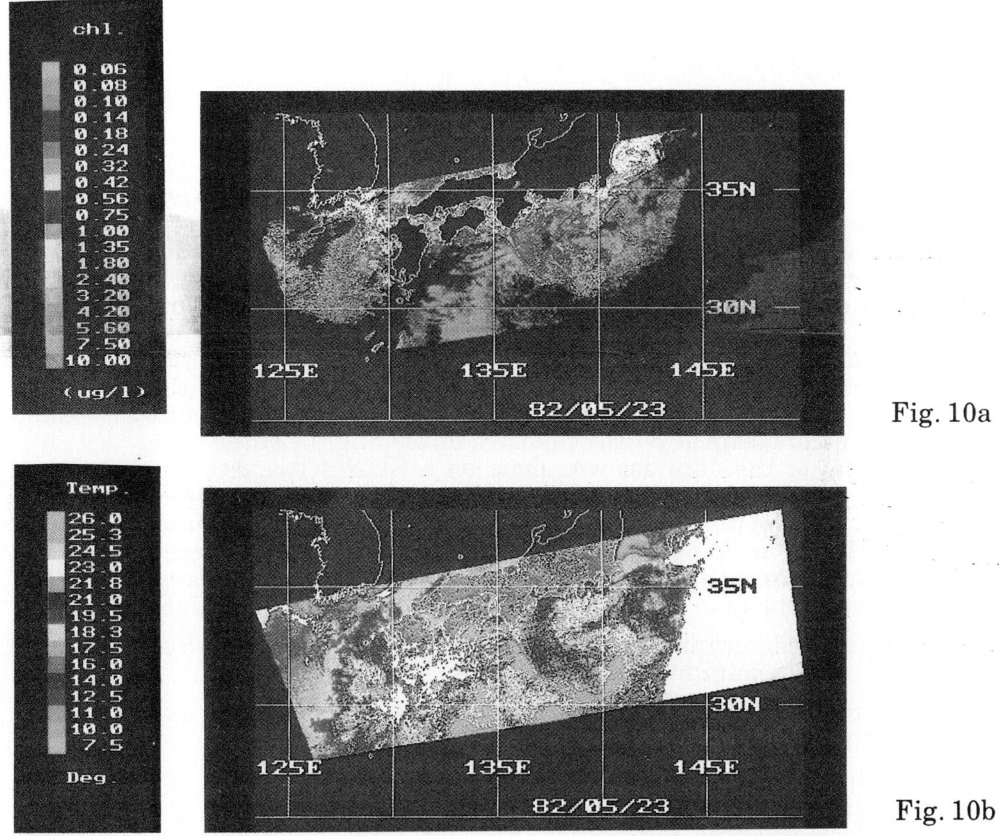

Figure 10 Pigment concentration (a) and temperature (b) distributions around southern Japan on May 23, 1982, obtained by CZCS (Ishizaka *et al.*, in press). Colorphotograph on p. 361

4. Coccolithophore blooms off the Sanriku area observed by CZCS

4.1 COCCOLITHOPHRES AND COCCOLITHS

Coccolithophres are a group of phytoplankton that produces external calcium carbonate plates, or coccoliths. Among the group, *Emiliania huxleyi* is known to produce vast amount of coccoliths, and is important in terms of inorganic carbon flux. The water appears "milky white" at high concentration of coccoliths, due to the strong backscattering of the plankton in detached form. Although this can be easily observed by eye from shipboard, the magnitude of the spatial and temporal scales of coccolithophore blooms have not been recognized until recently, when satellite remote sensing data became available for studying the phenomenon. With the use of LANDSAT Multi-Spectral Scanner (MSS), CZCS and/or AVHRR, the high reflectance waters are seen in the North Atlantic each summer in the open ocean (Holligan *et al.*, 1983) and in coastal/upwelling regions (Dupouy and Demarcq, 1987; Groom and Holligan, 1987) including the Gulf of Maine (Balch *et al.*, 1991). Fukushima *et al.* (1987) also reported one occurrence of the high-reflectivity water mass in a CZCS scene covering Japan.

4.2 SURVEY OF HIGH REFLECTANCE WATERS IN THE SANRIKU CZCS DATA SET

Fukushima *et al.* (in preparation) have been examining a selected CZCS data set for the area off Sanriku, which is off the northeastern coast of Honshu island. The area is known as a "highly perturbed area" due to contributions from the warm Kuroshio current, the cold nutrient rich Oyashio water, and the Tsugaru current (figure 1).

First, they checked the CZCS Video Browse System (Feldman *et al.*, 1989) to select 89 low cloud-coverage scenes, out of 301 totals covering the rectangular study area ranging from 35°N to 43°N and from 140°E to 151°E. The search was limited to data from 1979 to 1992, because of the paucity of data for other years. Most of the original 89 scenes were supplied by Gene Feldman and his team at NASA Goddard Space Flight Center.

Processing of the data set was done on a SUN 4 sparcstation-1, using a software system developed for this purpose. This uses the NASA standard atmospheric correction with ε parameters set to 0.95, 1.0, 1.0 for ch.1, 2 and 3, respectively. To avoid non- zero $L_W(670)$ which will cause underestimation of spectral reflectance at shorter wavelengths, the following simple spatial filtering was applied to $L_A(670)$ during the atmospheric correction:

(1) within a predetermined window along the scan line centered at the target pixel, look for the minimum $L_A(670)$ value;

(2) perform atmospheric correction taking the selected value as $L_A(670)$ for the target pixel.

After preliminary investigation, the window size was set to ± 20 pixels for 2 by 2 subsampled data. This corresponds to about ± 40 km on the ground.

Subsurface spectral reflectance $R(\lambda)$ for high reflectance waters were sampled to compare with the results reported in other papers. The method calculates $R(\lambda)$ from normalized water-leaving radiance $nL_W(\lambda)$, using the relation

$$nL_W(\lambda) = \frac{(1-\rho)(1-\rho')F_0'R(\lambda)}{n^2 Q} \tag{5}$$

where ρ is the Fresnel reflectance from the air to the water and ρ' from the water to the air), F_0' is the extraterrestrial solar irradiance with ozone absorption taken into consideration, and n is the refractive index of water. Q is a factor to convert subsurface upwelling radiance to upwelling irradiance and is assumed to be 5.0.

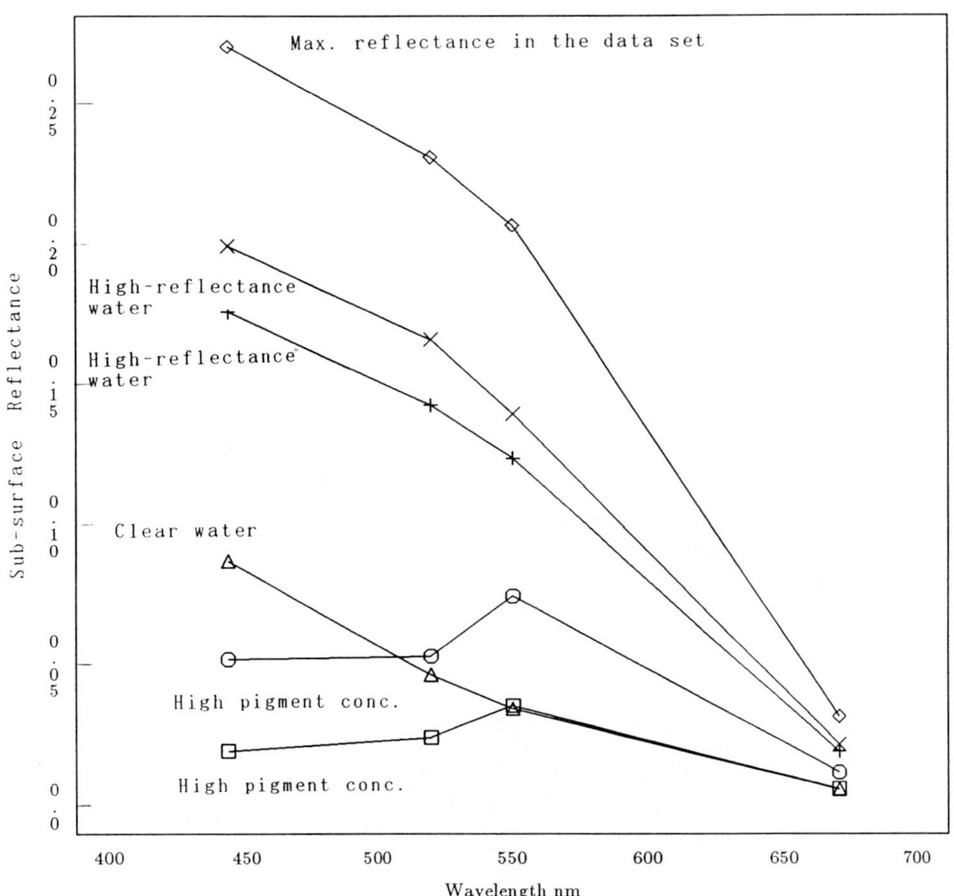

Figure 11 Typical sub-surface reflectances retrieved from the CZCS data set in the area off Sanriku area. The magnitude and spectral dependency of "high-reflectance" water agrees well with those reported by other authors, strongly suggesting that the high reflectance is due to detached coccoliths.

230

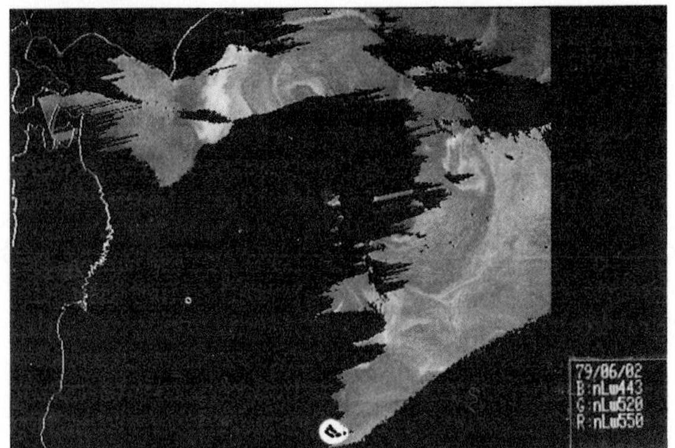

Fig. 12a

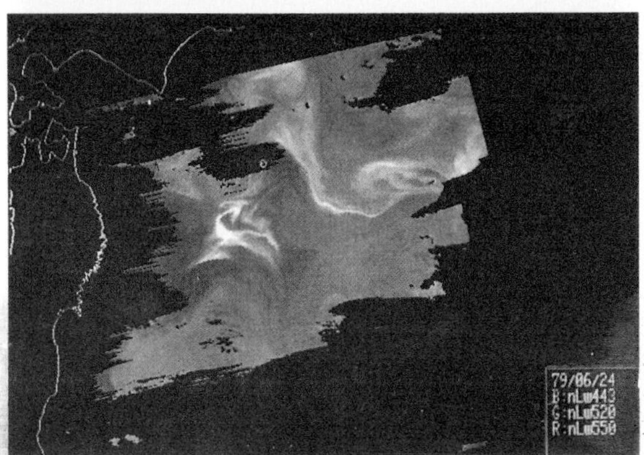

Fig. 12b

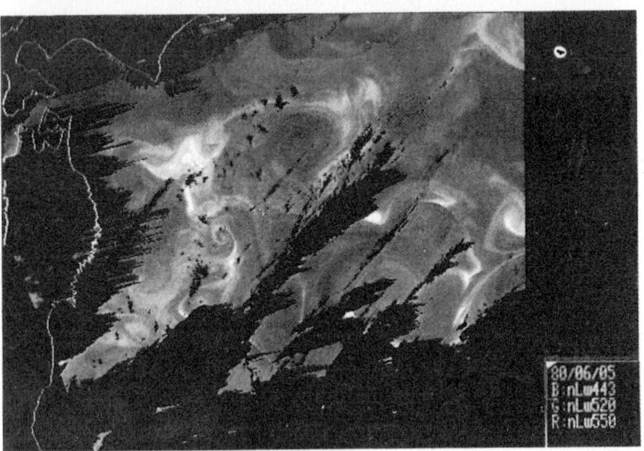

Fig. 12c

Figure 12 CZCS color composite image of the area off Sanriku with high concentration of coccolithophores. Normalized water-leaving rediances of channels 1, 2 and 3 are assigned to the blue, green and red color, respectively, so that high-reflectance waters appear in white. The scenes were collected on (a) June 2, 1979, (b) June 24, 1979, and on (c) June 5, 1980. A cloud ring masking algorithm (J.L. Mueller, personal communication) was applied.

Colorphotograph on p. 362

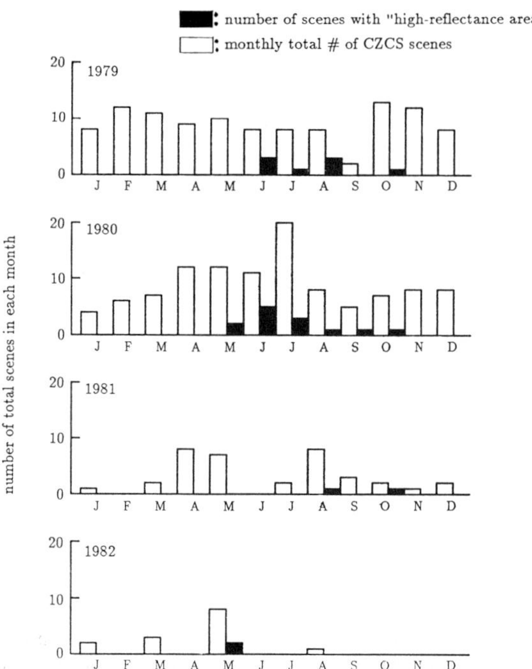

Figure 13 CZCS coverage of the area off Sanriku, showing the monthly total number of CZCS scenes (open bars) and the number of scenes with high-reflectance water appearance. Although the analysis was made by eye inspection, and is therefore subjective to some extent, this coccolithophore-like feature mostly occurs in early summer.

Figure 11 shows some of the spectral reflectance curves obtained. Samples from a clear water area with low pigment concentration and an area with high pigment concentration with little backscattering are also shown in the figure for comparison. As is shown, typical high-reflectance water has about 0.2-0.25 reflectivity at wavelength 443 nm with gradual decrease towards the longer wavelength region. These spectral shapes and the magnitude of the reflectances agree well with those reported elsewhere (Groom and Holligan, 1987; Balch *et al.*, 1991). This fact strongly supports the hypothesis that those high-reflectances are due to detached coccoliths of *Emiliania huxleyi*, with a concentration of at least 10^5 coccoliths per ml.

Figure 12 shows the resulting color-composite images with $nL_W(443)$, $nL_W(520)$ and $nL_W(550)$ loaded to the blue, green and red frame buffers of the display, so that water with high reflectance for all the three channels will appear as bright white. Due to the dynamical feature of the area, the coccolithophore-rich waters help to visualize the numerous eddies. As the three images show, high-reflectance water masses are frequently found in the area off Sanriku which is considered to be less eutrophic than the northern Oyashio waters extending to the southern coastal area of Hokkaido.

Fukushima *et al.* also investigated seasonal variability of high reflectivity. figure 13 shows monthly total number of CZCS scenes, covering the study area as well as the monthly total number of scenes where high reflectance areas were identified.

Although the identification was made by eye inspection and hence subjective is to some extent, it is obvious that the coccolithophore blooms typically occur during the early summer season. Also notable is the data paucity for the years '81 and '82.

5. High concentration of suspended solid in the Yellow Sea and the East China Sea

The East China Sea is a shallow sea (depth $<$ 200 m) which extends to the southwestern islands of Japan and Kyusyu. To the north is the shallow Yellow Sea, surrounded by the Korean Peninsula and the Chinese coast. The sea, particularly in the Chinese coastal area, is known for the high concentration of suspended solid (SS) attributable to the Yellow River discharge.

Figure 14 shows a CZCS image obtained on January 10, 1982, in the form of a color composite of channel 1 (blue), channel 3 (green) and channel 4 (red) with no atmospheric correction. The most striking feature of the image is a high suspended solids (SS) concentration area (brown pattern), about 300 km wide, along the Chinese coast, with its core part (bright brown) having saturated channel 4. Since the outer rim of this pattern corresponds well to the 50 m isobath in this region, the high sediment concentration could be interpreted as resuspension of the bottom sediment (as far as this particular image is concerned).

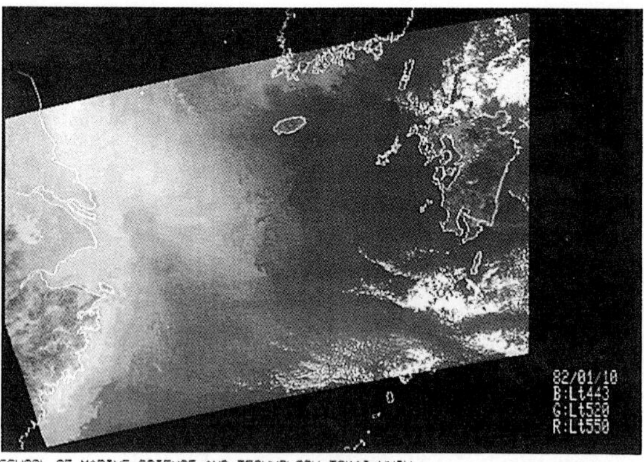

Figure 14 Suspended solid (SS) laden shelf waters in the Yellow Sea and the East China Sea. The image is color composite of original CZCS data; channel 1 assigned to blue, channel 3 to green and channel 4 to red.

Colorphotograph on p. 363

In addition to the importance as a fishing ground, the East China Sea has been drawing more attention recently in terms of ocean flux study: the Yellow River and the Yan-tze River are considered as sources of a vast amount of dissolved organic matter (DOM) for the area, apparently contributing to the high productivity there. Although there is no doubt that satellite observation is useful to monitor ocean productivity, this image clearly points out the necessity of developing algorithms that can handle such heavily SS-laden waters.

6. Conclusion

There have not been many CZCS applications in Japan for several reasons. One of them was lack of information about CZCS: there was no Japanese member in the Nimbus-7 Experimental Team, for example. Another strong reason was lack of resources from the stand point of data users: it took several years to establish the CZCS processing algorithms. It was also not easy to develop a dedicated data system, particularly for end-users such as marine biologists.

The situation has been steadily improving. As one of the activities of the Japanese community with regard to information exchange and international cooperation, a series of Joint Japan-US Workshop have been held in 1990 and in 1991 to discuss issues related to SeaWiFS and OCTS. The ISY'92 PGO activities will also provide a good opportunity in this respect. Regarding resource for data processing, the end-user oriented PC-Seapak system (McClain *et al.*,1990) has been introduced to seven sites in Japan. The Miami DSP software system, which is used as a standard CZCS data processing system by NASA/GSFC was also introduced to the School of High-Technology, Tokai University, and will be used for the large volume data processing required to the Japanese PGO activities. Under these circumstances, the number of Japanese satellite ocean color application studies will be much increased.

Looking now to the future, following the planned launch of SeaStar/SeaWiFS in 1993, NASDA is to launch ADEOS in early 1995. The satellite is to carry the Ocean Color and Temperature Scanner (OCTS), as well as seven other sensors including the NASA Scatterometer (NSCAT), the Total Ozone Mapping Spectrometer (TOMS) and the Polarization and Directionality of Reflectances (POLDER). With these sensors, ADEOS/OCTS will offer a unique opportunity to the science community. In order to establish a system that would allow end-users to extract full value of the data, the OCTS Mission Team was created in 1990 to plan validation, application and algorithm development, as well as to give advise with regard to data system operations and policy.

References

Ackleson, S., Balch, W. M., Holligan, P. M. (1988) 'White waters of the Gulf of Maine', Oceanography 1, 18-22.

Atkinson, L. P., Blanton, J. O., McClain, C., Lee, T. N., Takahashi, M., Ishimaru, T., Ishizaka, J. (1987) 'Observations of upwelling around the Izu Peninsula, Japan May 1982', J. Oceanogr. Soc. Japan 43, 89-103.

Balch, W. M., Holligan, P. M., Ackelson, S. G., Voss, K. J. (1991) 'Biological and optical properties of mesoscale coccolithophore blooms in the Gulf of Maine', Limnology and Oceanography 36, 629-643.

Betzer, P. R., Carder, K. L., Duce, R. A., Merrill, J. T., Tindale, N. W., Uematsu, M., Costello, D. K., Young, R. W., Feely, R. A., Breland, J. A., Bernstein, R. E, Greco, A. M. (1988) 'Long-range transport of giant mineral aerosol particles', Nature 336 (6199), 568-571.

Carder, K. L., Gregg, W. W., Costello, D. K., Haddad, K., Prospero, J. M. (1991) 'Determination of Saharan Dust radiance and chlorophyll from CZCS imagery', Journal of Geophysical Research 96 (D3), 5369-5378.

Dupouy, C., Demarcq, H. (1987) 'CZCS as an aid for understanding modalities of the phytoplankton productivity during upwelling off Senegal', Advances in Space Research 7, 63-71.

Feldman, G., Kuring, N., C. Ng, W. Esaias, McClain, C. , Elrod, J., Maynard, N., Endres, D., Evans, R., Brown, J., Walsh, S., Carle, M., and Podesta, G. (1989) 'Ocean color. Availability of the global data set', EOS 70, 634-641.

Fukushima, H., Hiramatsu, K., and Sugimori, Y. (1987) 'CZCS derived pigment concentration fields in Japanese coastal area', Advances in Space Research 7, 79-82.

Fukushima, H., Smith, R. C., Sugimori, Y., Toratani, M., Yasuda, Y. (1989) 'Aerosol anomalies in Nimbus-7 Coastal Zone Color scanner data obtained in Japan area', Proc. IGARSS '89, IEEE.

Gordon, H. R., Clark, D. K., Brown, J. W., Brown, O. B., Evans, R. H., Broenkow,W. W. (1983) 'Phytoplankton pigment concentrations in the Middle Atlantic Bight: Comparison of ship determinations and CZCS estimates', Applied Optics 22, 20-36.

Gordon, H. R., Castaño, D. J. (1987) 'The Coastal Zone Color Scanner atmospheric correction algorithm: multiple scattering effects', Applied Optics 26 (11), 2111.

Gordon, H. R., Brown, J. W., Evans, R. E. (1988) 'Exact rayleigh scattering calculations for use with the Nimbus-7 Coastal Zone Color Scanner', Applied Optics 27 (5), 862-871.

Gordon, H. R., Brown, O. B., Evans, R. H., Brown, J. W., Smith, R. C., Baker, K. S., Clark, D. K. (1988b) 'A semi-analytic radiance model of ocean color', Journal of Geophysical Research 93, 10909-10924.

Groom, S. B. and Holligan, P. M. (1987) 'Remote sensing of coccolithophore blooms', Advances in Space Research 7, 73-78.

Hiramatsu, K., Fukushima, H., Matsumura, S., Sugimori, Y. (1987) 'CZCS-derived pigment concentration fields in Japanese coastal area', (in Japanese with English Abstract), Sora to Umi 9, 1-7.

Holligan, P. M., Viollier, M., Harbour, D. S., Camus, P., and Champagne-Philippe, M. (1983) 'Satellite and ship studies of coccolithophore production along a continental shelf edge', Nature 304, 339-342.

Ishizaka, J. and Harashima, A. (1991) 'CZCS data number around Japan and surface chlorophyll data number simultaneously observed by ship', (in Japanese with English Abstract), Sora to Umi 13.

Ishizaka, J., Fukushima, H., Kishino, M., Saino, T., and Takahashi, M. (in press) 'Chlorophyll distributions in regional upwelling around Izu Peninsula detected by Coastal Zone Color Scanner on May 1982, J. Oceanogr. Soc. Japan.

Iwasaka, Y., Yamato, M., Imasu, R., and Ono, A.(1988) 'Transport of asian dust (KOSA) particles; importance of weak KOSA events on the geochemical cycle of soil particles', Tellus 40B, 494-503.

Matsumura, S. and Fukushima, H. (1988) 'Water mass analysis using ocean color map and sea surface temperature map obtained by NIMBUS-7/CZCS', (in Japanese with English Abstract), Sora to Umi 10, 27-39.

McClain, C. R., Fu, G., Darzi, M., and Firestone, J. K. (1990) 'PC-SEAPAK User's Guide', Version 3.0. NASA Goddard Space Flight Center.

Merrill, J. T., Uematsu, M., and Bleck, R. (1989) 'Meteorological analysis of long range transport of mineral aerosols over the North Pacific', Journal of Geophysical Research 94 (D6), 8584-8598.

Mueller, J. L. (1988) 'Nimbus-7 CZCS: electronic overshoot due to cloud reflectance', Applied Optics 27 (3), 438-440.

Nakajima, T., Tanaka, M., Yamano, M., and Shiobara, M. (1989) 'Aerosol optical characteristics in the yellow sand events observed in May, 1982 at Nagasaki - Part II Models, J. of the Meteorological Society of Japan' 67 (2), 279-291.

Ogishima, T., Fukushima, H., and. Sugimori, Y. (1986) 'Problems relating to atmospheric correction algorithm for CZCS data in Japanese coastal area', (in Japanese with English Abstract), Sora to Umi 8, 53-63.

Sasaki, Y., Asanuma, I., and Muneyama, K. (1983) 'A study on the atmospheric correction for the satellite observation for oceans', (in Japanese with English abstract), Journal Japan Society of Photogrammetry and Remote Sensing 22, 4-10.

236

Sugihara, S., Kishino, M., and Okami, N. (1985) 'Correlation of chlorophyll concentrations and suspended solids with near-surface upward irradiance within LANDSAT Bands 4, 5 and 6', J. Oceanogr. Soc. Japan 41, 81-88.

Takahashi, M. and Kishi, M. J. (1984) 'Phytoplankton growth response to wind induced regional upwelling occurring around the Izu Peninsula off Japan, J. Oceanogr. Soc. Japan. 40, 221-229.

Takahashi, M., Ishizaka, J., Ishimaru, T., Atkinson, L. P., Lee, T. N., Yamaguchi, Y., Fujita, Y., and Ichimura, S. (1986) 'Temporal change in nutrient concentrations and phytoplankton biomass in short time scale local upwelling around the Izu Peninsula, Japan', Journal of Plankton Research 8, 1039-1049.

Takahashi, M., Yasuoka, Y., Watanabe, M., Miyazaki, T., and. Ichimura, S. (1981) 'Local upwelling associated with vortex motion off Oshima Island, Japan', in F. A. Richards (ed.), Coastal Upwelling, American Geophysical Union, Washington, D.C., pp. 119-124.

Uematsu, M., Duce, R., Prospero, J. M., Chen, L., Merrill, J. T., and McDonald, R. L. (1983) 'Transport of mineral aerosol from Asia over the North Pacific Ocean', Journal of Geophysical Research 88 (C9), 5343-5352.

GLOBAL OCEANIC PRODUCTION AND CLIMATE CHANGE

G. P. HARRIS
CSIRO Office of Space Science and Applications
Cnr North & Daley Rds, ANU Campus Acton ACT
GPO Box 3023
Canberra 2601Australia

G. C. FELDMAN
Goddard Space Flight Centre
National Aeronautics and Space Administration
Greenbelt, MD 20771
USA

F. B. GRIFFITHS
CSIRO Division of Fisheries Research
GPO Box 1538
Hobart, Tasmania 7001 Australia

ABSTRACT. Even a cursory glance at the global Coastal Zone Color Scanner (CZCS) composites of oceanic phytoplankton biomass reveals that the regions of highest biomass are geographically separated and distinct. Three main regions can be distinguished - the northern hemisphere temperate oceans, the Equatorial divergences and the southern hemisphere temperate waters around 40-50°S. In the three regions the large scale mechanisms which control algal growth and primary production are quite different. Spring blooms dominate the northern hemisphere oceans, whereas divergence and interaction between the western boundary currents and the west wind drift dominate the processes in Equatorial and southern hemisphere regions respectively. Interactions between ocean biology and climate change occur at a variety of scales. The large scale circulation of the world ocean determines rates of subduction and upwelling and hence feeds back upon plankton growth, the global carbon cycle and climate change. CZCS composites show clear interannual variability in biomass in the Equatorial Pacific in response to El Niño Southern Oscillation (ENSO) events and there is some evidence of large interannual variability in the region of the Subtropical Convergence in the S Hemisphere. There are not yet sufficient data to make a clear statement about the large scale effects of climate change on oceanic production. In each of the three main reions it is the mesoscale interaction of physics and biology which controls the availability of nutrients and the increase in biomass at scales of a few kilometres. Given that it is the biology which is partly responsible for reducing the pCO_2 in surface waters and that exchange with deep waters is also important, subtle small scale interactions between the mesoscale physics and the stability of surface waters will have a large effect on algal growth and the atmospheric carbon flux. These interactions are likely to be influenced by climate change. Data from spacecraft sensors such as CZCS can provide a global monitoring capability, but there are problems with cloud cover and with the

237

V. Barale and P.M. Schlittenhardt (eds.),
Ocean Colour: Theory and Applications in a Decade of CZCS Experience, 237–270.
© 1993 *ECSC, EEC, EAEC, Brussels and Luxembourg. Printed in the Netherlands.*

compilation of large scale, long term, global mosaics. In the CZCS composites there is evidence of spatial and temporal aliasing because the dominant physical and biological interactions occur at mesoscales. Composites constructed from more than a few days data average out the ephemeral features of interest. Aliasing affects both the estimates of the global fluxes and the capability to observe the effects of climate change. In future missions new ways will have to be found to preserve the mesoscale dynamics and to prevent such spatial and temporal aliasing of the data. At present the required observation capability is lacking. It is also too early to say what the full impact of greenhouse warming might be: there might well be "surprises in the greenhouse". A recent example of interannual variability from Australian waters will be used as an example to illustrate the subtle effect of climate change or oceanic production.

1. The Oceanic Component of the Global Carbon Cycle

There is still some debate about the overall magnitude of oceanic production and the role the oceans play in the global carbon cycle (Tans *et al.*, 1990). Because of the high buffering factor in the oceans and the large disparity between the equilibrium carbon contents of the ocean and atmosphere it has always been known that over time scales of about 1000 years the oceans would be the dominant global carbon sink (Broecker, 1973). Under non steady state conditions (as now) however, and over shorter timescales, the portion of the anthropogenic carbon dioxide taken up by the oceans depends on a number of physical, chemical and biological processes: processes which determine the partial pressure of CO_2 in surface waters, the rates of exchange of CO_2 between the atmosphere and surface waters and the exchange between surface and deep waters. We can reasonably expect all these processes to be influenced by climate change.

The contribution of oceanic processes to the total annual global carbon flux has recently been revised downwards from about 4 Gt.y^{-1} to les than 2 Gt.y^{-1} (Tans *et al.*, 1990). The overall figure is, however, a composite of oceanic areas at temperate latitudes which are net sinks for atmospheric carbon and an equatorial area which is a strong source of carbon dioxide. Oceanic carbon sinks have surface pCO_2 values less than air equilibrium and are composed either of areas of cooling and subduction or areas of high biological productivity. Either process reduces the partial pressure of CO_2 in surface waters. Oceanic carbon sources arise from upwelling areas where cold, carbon rich, deep waters rise to the surface waters and are warmed. The CO_2 is consequently lost to the atmosphere. The subtropical gyres appear to be largely stable and unproductive and do not represent either strong sources or sinks.

There is still considerable uncertainty as to the magnitude of the oceanic and terrestrial carbon fluxes - the uncertainties are almost as big as the estimates of the mean fluxes (Table 1). Almost the only things known with any accuracy are the magnitudes of anthropogenic output and the amount of CO_2 remaining in the atmosphere. All the other fluxes must equal the difference between these two numbers. While the broad pattern of oceanic sources and sinks is known (Pearman and Hyson, 1980; Enting and Mansbridge, 1989; Tans *et al.*, 1990) the absolute magnitude, seasonality and the precise geographical distribution of sources and sinks in the ocean is not known with any certainty. Next to nothing is known about interannual

TABLE 1. Estimates of the componentsof the global C cycle [Gt C y-1]

Fossil fuel emissions	5.4 ± 0.5
Deforestation	1.6 ± 1.0
Increase in atmosphere	3.4 ± 0.2
Ocean "sink"	1.9 ± 0.9
Terrestrial "sink"	1.7 ± 1.4

variability and the possible response of ocean biology to climate change.

Problems have arisen with the determination of the balance of oceanic and terrestrial productivity in the Northern Hemisphere (Tans *et al.*, 1990, Enting and Mansbridge, 1987). Recent work has revealed that there is an enormous amount of small scale variability in the surface pCO_2 data in the North Atlantic (Watson *et al.*, 1991) and the uncertainty in that basin alone (caused by the ensuing sampling problems) is probably as big as the uncertainty in the global flux estimates. Part of the problem lies in the use of large scale averages for such numbers as the pCO_2 deficit in surface waters. Globally, a mean concentration difference of 1 µatm CO_2 between the ocean surface and the air gives a gloabl C flux of about 0.2 Gt y-1 (Watson *et al.* 1991). A mean deficit for the global ocean of 9 µatm would convert to a global oceanic carbon flux of 1.8 - 2.0 Gt y-1. The uncertainties in the North Atlantic data are themselves about 10 uatm - bigger than this mean global deficit (Watson *et al.*, 1991).

One of the major objectives of the Joint Global Ocean Flux Study (JGOFS) - a core project of the International Geosphere Biosphere Program - is to determine the magnitude of the oceanic fluxes of carbon with greater accuracy. Given that the total oceanic productivity is of the order of 45 Gt y-1; export from surface waters may be of the order of 10-20% of this. As the increase in atmospheric carbon dioxide represents about 2 Gt y-1, it will be necessary to determine the oceanic fluxes with at least a 10% accuracy in order to monitor and detect change. This is a tall order. If this is to become possible it will be necessary to use data from a number of sources: atmospheric chemistry, remote sensing, physical, chemical and biological oceanography and modelling.

Another challenge facing programs such as JGOFS is not only to attempt to obtain better estimates of global sources and sinks but also to account for the way in which small scale processes scale up to drive phenomena of global significance. In addition, the coupling between satellite derived phytoplankton biomass estimates, primary productivity and the so-called "*J*" flux (the vertical carbon flux out of the photic zone) needs to be much better understood if synoptic biomass surveys from spacecraft are to be used for process studies and an understanding of the global carbon cycle.

2. The Subtropical Convergence (STC) of the Southern Ocean

Global composite images of the phytoplankton biomass of the world ocean (Feldman *et al.*, 1989; Lewis, 1989; Post *et al.*, 1990) have shown that the highest biomass is to be found in northern temperate regions - particularly the North Atlantic Ocean. Esaias (unpublished) has estimated the global oceanic production from global ocean CZCS composites and has shown that even though the North Atlantic has high seasonal blooms of phytoplankton overall, when the biomass data is converted to primary production and the basin scale data is corrected for ocean area, the production of the Southern Oceans around 40-50°S is the highest single CO_2 sink. It is becoming clear that the region of the Subtropical Convergence (STC) in the southern hemisphere between 40° and 50° South latitude is an atmospheric carbon dioxide sink of global significance (Tans *et al.*, 1990) with a carbon flux as high as 1-2 Gty-1.

This paper will discuss the interactions between climate and the productivity of the southern hemisphere STC region in order to describe some of the possible interactions between climate change and the ocean carbon sink.

2.1 DESCRIPTIVE OCEANOGRAPHY

The STC extends right around the Southern Ocean at about 40°S latitude (Haedrich and Judkins 1979) and is therefore one of the major features of southern waters: the 15°C isotherm has long been known as "Ortmann's line" (Ortmann, 1896). The STC is effectively the northern edge of the Southern Ocean and is a major biogeographical boundary for many pelagic and planktonic organisms (Van der Spoel and Pierrot-Bults, 1979). What was thought to be a single "line" (Deacon, 1945) is, in reality, a region which is highly variable in space and time. Eddy activity blurs the "line" and makes the sea surface temperature (SST) distributions show great variance in certain areas.

The STC is the site of active subduction where warm, saline, subtropical, surface waters cool and sink beneath the cold, fresher, subantarctic waters. Temperature and salinity inversions are found frequently in the top 200 to 300 m in depth profiles. The extensive eddy fields associated zith the STC ensure that most of the subduction takes place in eddies and other mesoscale features.

Satellite measurements of sea surface temperature in the Southern Ocean show a banded appearance from north of the STC to the Polar Front (Anon, 1989). Sea surface temperature decreases from 18-20°C to 9-10°C in a series of steps from north to south across the northern boundary of the Southern Ocean. In broad terms, the STC lies at the point where the thermocline of the tropical and subtropical oceans breaks the surface. This leads to the presence of a series of step-like frontal zones separating warm, stratified, nutrient depleted waters to the north from cool, vertically mixed nutrient rich waters to the south. It is not so much a single convergence as a region of convergence marked by strong mesoscale activity.

The position of these frontal zones is largely set by the distribution of land and sea and the strength of the zonal westerlies, so that the temperature gradients and the spatial variances in SST are not evenly distributed around the hemisphere. The sharpest temperature gradients, the greatest amount of mesoscale eddy activity and the greatest spatial variances in SST are found on the "downstream" sides of the tips of the southern continents where the western boundary currents of the Atlantic, Indian and Pacific Oceans merge with the easterly flow of the Circumpolar Current (figure 1; Anon, 1989). These frontal and eddy regions appear to be highly variable both within and between years and the width of the frontal zones may, themselves, also vary.

Details of the oceanography of the southeastern Australian region were published in Harris *et al.* (1987a) as were details of the regional oceanography and seasonal and interannual variability observed at coastal stations (Harris *et al.*, 1987b, 1991)

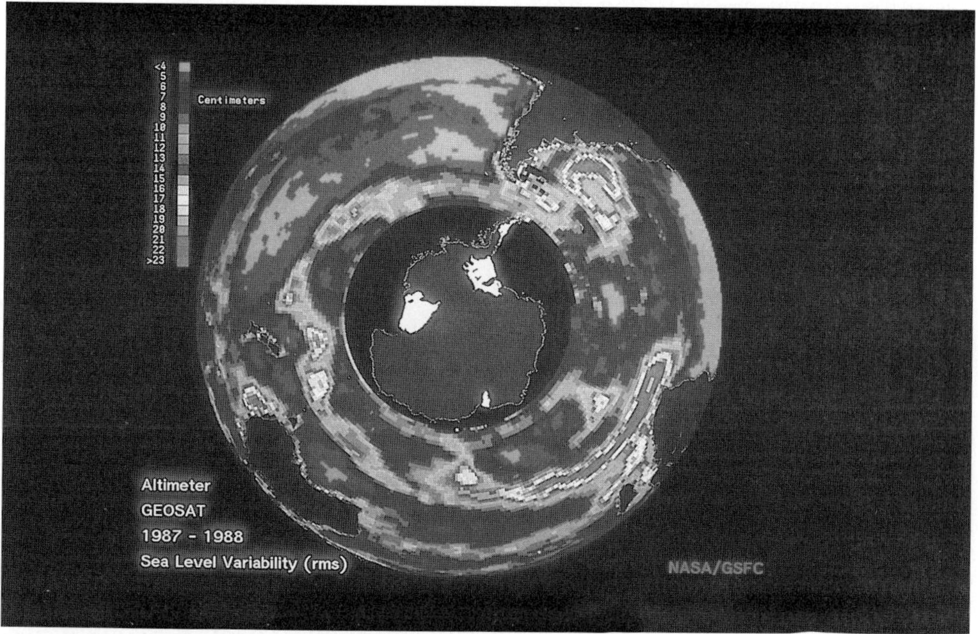

Figure 1. GEOSAT altimeter from the S hemisphere showing regions of high mesoscale variability. Colorphotograph on p. 363

2.2 ATMOSPHERIC CHEMISTRY AND MODELLING

The mean concentration of carbon dioxide in the atmosphere at the CapeGrim monitoring station in Tasmania (44°S latitude) is the lowest of any of the atmospheric monitoring stations (Tans *et al.*, 1990). There is little land at these Southern latitudes (40-50°S) and very little anthropogenic CO_2 production, so the CO_2 concentration in the atmosphere must be a function of long range, interhemispheric transport of anthropogenic CO_2 and a nearby sink which must be predominantly oceanic. Not only is the mean CO_2 concentration at Cape Grim the lowest observed at any monitoring station, but also the seasonal fluctuation in the data is weak. The aseasonal atmospheric signal at Cape Grim is quite unlike the strong seasonal signal seen in CO_2 data from the Northern Hemisphere (Keeling and Whorf, 1985; Bacastow *et al.*, 1985), where the carbon sources and sinks are dominated by large anthropogenic inputs and a considerable seasonal fluctuation in the terrestrial biota. The Cape Grim data point to a CO_2 sink due to oceanic phytoplankton production with weak seasonality or a CO_2 sink due to physical processes such as surface water cooling and subduction. Whatever the reason, the sink is apparently large and relatively non seasonal.

A number of inverse models of global carbon fluxes have used the latitudinal variation of atmospheric CO_2 and an atmospheric transport model to determine the surface fluxes (Pearman and Hyson, 1980; Tans *et al.*, 1990). Pearman and Hyson (1980) were the first to use a relatively unsophisticated two dimensional model which, nevertheless, gave strong evidence for a large oceanic sink between 40-50°S sink. More recent three dimensional global circulation models (Tans *et al.*, 1990) have given much the same result. While there is clearly uncertainty in the norhern hemisphere fluxes (Enting and Mansbridge, 1989) there is good agreement between all the models as to the magnitude of the 40-50°S. Converting the results of the Pearman and Hyson model to common units and knowing the area of the 40-50°S latitude band ($31.5 \ 10^{12} \ m^2$) gives an estimated carbon flux into the ocean of approximately $62 \ gCm^{-2}.y^{-1}$ or a total of $1.95 \ Gt \ y^{-1}$.

2.3 CHEMICAL OCEANOGRAPHY - SURFACE PCO_2 DATA.

Shipboard observations from the Indian and Pacific Oceans indicate a broad zone of low pCO_2 in surface waters in parts of the STC region but the data are sparse (Miyake and Sugimura 1969, Miyake *et al.*, 1974, Inoue and Sugimura 1986, 1988). Minimum pCO_2 values (approximately -50 uatm) were found in surface waters on the western sides of the ocean basins with higher values (occasionally above air equilibrium) on the eastern margins. Data published from the Australian region (Miyake *et al.*, 1974, Inoue and Sugimura, 1988) shows a similar pattern to that found in the South Atlantic with pCO_2 decreasing with increasing latitude until about 50°S. The lowest pCO_2 values (-50 uatm) were found in the region where the East Australian Current (EAC) flowed eastwards into the Tasman Sea. Values in the Leeuwin Current off western Australia were close to air equilibrium at 40°S (Inoue and Sugimura, 1988).

Mackey and Butler (1991) give a minimum pCO_2 of -40 uatm in the STC region south of Tasmania which, if zonally averaged, would give a total C flux of $77 \ gCm^{-2}y^{-1}$ or $2.42 \ Gty^{-1}$. A minimum of -50 uatm would equate to a flux of $3.02 \ Gty^{-1}$. This flux estimate assumes that the reduced pCO_2 in surface waters is a seasonal drawdown - almost certainly not a valid assumption - and it also includes both surface cooling and biological drawdown. Mixing of water, both laterally and vertically, would reduce the concentration. Furthermore, it is probably not valid to use the minimum pCO_2 values for a zonal average, given that the minimum values only occur on the western sides of the ocean basins. A better zonally averaged annual estimate (until there are better spatially averaged data) would be to use a mean value of about 25 uatm ($48 \ gCm^{-2}y^{-1}$) or a total flux of $1.5 \ Gty^{-1}$. Tans *et al.*, (1990) used a mean pCO_2 deficit of 17-23 uatm or a total flux of 1.03 to $1.39 \ Gty^{-1}$.

In all cases these estimates of the regional flux represent heavily spatially and temporally averaged estimates which will be subject to the same types of errors discussed by Watson *et al.* (1991). The uncertainties must be large even though, in the case of the southern STC the pattern of low pCO_2 is fixed in space and, apparently, relatively consistent in time.

These data give some evidence for an important role for physical and biological interactions in the maintenance of a carbon sink in the regions where the western boundary currents interact with the West Wind Drift (WWD) of the Southern Ocean. These regions of the STC are characterized by large eddy fields concentrated in the Falklands/Drake Passage region of the South Atlantic, the area of the Agulhas retroflection in the Indian Ocean and the region where the EAC interacts with the easterly flow in the Tasman Sea. These and other data confirm that there are distinct zonal patterns in the processes which control the magnitude of the CO_2 sink.

The data from the southern hemisphere is quite unlike that from the Northern Hemisphere oceans. Data from the North Atlantic and North Pacific show a broad region of low pCO_2 values extending to high latitudes - a result of the depletion of pCO_2 in surface waters by the spring bloom (Miyake *et al.*, 1974). Ship and remote sensing data shows a large seasonal bloom in the North Atlantic the North Pacific whereas there is little or no spring bloom at high latitudes in the Southern Ocean. This is probably due to the deep wind induced mixing in such waters (Mitchell *et al.*, 1991). Whatever the cause of the difference between the two hemispheres, it means that intermediate and deep water formation in the Southern Ocean will not (unlike the North Atlantic) be a strong sink for atmospheric CO_2.

2.4. REMOTE SENSING

The methods used in the production of the images from the Coastal Zone Colour Scanner (CZCS) on NIMBUS-7 have been described in Feldman *et al.* (1989). The satellite data analysis used a number full resolution and subsampled, mosaic images. Full resolution (1 km) images are subsampled to 4 km resolution before binning and averaging to produce cloud free mosaics (Feldman *et al.*, 1989). Subsampling does not affect the statistical properties of the CZCS data but binning and averaging tends to average out extreme values. Denman and Abbott (1988) showed that spatial and temporal aliasing resulted if composites were made over periods longer than 5 days. To determine the full extent of the aliasing problem the 16 km data was compared to monthly full resolution (1 km) composites and composites made from 1 km data over a period of a few days.

2.4.1. CZCS images of the Tasman Sea region. This project began with a search of the entire CZCS archive for a full resolution set of images of the Tasman Sea. During the period 1978-86 only a very few periods were sufficiently cloud free to produce either a full cloud free Tasman Sea image or to produce a set of seasonal monthly composites. The period from November 1981 to March 1982 was found to contain sufficient recorded images to produce the required data. The final data set comprised a full resolution Tasmanian image from 27 November 1981 at the height of the spring bloom (figure 2), a five day composite of the entire Tasman Sea from Tasmania to New Zealand (14-19 January 1982, figure 3), a similar eight day composite

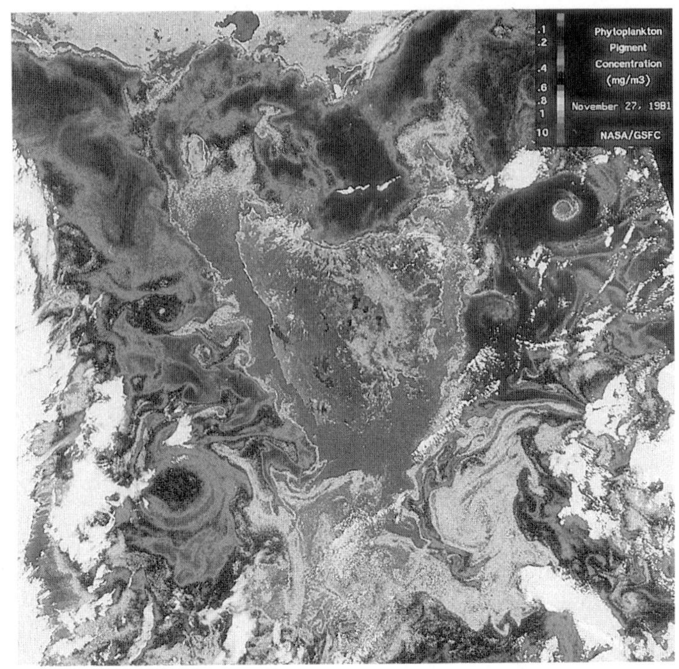

Figure 2. CZCS Tasmania 27 November 1981.   Colorphotograph on p. 364

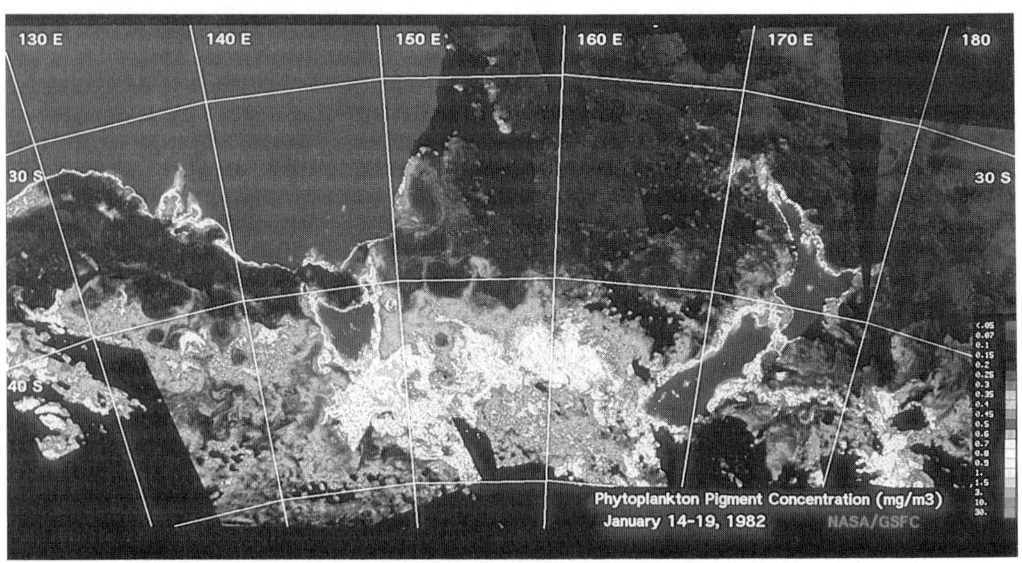

Figure 3. CZCS Tasman Sea 14-19 January 1982.   Colorphotograph on p. 364

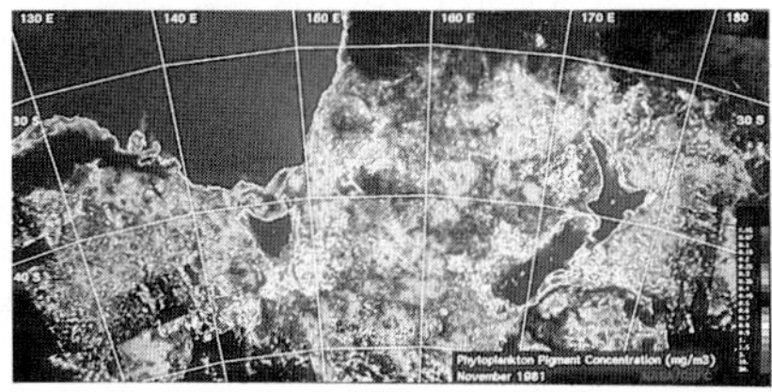

Fig. 4a November

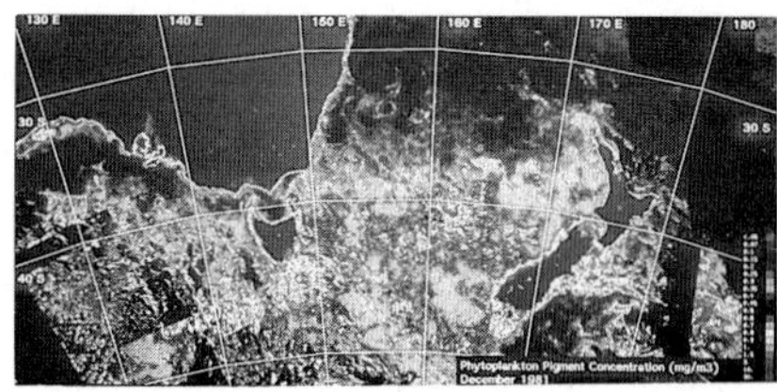

Fig. 4b December

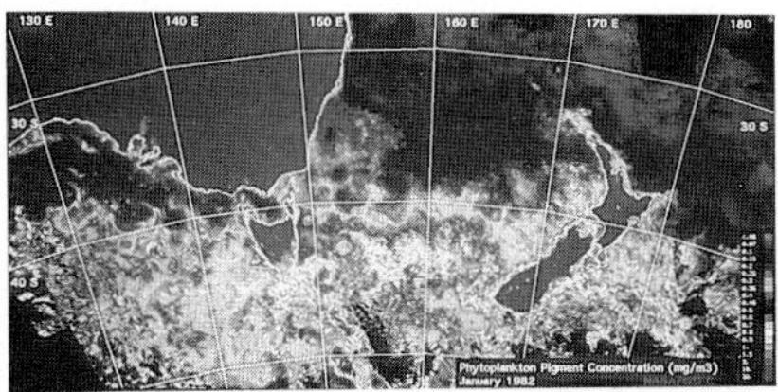

Fig. 4c January

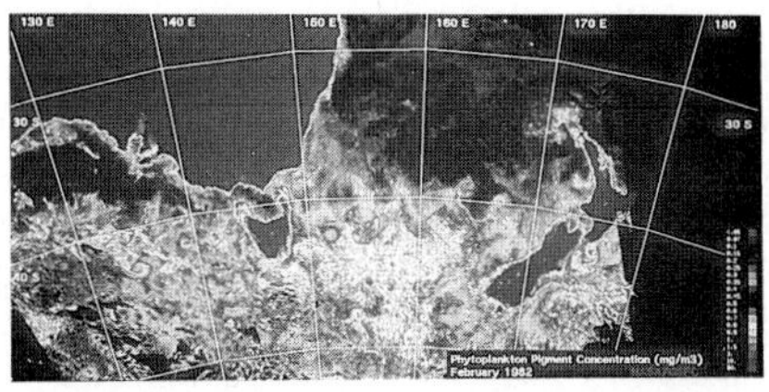

Fig. 4d February

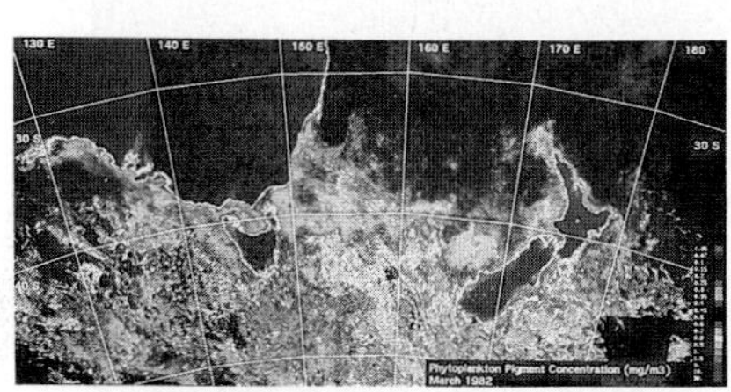

Fig. 4e March

Figure 4. CZCS Monthly composites November 1981 - March 1982.

Colorphotograph on p. 365

from the same period, and a set of monthly composites spanning the period from November 1981 to March 1982 (figure 4). These monthly composites were combined into a final five month composite (figure 5).

The spring bloom image from November 1981 (figure 2) showed an enormous amount of mesoscale activity with the phytoplankton biomass forming "edge enhancements" of fronts and eddies. The highest biomass occurred on the shelf on all Tasmanian coasts except the Bass Strait region. The centre of Bass Strait was particularly unproductive at this time. The eddy patterns observed were consistent with those frequently seen in National Oceanic and Atmospheric Administration (NOAA) / Advanced Very High Resolution Radiometer (AVHRR) images - particularly the large eddy off southeast Tasmania, "downstream" from the island in the wake of the West Wind Drift and the mushroom shaped eddy pair (modon) off northeastern Tasmania with cold and warm core features. Such features can be frequently seen in Tasmanian AVHRR imagery (figure 6).

The Tasman Sea image from January 1982 (figure 3) revealed similar features around Tasmania, including the same topographically locked "downstream" eddy southeast of the Island, a large pair of warm core eddies to the east of Tasmania and a series of wave-like features all along the northern edge of the STC at about 40°S. Phytoplankton biomass was low both to the north of 40°S and to the south of 46°S. A large warm core eddy was visible on the coast of New South Wales at about 36°S. Around New Zealand phytoplankton biomass occurred in the shelf waters of South Island with a streamer extending to the east across Chatham Rise. Biomass appeared to be higher

Figure 5. CZCS 5 month composite November-March 1981-82.

Colorphotograph on p. 365

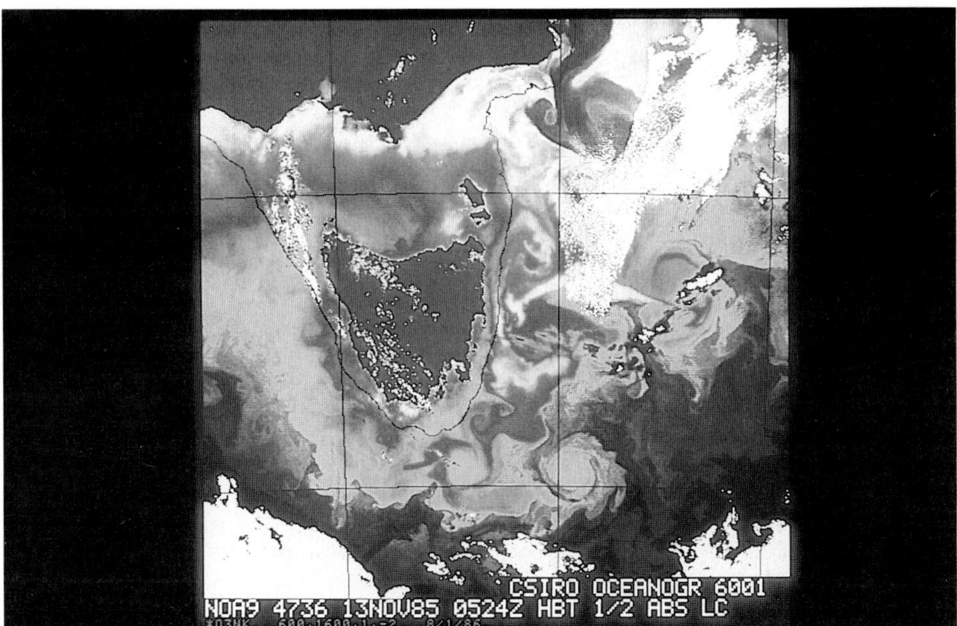

Figure 6. AVHRR image of Tasmania showing mesoscale eddies. Colorphotograph on p. 366

"downstream" of Tasmania and New Zealand as compared to the region to the west of Tasmania.

The set of seasonal composites (figure 4) showed that a weak spring bloom in 1981 was widespread as far north as 30°S and that as the season progressed the phytoplankton biomass became restricted to the STC frontal regions. By February 1982 warming of southern waters produced some plankton growth to the south of the STC. The monthly composites showed the biomass distributions around New Zealand particularly well. The Southland Front - the sharp discontinuity between shelf waters and the deep Southern Ocean - was particlarly well marked southeast of the south Island and mushroom eddies could be seen in the outflow from Cook Strait.

What was particularly striking was the fact that many of the mesoscale physical features associated with the STC region were clearly visible even on the monthly composites (figures 3 and 4). The monthly composites showed some considerable blurring of small scale features - as would be expected from a process which averaged data over 30 day periods - but some frontal features like the waves on the northern edge of the STC and the topographically locked eddies were preserved. The warm core eddy off eastern Tasmania moved little between January and February 1982. Even the five month composite (figure 5) still clearly showed the position of the Southland Front off southeastern New Zealand and the region of high biomass off the southern tip of Tasmania where the Southern Ocean flows to the east.

2.4.2. *Hemispheric composites.* Composite cloud-free images from CZCS in South Polar projection (figure 7) showed a permanent non seasonal circumpolar band of phytoplankton biomass (chlorophyll) at the latitude of

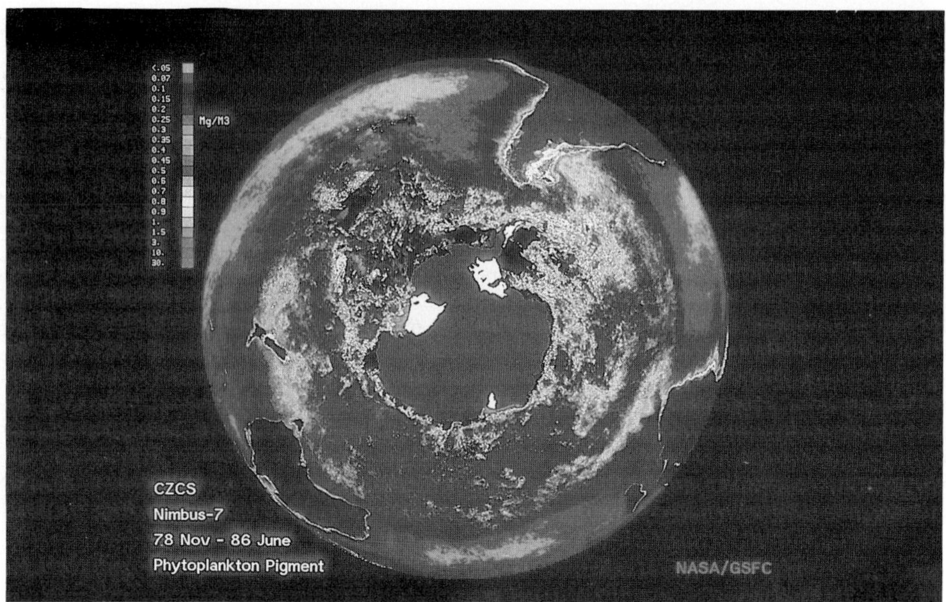

Figure 7. CZCS polar composite. All available CZCS data combined.

Colorphotograph on p. 366

the STC with the areas of highest biomass associated with areas of active mesoscale physics. These areas of highest biological activity are associated with the mesoscale eddy fields of the "downstream" ends of the western boundary currents where they merge into the West Wind Drift of the Southern Ocean. There was a striking similarity between the spatial patterns of the CZCS data and the data on the variance of sea surface elevation produced by the GEOSAT altimeter (figure 1). The composite CZCS images show that many of the mesoscale features of the STC are semipermanent, topographically locked features. In particular the wave motions in the region of the Agulhas retroflection appear very clearly in this long term (8 year) composite. Also some of the eddies in the "downstream" region to the east of the EAC extension in the Tasman Sea appear very clearly in the monthly composites as well as in the five day composites; an indication of little or no motion over long periods.

An analysis of longitudinal sections of CZCS data abstracted from the global 16 km cloud free data base by Esaias (unpublished) showed that there was (by global standards) little phytoplankton biomass in the 40-50°S band. The biomass in the North Atlantic rose to approximately 2 mgm-3 in the spring and declined in summer and winter, whereas the biomass in the STC band in the southern hemisphere averaged about 0.4 mgm-3 with little seasonal variance. The highest biomass occurred around 50°S and was lower at higher latitudes in the Southern Ocean. Higher values (< 2 mgm-3) were found at the southern ice edge. When the total global oceanic production was calculated the southern STC band showed the highest and most consistent production of any oceanic area. The result was explained by the very large oceanic area of the southern STC region and the lack of seasonality in the pattern of production.

Phytoplankton biomass in the main part of the Southern Ocean is low and appears, from the composite images, to be tightly coupled to a number of physical processes. The striking similarities between the phytoplankton biomass distribution and the mesoscale eddy patterns revealed by the GEOSAT altimeter (figure 1) has already been noted. There was also a clear relationship between the position of the biomass band and the northern edge of the hemispheric cloud cover band (figure 8) The region between 40-50°S is the band of the "roaring forties" with high winds and high significant wave heights. It is, however, on the northern edge of the region of nearly continuous cloud. Waters to the south of the STC are deeply mixed and it is very cloudy year round: this will limit phytoplankton production (Mitchell *et al.*, 1991). Some subtle effects of bottom topography can also be identified in regions of modifications in the flow patterns of the Antarctic Polar Current. The phytoplankton biomass is concentrated in regions of strong mesoscale eddy activity but only where stability is produced by the interaction of deeply mixed subantarctic waters with saline subtropical waters. There is very little plant biomass in the deep eddies of the core of the Polar Current (figure 7).

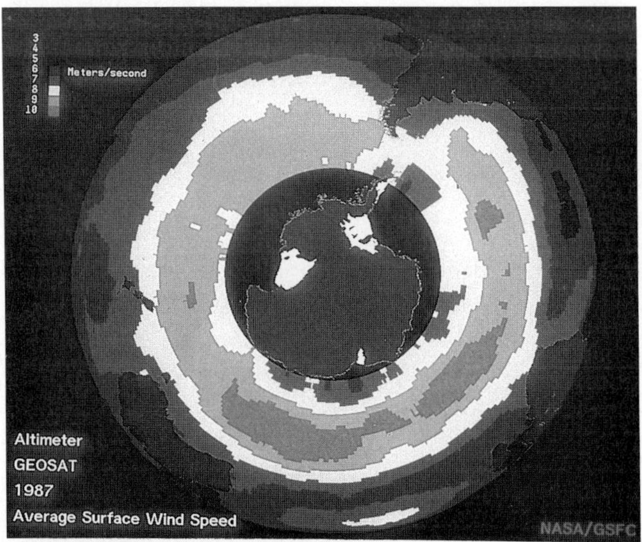

Figure 8. Polar cloud and wind composites.   Colorphotograph on p. 367

2.4.3. Data from sections along CZCS images.   The links between phytoplankton biomass and mesoscale eddy activity can strongly seen in figure 9 where zonal averages of phytoplankton biomass (between 40-50°S) are plotted against longitude. These data were obtained from the hemispheric CZCS composite. Biomass was highest on the western sides of the Atlantic, Indian and Pacific basins and there was no indication of consistent seasonality. It was immediately evident that the zonal distribution of

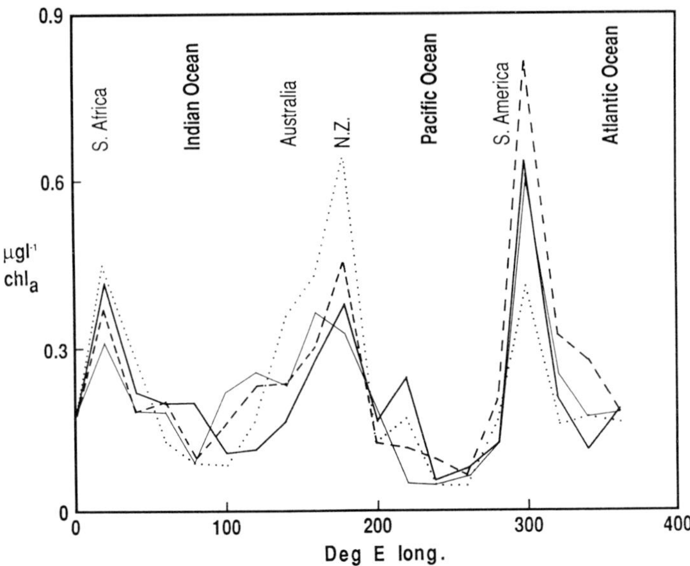

Figure 9. CZCS zonal averages. Data obtained from polar composites.

phytoplankton biomass was spatially correlated with the regions of subduction and lowest pCO_2 (see above).

Phytoplankton biomass data representing a section south along 155°E were recovered from the global data base. Comparison of the spatial extent of the areas of high phytoplankton biomass on this section with the areas of low pCO_2 in the published cruise data of Miyake *et al.* (1974) indicated the areas of low pCO_2 corresponded to the areas of highest biomass only around 45°S (figure 10).

While biological activity is of clearly of great significance in the subantarctic waters on the southern edge of the STC region, physical processes (cooling and subduction) appear to dominate the pCO_2 signal in the subtropical waters on the equatorial side of the STC. A comparison of the mean position of the phytoplankton biomass peak derived from CZCS composites with the surface pCO_2 data from 155°E (figure 10) indicates strong depletion in pCO_2 between 30°S and 40°S where cooling occurrs. Biomass levels in these waters are too low to solely account for the reduction in pCO_2.

Full resolution images (figure 2) revealed the close spatial correspondence of mesoscale eddy activity and phytoplankton biomass. Indeed the plant biomass appeared to be an "edge enhancement" for mesoscale physics. The band of productivity associated with the STC is therefore somewhat different from the productivity of the northern hemisphere oceans. The CZCS global composites reveal a large seasonal spring bloom in the northen hemisphere whereas the spring bloom in the region of the STC is restricted to a narrow band. The predominant mechanism of productivity in the STC region appears to be a semipermanent series of interactions between the cool, deeply mixed, nutrient rich subantarctic waters and the warm, stable, nutrient poor waters

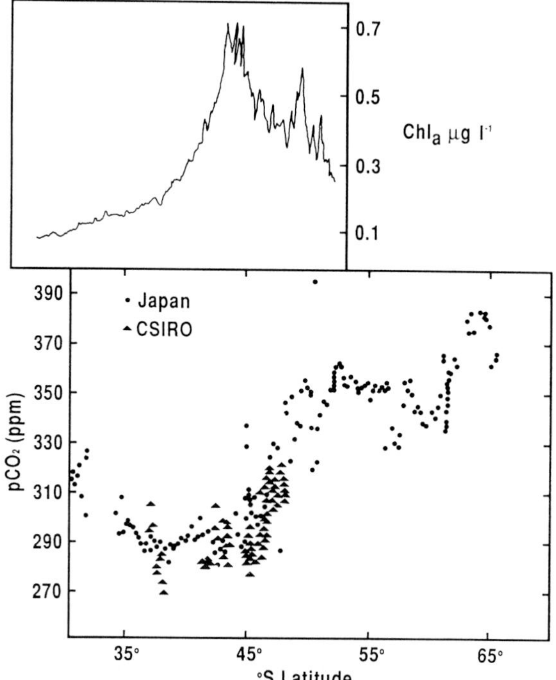

Figure 10. pCO$_2$ and 155 E biomass. pCO$_2$ data from Miyake et al. (1974) and CSIRO surveys. Biomass data from CZCS.

of the subtropics. These interactions are almost all in sharp frontal structures where the western boundary currents interact with the waters of the Southern Ocean.

2.4.4. *Spatial Spectra.* Spatial spectra were computed from transects across CZCS images. The data were obtained by recording the biomass estimates from contiguous pixels across an image. Transects across single images of Tasmania (figure 2) revealed strong mesoscale variability and, as expected, the spatial spectra showed that the maximum variance occurred at scales of 50 to 80 km with an exponential decrease in variance to scales below 20 km (figure 11). Transects across the STC at 155°E and 160°E gave similar results; confirming that the phytoplankton biomass was predominantly associated with mesoscale physical features of similar scale (figure 6). All spectra showed considerable structure in the 2 to 20 km region indicating that the patches of biomass were small and/or narrow and associated with the edges of fronts and eddies.

Spectra from composite images frequently showed little or no decrease in variance at scales below 10 to 20 km and a blurring of the 80 km variance peak. Some spectra were flat or "white" from 20 km to the smallest scales (order 2 km, figure 11). These spectra were the result of the image mosaicing process and resulted from the process of combining and averaging the pixel data from consecutive images. The averaging and binning process involved in the production of mosaic images was therefore smearing the images and reducing the true spatial pattern to "white noise" at scales below 20 km.

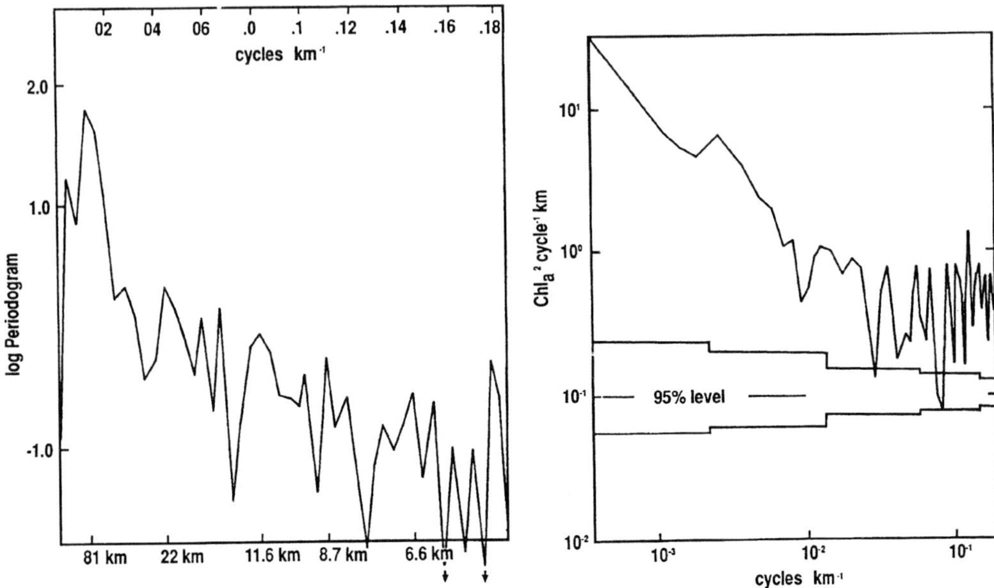

Figure 11. Autospectrum of data from single and mosaic CZCS images.

2.4.5. *Statistical distributions of biomass data from CZCS.* Data from discrete areas of the November and January images (figures 2 and 3) were converted to frequency distributions. These data were plotted as probits to analyse the underlying statistical distributions of the data (figure 12). The data from oligotrophic, mesotrophic and eutrophic regions of the January image revealed regular lognormal distributions. Comparison with data from the November Tasmanian image (figure 2) revealed that a regular pattern of statistical behaviour existed: higher mean biomass being associated with increased standard deviations and the inflexion of the probit plots.

Data from the single, full resolution image, when compared to data from the four day, eight day and monthly composite images showed an obvious blurring of small scale features; as would be expected from a process which averaged data in each pixel from a number of images (Denman and Abbott, 1988). The highest biomass values occurred in small regions of the full resolution images, were of small spatial extent and moved slightly from image to image over a period of days. The net result of the compositing system was to reduce the apparent frequency of small, high biomass patches in the composite image (figure 13). Any estimate of production based on composite image chlorophyll data must therefore be suspect for the same reason as the averaged pCO_2 data discussed above.

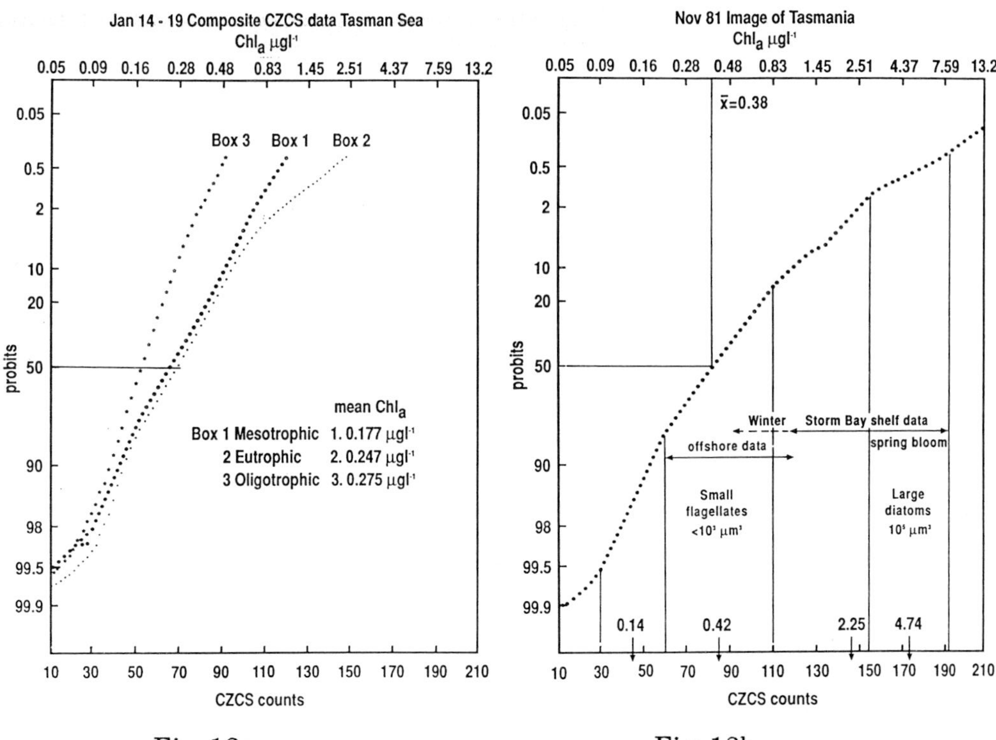

Fig. 12a   Fig. 12b

Figure 12. Probit plots from CZCS image data.

2.5. BIOLOGICAL OCEANOGRAPHY - INTEGRAL PRODUCTION

The high productivity of the STC region in the southern hemisphere has been known for some time, ever since the first global plots of ^{14}C productivity were produced. A number of maps of global ^{14}C productivity have been produced with the number of data points and the contouring dependent on the quality control employed. All the maps show an increase in productivity south of 40°S but opinions have been divided about the magnitude of productivity in the Southern Ocean. A compilation of data by Bunt (1975) showed a clear band of primary productivity between 40° and 50°S. How do we convert a satellite map of biomass to a map of productivity? Platt (1986) and Platt and Sathyendranath (1988) have discussed the problem of the calculation of oceanic production from the surface biomass data obtained from satellites. The calculation requires a number of assumptions to be made, most of which have been known since Talling (1957) produced the first one dimensional model of aquatic productivity. Parslow and Harris (1990) have recently reviewed the topic and have assessed the major errors. The two major sources

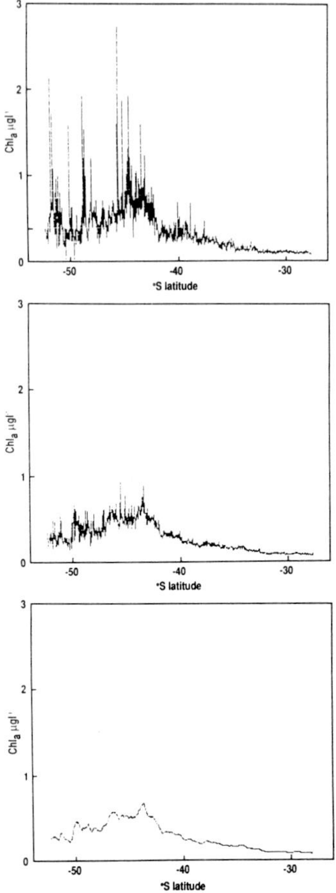

Figure 13. Comparison of raw and binnedCZCS data along the 155E section. Averaging and binning removes the peak biomass values.

of error are the physiological state of the phytoplankton and the vertical distribution of the biomass. A simple model of the form as in Talling (1957),

$$Sum\ P = B\ Pmax.e^{-1} \qquad ln\ f\!\left(I_0\right)$$

is sufficient to encapsulate the problem. The calculation of integral production requires a knowledge of the near-surface biomass (B), the photosynthetic capacity $(Pmax)$, the vertical attenuation coefficient (e), and the saturation light intensity at $Pmax$ as a function of Io (this is equivalent to knowing the photosynthetic efficiency or the slope of the $P\ v\ I$ curve at low light).

Calculation of integral production knowing only B and assuming the physiological parameters yields estimates accurate to about an order of magnitude (Eppley *et al.*, 1985, 1987). More site specific information clearly is required. Banse and Yong (1990) have recently performed an analysis of the greatest sources of error in the estimation of integral production and have shown that a knowledge of *Pmax* is very important. Parslow and Harris (1990) came to much the same conclusion by showing that *ln f(Io)* was more or less constant due to light adaptation by the plankton and the logarithmic form of the parameter. *Pmax* itself is very sensitive to the light history of the phytoplankton community and changes over a period of about 5 days in response to surface light fluctuations and vertical mixing (Harris *et al.*, 1980, Parslow and Harris, 1990).

Cruise data from CSIRO cruises in the STC region has shown (Harris *et al.* 1987b) that the value of *Pmax* rarely exceeded 2 mgCmg Chl-1h-1. This low value was itself an indication of a history of variations in the underwater light climate prior to measurement either due to surface light fluctuations or to changes in vertical stability in the margins of mesoscale eddies (Harris and Parslow, 1990). This *Pmax* value, together with observed values of e and *f(Io)*, was used to estimate integral production in these waters as follows:

Using an observed euphotic zone depth of 65m, a *Pmax* value of 2 mgCmg Chl-1h-1 and *ln f(Io)* of 2.5 (Harris, 1978) and a mean chlorophyll biomass from 40-50°S of 0.47 mgm-3 gave an integral gross production of 335 mgCm-2d-1 (assuming a 10h day) and an annual integral gross production of 122 gCm-2y-1. This would be equivalent to a gross carbon flux of 3.84 Gty-1. Between 1958 and 1962, Jitts (1966) carried out an extensive survey of the productivity of the Coral and Tasman Seas and obtained a mean integral productivity of 49 mgCm-2h-1, equivalent to 178 gCm-2y-1. Jitts' data indicates only weak seasonality and low standard deviations.

The methods used to obtain the oceanographic, primary productivity and particle size data were published in Harris *et al.* (1987b, 1991). Cruise data was obtained from cruises on the CSIRO vessels FRV "Soela" and RV "Franklin" in the Southern Ocean and the Tasman Sea in the period 1986-90.

2.6. BIOLOGICAL - OCEANOGRAPHY NEW PRODUCTION AND THE "J" FLUX

In order to calculate the downward flux of carbon from integral production is it necessary to know the proportion of the total production that is "new" production (Eppley and Peterson, 1979). This is best done by using 15N or distinguishing between the uptake of nitrate and ammonia. In the absence of such data from these waters the only way to estimate the "f" number (or the fraction of total production that is new production) is to use published empirical relationships between integral P and "f" (Eppley and Peterson 1979). Spatial and temporal variability in "f" and the different scales of the various food chain processes means that some inaccuracy will be inevitably be introduced by the use of large scale averages (Vezina and Platt 1987).

Eppley and Peterson (1979) estimated an "f" value of 0.31 for a gross production of 122 gCm-2y-1 using data from the Southern California Bight. If Jitts' (1966) estimates of the gross production are used the "f" number would be between 0.27 and 0.35. Vezina and Platt (1987) modelled the relationship between production and "f" and showed that there was a nonlinear relationhip between "f" and the biomass in the photic zone. For the total production calculated above their estimate of "f" was about 0.4.

Assuming that the total annual net production is sustained (at equilibrium) by an equivalent flux of new nutrients these data can also be used to estimate the "f" value - or the fraction of gross production that is new production. Using the technique of Sathyendranath *et al.* (1991) it is possible to derive estimates of the 'f' ratio from CZCS and AVHRR images. CZCS gives the total production and the AVHRR images (from the relationship between temperature and nitrate in surface waters and a C:N ratio) gives "new" production. In these waters, if the net production is assumed to be between 48 and 62 gCm-2y-1 (see Section 2.7 below) then "f" lies between 0.39 and 0.51. These figures compare well with published data for similar values of total gross production (Platt and Harrison, 1985, Vezina and Platt, 1987), but are slightly higher than those quoted in Eppley and Peterson (1979).

Estimates of new production in the STC region based on these empirical relationships between new production and integral production data give values for the total net annual carbon flux through the STC region of the order of 1.49 to 1.95 Gt Cy-1.

There is considerable indirect evidence that these estimates are reasonable and that the region of 40-50°S is a region of high new production. For example, Prakash *et al.* (1991) examined the rate of dark uptake of ^{14}C as a proportion of total ^{14}C production in an analysis of data from the Atlantic and Pacific Oceans. Those data indicated that the high latitude regions of the Northern and Southern Oceans and the region of the equatorial divergence were areas where dark ^{14}C uptake was a small fraction of the total. Harris *et al.* (1989a) showed that regions of low dark ^{14}C uptake were regions of active new production in the Tasman Sea. There is a fundamental difference in the food chain between regions of high and low new production (Vezina and Platt, 1987; Cushing, 1989).

There appears to be a fundamental difference between the high latitude regions of the North Atlantic and the Southern Ocean. Dark ^{14}C uptake increases with latitude beyond 50°S whereas it remains low in the subarctic regions of the North Atlantic. This would indicate higher new production at high latitudes in the Northern Oceans. This is consistent with del ^{13}C data (Rau *et al.*, 1989), pCO$_2$ distributions (see above) and the CZCS composite images.

The derivation of the vertical flux of carbon (the "J" flux) from integral production is not a simple matter. One major problem is how to couple the time and space scales of production and sedimentation. Satellites and sediment traps measure processes at difference scales (Bishop and Marra 1984). All the available evidence points to a non linear relationship between integral P and "J" (Vezina and Platt, 1987), although the spatial and temporal averaging required leads to error bars of the order of at least $\pm$ 30%. Betzer *et al.* (1984), for example, recorded a power relation between *sumP* and "J" with an exponent of 1.4. Vezina and Platt, (1987) concluded that the relationship between the vertical carbon flux and sum P was a function of sum P to a power between 1.7 and 1.8. Betzer's empirical relation would give a "J" flux of 13 gCm-2y-1 for a sumP of 122 gCm-2y-1 at 100m whereas Pace *et al.*, (1987), using the VERTEX data, would have estimated a "J" flux of 14.6 gCm-2y-1.

These values, when compared to the net fluxes through the surface and the nutrient depletion data, appear to be on the low side. This would be consistent with the recent realisation that sediment traps underestimate the "J" flux at depths less than 1000 m. In the JGOFS North Atlantic Bloom Experiment [234]Thorium data gave estimates of the vertical flux of particulate matter as much as three times higher than the estimates from shallow sediment traps (Buesseler *et al.*, 1991). If the "J" flux in the STC is estimated using data and relationships from from VERTEX, the vertical carbon flux can be estimated to be 0.46 Gt Cy[-1] - about 30% of the apparent flux calculated from other sources.

2.6.1. *Particle size data and the structure of the food chain.* Particle size data from inshore and offshore waters and from regions of different phytoplankton biomass in different seasons revealed systematic differences between high and low biomass regions. In late summer 1989 phytoplankton biomass ranged from 0.2 to 1.2 mg Chlm[-3] offshore from the east coast of Tasmania - the same range of values as those seen in the images from 1981-82. Mean slopes (log number *vs* log size) were consistently about -1.0. Particle size distributions showed no changes in slope with increased biomass; (if anything there was a slight decrease in slope with increased biomass, figure 14). There was, however, a significant correlation between increases in intercept with increases in biomass (figure 14). The highest correlation between particle number and total biomass (as chlorophyll) was that between chlorophyll and particles in the size range 100 μ^3 to 1000 μ^3. Mean particle volume was constant around 500 μ^3. This indicated that increases in phytoplankton biomass were associated with changes in the abundance of a single size class of small diatoms and flagellates.

This contrasted sharply with data published by Harris *et al.* (1987b) from the spring period in these same waters where increases in total biomass were correlated solely with the abundance of large diatoms. Biomass ranged up to 5 mgm[-3] in spring. A similar pattern was found in shelf waters (Harris *et al.* 1991) and during transient blooms. Thus the community structure and the nature of the food chain offshore in summer was different from that in spring blooms, in shelf waters, in eddy margins and in sporadic regions of high "f" (Vezina and Platt, 1987; Cushing, 1989).

These differences in the food chain structure are significant because periods of high new production in these waters tend to be dominated by large diatoms and salps (Cushing, 1989; Harris *et al.*, 1991) and are times and places of high "f" and non linear coupling between biomass and carbon export (Vezina and Platt, 1987). Salps are responsible for massive exports of carbon from the euphotic zone (Michaels and Silver, 1988) and occur as massive blooms in Tasman Sea waters (Harris *et al.*, 1991). Thus the times and places when phytoplankton biomass was above about 1.2 - 1.5 mg.m[-3] were times and places of high "J" fluxes. These were almost exclusively eddy margins and small frontal zones. These events were sporadic in space and time (Harris *et al.*, 1991) and were precisely the events which were averaged out by the CZCS compositing process.

PARTICLE VOLUME VS CHLOROPHYLL
SIZE CLASS 10^{1.5} μm³
VOLUME OF PARTICLES, μ3/ML IN SIZE CLASS
(Thousands)
30
20
10
0
0
0.4
0.8
1.2
chlorophyll a μgl⁻¹

PARTICLE VOLUME VS CHLOROPHYLL
SIZE CLASS 10^{2.5} μm³
VOLUME OF PARTICLES, μ3/ML IN SIZE CLASS
(Thousands)
150
100
50
0
0
0.4
0.8
1.2
chlorophyll a μgl⁻¹

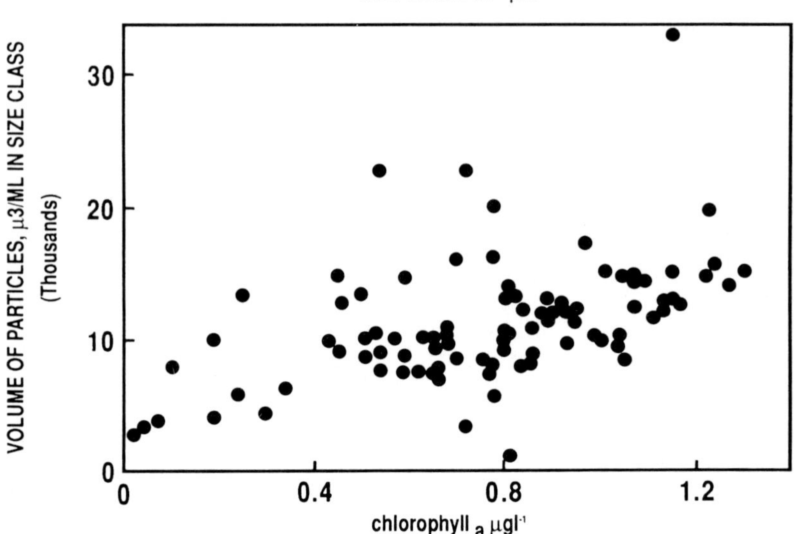

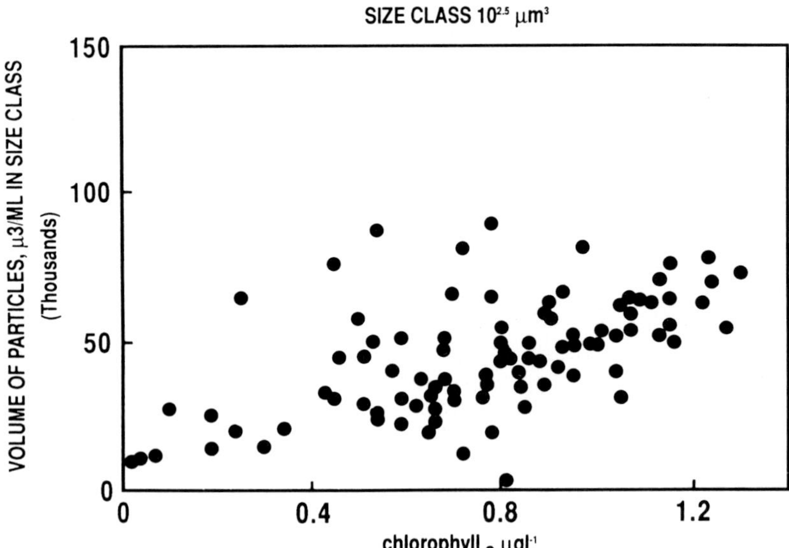

Figure 14a

SLOPE vs CHLOROPHYLL

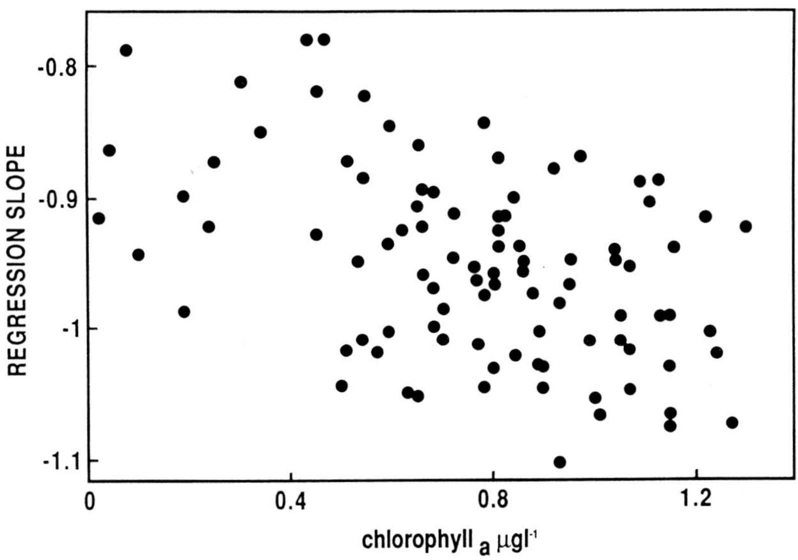

CONSTANT vs CHLOROPHYLL

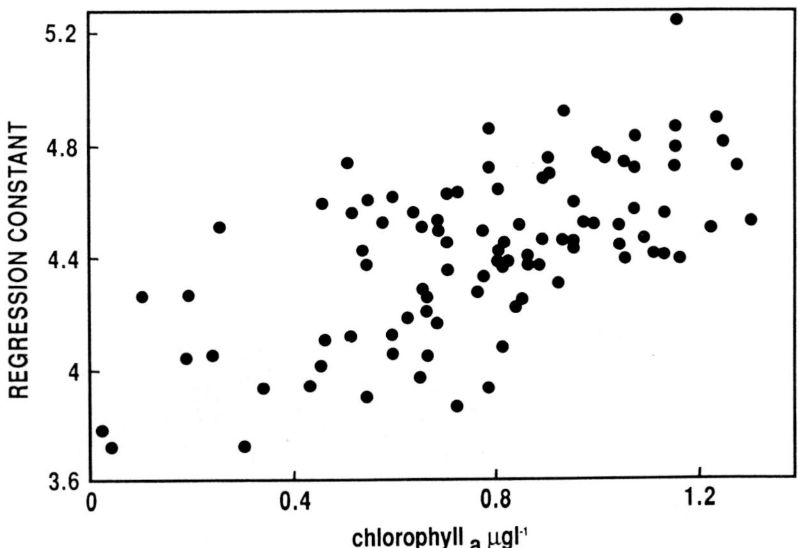

Figure 14b

Figure 14. Particle size - relationships between slope and intercept with biomass as measured by chlorophyll.

2.7. CHEMICAL OCEANOGRAPHY - NUTRIENT DEPLETION DATA

It is possible to estimate new production by measuring the depletion of nitrate and phosphate during spring blooms and assuming standard stoichiometries between C:N:P. Data in Harris *et al.* (1991) and other previously unpublished data give an estimate of the depletion of NO_3 and PO_4 during the spring bloom in Tasmanian latitudes. This data can be converted to a C flux using a Redfield ratio and hence an estimate of total "new" production.

Nutrient data relating temperature to nitrogen in Tasmanian waters shows a seasonal depletion of approximately 10 µM total nitrogen (figure 15). Using a C:N ratio of 7 by weight gives an annual net production of 63.6 gCm-2y-1 or 2.00 Gty-1.

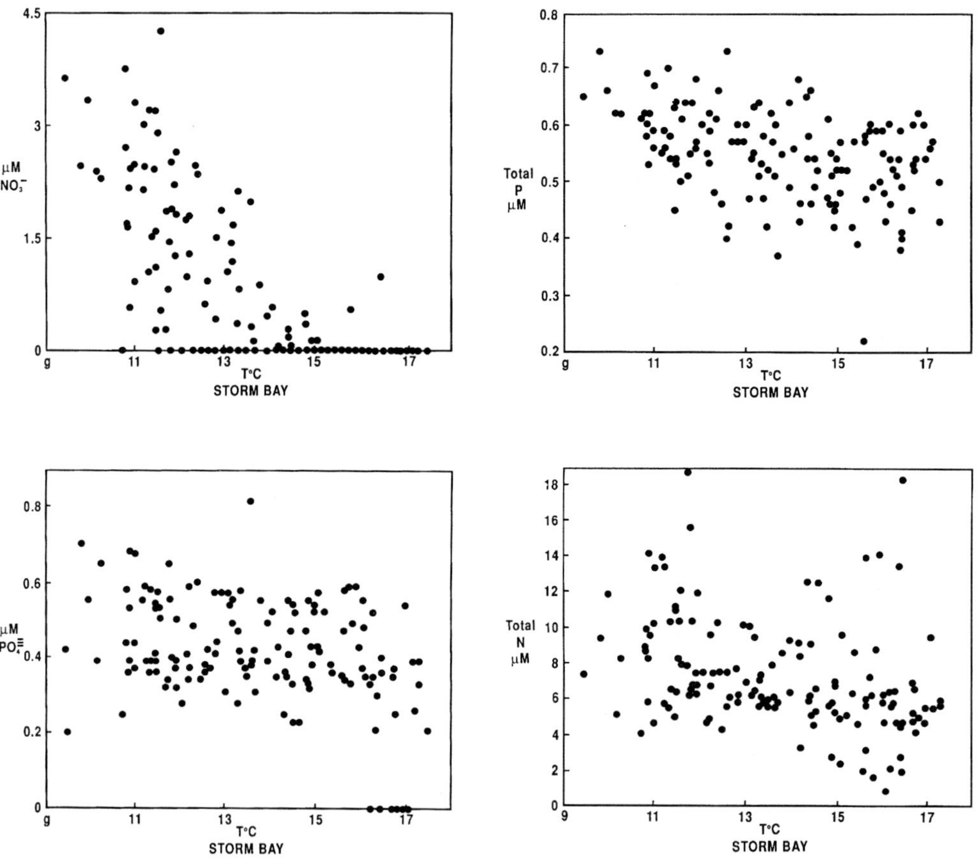

Figure 15. Relationship between temperature and nutrient concentrations in Tasmanian waters.

Simultaneously, total phosphorus declined by approximately 0.4 µM. Using Redfield stoichiometries, the TP depletion equates to a very similar C flux of 62.3 gCm-2y-1 or 1.95 Gty-1.

2.8. A SUMMARY OF THE CALCULATED FLUXES

These calculations, using a variety of data, are summarised in figure 16. In general terms there is good agreement between the different methods of calculation if an "f" number of about 0.5 is assumed - despite all the assumptions and known problems with the methods. In that case the gross production (averaged over the zonal band from 40-50°S) is about 122 gCm-2y-1 and the export or new production is about 65 gCm-2y-1 equalling a carbon flux of about 1.6 Gt Cy-1. Estimates of "f" based on empirical correlations between 14C productivity measurements and "f" appear to underestimate the true export of carbon from the photic zone. The true value of "f" appears to lie closer to 0.5 rather than the value of 0.3 estimated from published data - both of these estimates are higher than the data from sediment traps would suggest. Given the evident non-linearity between the "J" flux and B and sum P is it not surprising that the true spatial and temporal integral of the flux obtained from pCO$_2$ and nutrient data is higher than simple scaling up of point estimates.

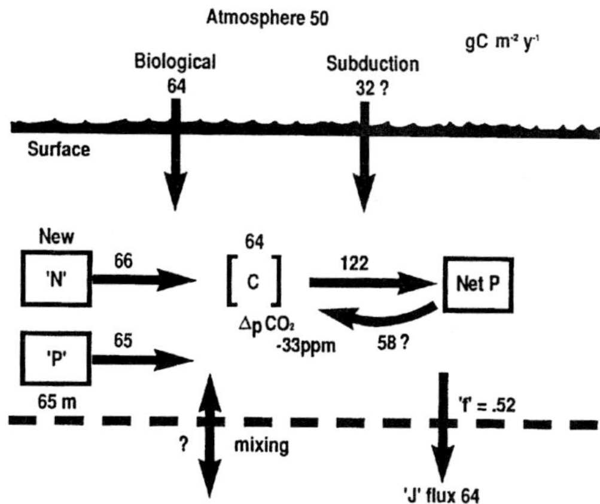

Figure 16. Summary of the calculated fluxes.

3. The Impact of Climatic Variability

A number of detailed oceanographic studies have been carried out in the STC region in the Tasman Sea over the last few years. Physical forcing of biological processes is a notable feature of the oceanography of the region. Harris and his co-workers (Harris *et al.*, 1988, 1991) have published a series of papers showing that the region of the Tasman Sea between Tasmania and

262

New Zealand responded to strong interannual variability in climate. Earlier, Harris *et al.* (1987a) showed that the interannual variability in coastal waters was coupled to offshore events and presented evidence to show that changes in the oceanic circulation influenced the biological productivity of the entire Tasman Sea region. These events appear to be driven by changes in the climate and oceanography of the region on the northern edge of the STC.

The region of the STC is known to be a region of both high interannual climate variability and long term change. Harris *et al.* (1988) documented the changes in the temperature, nutrient status and productivity of waters in the west of the Tasman Sea resulting from El Niño Southern Oscillation (ENSO) events.

In ENSO years (warm events) the westerly winds were much stronger at 43°S, water temperatures were colder and the spring bloom was extended by as much as 3 months (figure 17). Continual disruption of the small region of

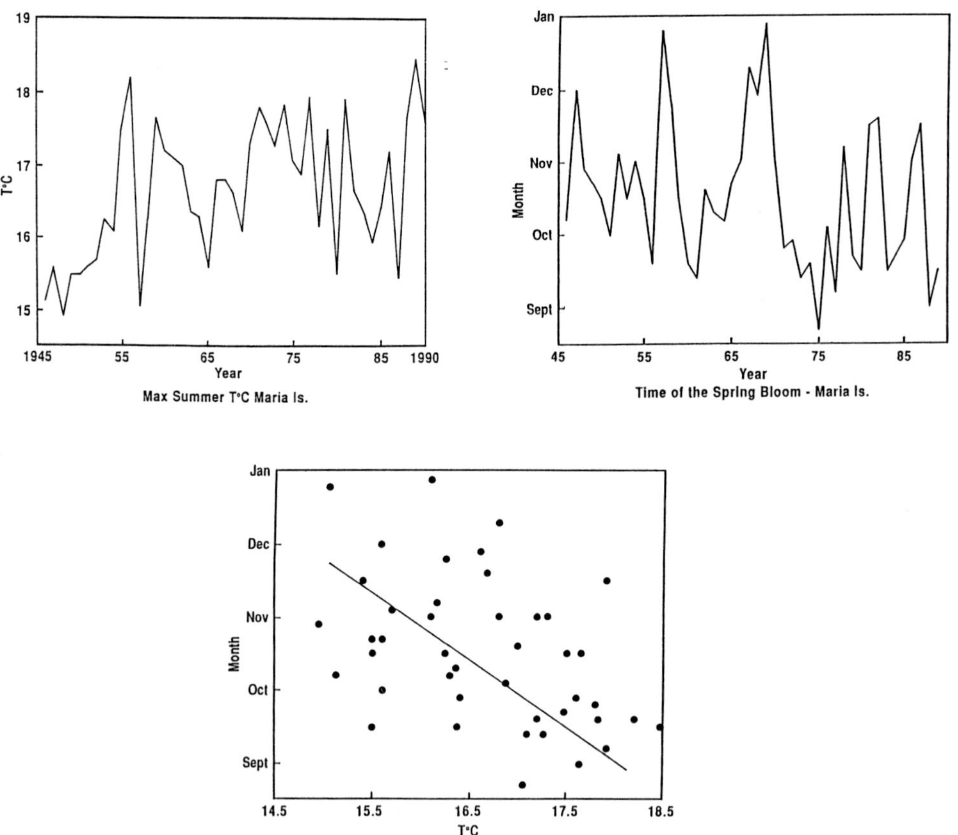

Figure 17. Long term changes in the timing of the, spring bloom at Maria Island, Tasmania.

seasonal stratification occurred. Judging by the productivity, phytoplankton species composition and zooplankton biomass in these years new production was greatly increased (Harris *et al.*, 1991). In anti-ENSO years (cold events) the westerlies weakened around Tasmania and the water warmed. These events occurred throughout the Tasman Sea. In these years the productivity of surface waters around Tasmania declined dramatically, zooplankton biomass decreased by an order of magnitude and fisheries failed (Harris *et al.*, 1989b, 1991).

The interannual variability in sea surface temperatures off Tasmania is more than 3°C; temperatures are colder in windy years because of the disruption of the seasonal stratification. Buried in the interannual variability is evidence for a long term warming of approximately 0.5°C (Harris *et al.*, 1988, figure 17). These long and short term variations in SST are driven by changes in the strength of the westerlies over the Tasman Sea. In ENSO years the westerlies strengthen and there is evidence for a strengthening of the circumpolar flow in the Tasman Sea sector. CZCS composites from 1978-83 indicate changes in the phytoplankton biomass in the southern Tasman Sea. Biomass increased over the period 1978-83 as the westerlies strengthened and peaked in 1980-81: the westerlies fluctuate with an eleven year cycle at these latitudes (Harris *et al.*, 1988).

3.1. THE EFFECT OF CLIMATE ON NUTRIENT AVAILABILITY

There was a strong effect of sea surface temperature on nutrient availability. In these waters both nitrogen and phosphorus concentrations are strongly correlated with SST (Section 2.7). Total nitrogen concentration decrease by approximately 1 µM N.°C^{-1} or about 12-15%.°C^{-1} (figure 15). A further effect of warming was observed in 1987-9 when warmer sea surface temperatures (SST) led to a reduction in the depth of winter mixing in Southern Ocean waters off Tasmania. This led to a decrease in winter and early nitrate levels to one half the normal concentration.

It is noticeable that the consistent warming observed in the last forty-five years over Tasmania (figure 17), coupled to the increase in SST over the same period, has led to an earlier, and presumably weaker, spring bloom. Over all, in the last forty-five years the spring bloom has been tended to occur about one month earlier (figure 17) and has weakened by up to 7-10% if nitrate concentrations in the water in early spring can be used as a guide. The effect of the 1988 event was to bring about a decrease in the westerlies, a warming of SST over a wide area, a 50% decrease in nitrate and a reduction in new production.

A similar long term warming trend has been noticed in New Zealand and throughout the subantarctic region (Jones *et al.*, 1986, Karoly 1990). It is particularly noticeable on all the major southern hemisphere continents and on the subantarctic islands (figure 18).

3.2. EFFECTS OF CLIMATE VARIABILITY ON THE PELAGIC FOOD CHAIN

The combined effect of ENSO and long term warming led, in the years from 1987-89, to a major reduction in nutrient availability, an order of magnitude reduction in phytoplankton and zooplankton biomass, and a change in the structure of the food chain (Cushing, 1989; Harris *et al.*, 1991). Diatoms were replaced by small flagellates as the dominant phytoplankton in warm years. The interannual variability in the regional oceanography also had a strong effect on the geographical distribution of, and recruitment to, commercial fish stocks. In particular, the commercial jack mackerel fishery in this region was subject to strong interannual variability driven by changes in climate and consequent changes in the coastal food chain (Harris *et al.*, 1989b, 1991; figure 19).

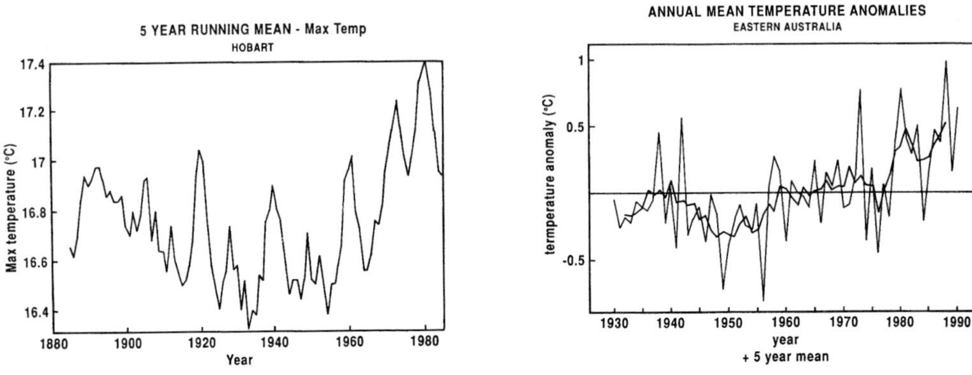

Figure 18. Long term changes in land surface temperatures. Hobart and SE Australia.

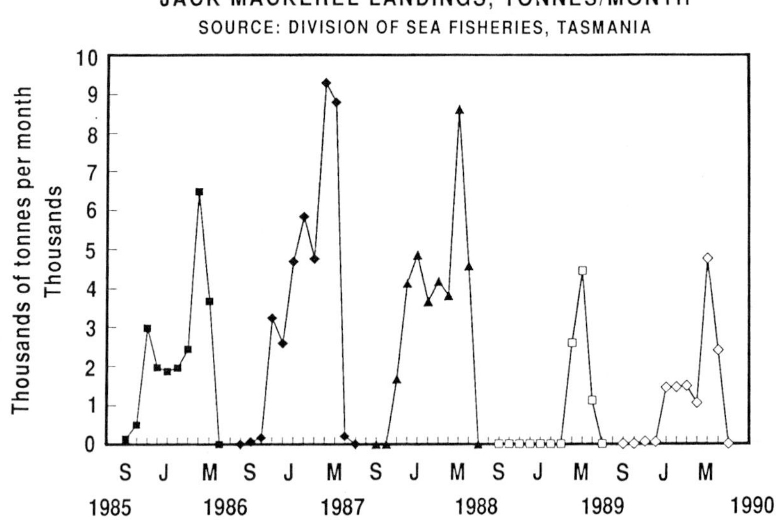

Figure 19. Mackerel catches Tasmania 1985-90.

It is not known to what extent these regional events are more widespread. The composite CZCS images for each year show strong interannual variability in phytoplankton biomass in the region of the STC in the Tasman Sea. Certainly there are ENSO events visible in the ocean circulation and recruitment to commercial fish and crustacean stocks all the way from from South Africa to New Zealand (Harris *et al.*, 1988) and from Australia south to at least Macquarie Island; but we do not know the full spatial extent of these large scale changes in productivity and the timing of events.

How ENSO and longer term events interact with the global carbon cycle in these waters is still a mystery. Clearly the main areas of productivity are coupled with the basin scale circulation patterns of the oceans. From changes in the recruitment of fish and crustacea with pelagic larvae it is possible to infer that ENSO events have a widespread effect on the patterns of oceanic transport in the entire STC region (Harris *et al.*, 1988).

There is strong circumstantial evidence for long term changes in the carbon sink in southern waters by virtue of fact that there are noticeable interactions between ENSO events and the atmospheric carbon dioxide concentration which appear to be mediated by biological events in southern waters (Bacastow,1976). Also there is evidence for a long term decrease in the difference in carbon dioxide concentration between the South Pole and the Equator between 1958 and 1974 (Enting, 1990; Francey and Enting, 1990). This might be taken as evidence for a weakening of the strength of the carbon sink in the southern oceans since the 1950s - it might equally be due to poor calibration in early data. The strength of the southern oceanic sink may have increased again in the early 1980s.

Such a scenario would be consistent with the arguments presented above which show a complex modulation between the eleven year cycle in the westerlies and the effects of a general, large scale warming in the subantarctic region. Certainly there is no evidence to suggest that the oceanic carbon dioxide sink is constant, nor can a steady state be assumed.

4. Conclusions

In order to adequately describe the sources and sinks of carbon in the global carbon cycle and to monitor the interannual variability in the critical processes, it will be necessary to have both synoptic biomass surveys from orbiting spacecraft and detailed *in situ* data from ships. This paper has shown that it is possible to combine data from a number of sources to obtain estimates of fluxes which are important for the global C cycle. The data quoted above give an estimate of the mean net carbon sink in the southern STC region of 1.6 ± 0.22 Gt Cy^{-1} ($\pm$ 2xSEM) - or a precision of about $\pm$ 30%. The available data give no indication of long term changes in seasonality or interannual variability. Even using these sources the accuracy is, at present, not sufficient to detect changes due to interannual variability (estimated at $\pm$ 30%) or long term warming due to anthropogenic effects which can be estimated to require a precision of better than $\pm$ 10%.

One modest goal of research in the region of the STC must be to obtain better data on the temporal and spatial extent of the regions of reduced pCO$_2$ in surface waters and on nutrient mass balances. The first global composites of CZCS data have given a synoptic view of the phytoplankton biomass data but incomplete orbital coverage and cloud cover have placed restrictions on our ability to detect changes from year to year. An ability to produce spatial and temporal composites from satellite data without compromising the required high frequency resolution will be essential.

Without more data from missions such as the forthcoming SeaWiFS mission it will not be possible to study the interactions between mesoscale physics and biology. Even then it will require a concerted effort to provide "sea truth" to complement the satellite data. In situ estimates of Pmax and measurements of optical properties will also be required to validate the models of integral production. A search for empirical relationships showing large number "system" properties (such a nutrient mass balance approaches) may be one way to attempt to scale up from small scale, high frequency processes to estimates of large scale, global fluxes.

References

Anon (1989) 'Climatic charts of sea surface temperature of the Western North Pacific and the global ocean', Marine Dept, Japan Meteorological Agency, 75p.

Bacastow, R.B. (1976) 'Modulation of atmospheric carbon dioxide by the Southern Oscillation', Nature 261, 116-118.

Bacastow, R.B., Keeling, C.D. and Whorf, T.P. (1985) 'Seasonal amplitude increase in atmospheric CO_2 concentration at Mauna Loa, Hawaii, 1959-82', Journal of Geophysical Research 90 (D6), 10529-10540.

Broecker, W.S. (1973) 'Factors controlling the CO_2 content of the oceans and the atmosphere', in G.M. Woodwell and E.V. Pecan (eds.), Carbon and the biosphere, Proc. 24th Brookhaven Symp. US AEC NTIS CONF-720510, Springfield, Va., 32-50.

Banse, K. and Yong, M. (1990) 'Sources of variability in satellite derived estimates of phytoplankton production in the eastern tropical Pacific', Journal of Geophysical Research 95, (C5), 7201-15.

Betzer, P.R., Showers, W.J., Laws, E.A., Winn, C.D., DiTullio, G.R., and Kroopnik, P.M. (1984) 'Primary productivity and particle fluxes on a transect of the equator at 153°W in the Pacific Ocean', Deep-Sea Research 31, 1-11.

Bishop, J.K.B. and Marra, J. (1984) 'Variations in primary production and particulate carbon flux through the base of the euphotic zone at the site of the sediment trap intercomparison experiment (Panama Basin)', Journal of Marine Research 42, 189-206.

Buesseler, K.O., Bacon, M.P., Cochran, J.K., and Livingston, H.D. (1990) '234Th as a tracer of expert flux and new production during the North Atlantic Spring Bloom Experiment', p 29 in JGOFS Rep No 7, Proc. JGOFS N Atlantic Bloom Exp. Int. Sci. Symp. Washington, DC, Nov 1990.

Bunt, J.S. (1975) 'Primary productivity of marine ecosystems', in H.Leith and R.H. Whittaker (eds.), Primary productivity of the biosphere, Springer, Berlin, pp. 169-183.

Cushing, D.H. (1989) 'A difference in structure between ecosystems in strongly stratified waters and in those that are only weakly stratified', Journal of Plankton Research 11, 1-13.

Deacon, G.E.R. (1945) 'Water circulation and surface boundaries in the oceans', Quart. J. Roy. Met. Soc. 71, 11-25.

Denman, K.L. and Abbott, M.R. (1988) 'Time evolution of surface chlorophyll patterns from cross spectrum analysis of satellite color images', Journal of Geophysical Research 93 (C6), 6789-98.

Enting, I.G. (1990) 'Carbon cycle modelling: illustrations of modelling problems in IGBP studies', in G.A. Latham and Taylor, J.A. (eds.), Proc. IGBP workshop 13, Mathematical and statistical modelling of global change processes, Proc. Centre for Math. Analysis, Australian National Uni Vol 25, pp. 212-234.

Enting, I.G. and Mansbridge, J.V. (1989) 'Seasonal sources and sinks of atmospheric CO_2: direct inversion of filtered data', Tellus 41B, 111-126.

Eppley, R.W. and Peterson, B.J. (1979) 'Particulate organic matter flux and planktonic new production in the deep ocean', Nature 282, 677-80.

Eppley, R.W., Stewart, E., Abbott, M.R., and Heyman, U. (1985) 'Estimating ocean primary production from satellite chlorophyll. introduction to regional differences and statisitcs for the Southern California Bight', Journal of Plankton Research 7, 57-70.

Eppley, R.W., Stewart, E., Abbott, M.R., and Owen, R.W. (1987) 'Estimating ocean production from satellite derived chlorophyll: insights from the Eastropac data set', Oceanologica Acta 1987, Proc. Int. Symp. Equatorial Vertical Motion, Paris, May 1985, 109-113.

Francey, R.J. and Enting, I.G. (1990) 'The role of terrestrial biota in the atmospheric carbon budget', in G.A. Latham and J.A. Taylor (eds.), Proc. IGBP workshop 13, Mathematical and statistical modelling of global change processes, Proc. Centre for Math. Analysis, Australian National Uni. Vol 25, pp. 235-245.

Harris, G.P. (1978) 'Photosynthesis, productivity and growth: the physiological ecology of phytoplankton', Erge. Limnology 10, 1-171.

Harris, G.P., Haffner, G.D., and Piccinin, B.B. (1980) 'Physical variability and phytoplankton communities II. Primary productivity by phytoplankton in a physically variable environment', Archiv fuer Hydrobiologie 88, 393-425.

Harris, G.P., Nilsson, C., Clementson, L.A., and Thomas, D.P. (1987a) 'The water masses of the east coast of Tasmania: Seasonal and interannual variability and the influence on phytoplankton biomass and productivity', Australian Journal of Marine and Freshwater Research 38, 469-490.

Harris, G.P., Ganf, G.G., and Thomas, D.P. (1987b) 'Productivity, growth rates and cell size distributions of phytoplankton in the SW Tasman Sea: implications for carbon metabolism in the photic zone', Journal of Plankton Research 9, 1003-1030.

Harris, G.P., Griffiths, F.B., and Thomas, D.P.(1989a) Light and dark uptake and loss of ^{14}C: methodological problems with productivity measurements in oceanic waters', Hydrobiologia 173, 95-105.

Harris, G.P., Nilsson, C.S., and Booth, B. (1989b) 'CSIRO harnesses satellite power to search out fish', Australian Fisheries 48, 22-25 February 1989.

Harris, G.P., Griffiths, F.B, Clementson, L.A., Lyne, V., and Van Der Doe, H. (1991) 'Seasonal and interannual variability in physical processes, nutrient cycling and the structure of the food chain in Tasmanian shelf waters', Journal of Plankton Research 13, (Suppl) 109-131.

Inoue, H. and Sugimura, Y (1986) 'Distribution of pCO_2 and del^{13}C in the air and surface sea water in the Southern Ocean, South of Australia', Mem. Natl. Inst. Polar Res. Spec. Issue 40, 454-461.

Inoue, H. and Sugimura, Y. (1988) 'Distribution and variations of oceanic CO_2 in western North Pacific, eastern Indian, and Southern Ocean south of Australia', Tellus 40B, 308-320.

Jitts, H.R. (1966) 'The summer characteristics of primary productivity in the Tasman and Coral Seas', Australian Journal of Marine and Freshwater Research 16, 151-162.

Jones, P.D., Raper, S.C., and Wigley, T.M.L. (1986) 'Southern hemisphere surface air temperature variations 1851-1984', Journal of Climate and Applied Meteorology 25, 1213-30.

Karoly, D.J. (1990) 'Evidence of recent temperature trends in the southern hemipshere', in G.I. Pearman (ed.), Greenhouse CSIRO Australia, pp.52-59.

Keeling, C.D. and Whorf, T.P. (1985) 'The concentration of atmospheric CO_2 at ocean weather station P from 1969 to 1981', Journal of Geophysical Research 90 (D6), 10511-28.

Lewis, M.R. (1989) 'The variegated ocean: a view from space', New Sci. 7 Oct 1989 (1685), 37-40.

Mackey, D.J. and Butler, E.C.V. (1991) 'Continuous surface measurements of pH in the south Pacific: the carbon dioxide system', Proc. Brest Symp. Chem. Southern Ocean (in press).

Michaels, A.F. and Silver, M.W. (1988) Primary production, sinking fluxes and the microbial food web. Deep-Sea Research 45, 473-490.

Mitchell, B.G., Brody, E.A., Holm-Hansen, O., McClain, C., and Bishop, J. (1991) 'Light limitation of phytoplankton biomass and macro-nutrient utilization in the Southern Ocean', Limnology and Oceanography, 36 (8), 1662-1677.

Miyake, Y. and Sugimura, Y. (1969) 'Carbon dioxide in the surface water and the atmosphere in the Pacific, the Indian and the Antarctic Ocean areas', Rec. Oceanogr. Works Japan 10, 23-28.

Miyake, Y., Sugimura, Y., and Saruhashi, K. (1974) 'The carbon dioxide content in the surface waters in the Pacific Ocean', Rec. Oceanogr. Works Japan. 12, 45-52.

Ortmann, A. (1896) 'Grundzuge der marinen Tiergeographie', G. Fischer, Jena.

Pace, M.L., Knauer, G.A., Karl, D.M., and Martin, J.M. (1987) 'Primary production, new production and the vertical flux in the eastern Pacific Ocean', Nature 325, 803-804.

Parslow, J.S. and Harris, G.P. (1990) 'Remote sensing of marine photosynthesis', in R.J. Hobbs and H.A. Mooney (eds.), Remote Sensing of Biosphere Functioning (Ecological Studies 79), Springer-Verlag, New York, pp. 269-290.

Pearman, G.I. and Hyson, P. (1980) 'activities of the global biosphere as reflected in atmospheric CO_2 records', Journal of Geophysical Research 85 (C8), 4468-4474.

Platt, T. (1986) 'Primary production of the ocean water column as a function of surface light intensity: algorithms for remote sensing', Deep-Sea Research 33, 149-163.

Platt, T. and Harrison, W.G. (1985) 'Biogenic fluxes of carbon and oxygen in the ocean', Nature 318, 55-58.

Platt, T. and Sathyendranath, S. (1988) 'Oceanic primary production: estimation by remote sensing at local and regional scales', Science 241, 1613-1620.

Post, W.M., Peng, T-H., Emanuel, W.R., King, A.W., Dale, V.H. and De Angelis, D.L. (1990) 'The global carbon cycle', American Scientist 78, 310-326.

Prakash, A., Sheldon, R.W., and Sutcliffe, W.H. (1991) 'Geographic variation of oceanic ^{14}C dark uptake', Limnology and Oceanography 36, 30-39.

Rau, G.H., Takahashi, T. and Des Marais, D.J. (1989) 'latitudinal variations in plankton del 13C: implications for CO_2 and productivity in past oceans', Nature 341, 516-518.

Talling, J.F. (1957) 'The phytoplankton population as a compound photosynthetic system', New Phytologist 56, 133-49.

Tans, P.P., Fung, I.Y., and Takahashi, T. (1990) 'Observational constraints on the global atmospheric CO_2 budget', Science 247, 1431-1438.

van der Spoel, S. and Pierrot-Bults, A.C. (eds) (1979) 'Zoogeography and diversity in plankton', Edward Arnold, London.

Vezina, A.F. and Platt, T. (1987) 'Small-scale variability of new production and particulate fluxes in the ocean', Canadian Journal of Fisheries and Aquatic Science 44, 198-205.

Watson, A.J., Robinson, C., Robinson, J.E., Williams, P.J. Le B., Fasham, M.J.R. (1991) 'Spatial variability in the sink for atmospheric CO_2 in the N. Atlantic', Nature 350, 50-53.

COUPLING OF OCEAN COLOR DATA
TO PHYSICAL-BIOLOGICAL MODELS

J. ISHIZAKA
National Institute for Resources and Environment
16-3 Onogawa, Tsukuba, Ibaraki
305 Japan

E. E. HOFMANN
Center for Coastal Physical Oceanography
Crittenton Hall, Old Dominion University,
Norfolk, VA 23529, USA.

ABSTRACT. Satellite-derived ocean color distributions provide synoptic views of biological fields that cover wide areas of the ocean. Physical-biological models are useful for quantitative analyses of ocean color data and satellite-derived phytoplankton fields can be used for initialization and verification of physical-biological models. Also assimilation of phytoplankton fields, derived from ocean color measurements, into physical-biological models represents on approach for improving the predictive capability of these models. Five examples of using ocean color data with physical-biological models are described. These include using ocean color data to compare the patterns in temperature and flow fields, to analyze numerical Langrangian particle tracing experiments, evaluation of Eulerian models, to provide input for data assimilation, and to estimate the magnitude of physical and biological processes. Approaches for more sophisticated coupling of ocean color data with physical-biological models are discussed.

1. Introduction

Satellite-borne ocean color sensors are presently the only method for obtaining synoptic views of biological fields over a wide area of the ocean surface. The good space and time resolution provided by these data have supported a large variety of research topics (Abbott and Chelton, 1992; see also lectures in this course). The first studies that used Coastal Zone Scanner (CZCS) measurements tended to emphasize comparisons between CZCS and ship-observed chlorophyll distributions and were rather descriptive in nature. However, recent studies have applied statistical and quantitative analyses to CZCS data.

Physical-biological models are a developing tool in biological oceanography. Physical models reproduce environmental forcing, and biological models reproduce the response of the biological system to this forcing as well as the response to biological interactions. Combined physical and biological models can then be used to understand the various responses of marine ecosystems. It is also possible to use a combined model to precast/nowcast/forecast ecosystem

271

V. Barale and P.M. Schlittenhardt (eds.),
Ocean Colour: Theory and Applications in a Decade of CZCS Experience, 271–288.

behaviour and to estimate biological production and associated material fluxes. However, this requires that physical-biological models be verified with space- and time-dependent biological distributions.

Most of the physical-biological models developed previously were verified by comparison with biological distributions sparsely sampled in time and space. Nihoul (1984) suggested that satellite observations are useful for initialization and verification of numerical models because of the good time and space coverage provided by these data. Thus, a natural use for satellite-derived chlorophyll distributions is initialization and verification of physical-biological models. This also provides a method for understanding the vast amount of information provided by satellite ocean color sensors.

Ishizaka (1990 a,b,c,) and McClain *et al.* (1990) undertook a series of studies that were attempts to couple CZCS data with a physical-biological model of the southeastern U.S. continental shelf (figure 1) ecosystem. The first study (Ishizaka, 1990 a), was a comparison of 9 CZCS images with optimally-interpolated temperature and flow fields. This study also used numerical Lagrangian particle tracing experiments to follow and identify features seen in the CZCS images. The second study (Ishizaka, 1990 b) provided statistical comparisons between CZCS data and the output of an Eulerian model. The third study (Ishizaka, 1990 c) presented approaches for assimilation of CZCS data into an Eulerian model to improve estimates of across-shelf phytoplankton carbon flux. The final study (McClain *et al.*, 1990) used an

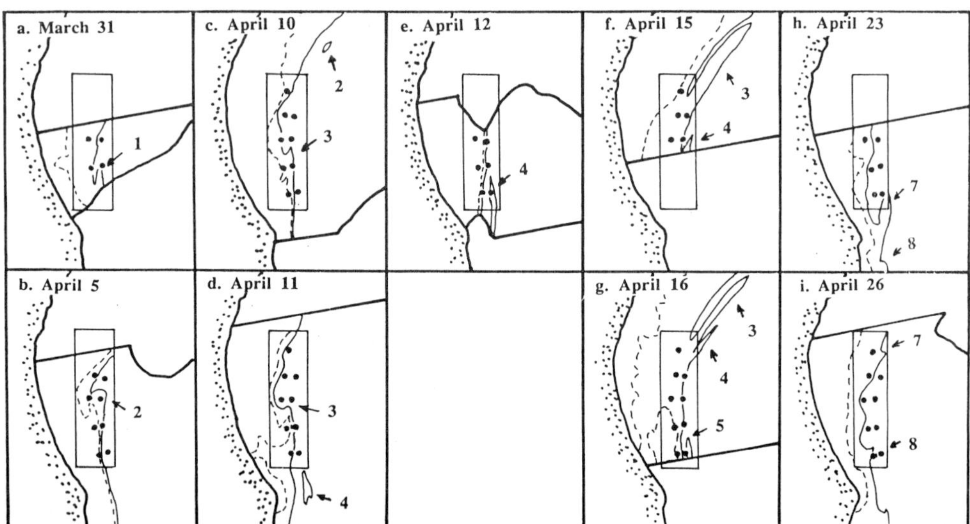

Figure 1. CZCS-derived chlorophyll distributions on the southeastern U.S. continental shelf from spring 1982. The approximate location of the 0.5 µg l-1 and 1.5 µg l-1 chlorophyll concentration are shown by the solid and broken lines, respectively. The region used for the model domain is shown by the box. Numbers indicate frontal eddy events identified from the optimally interpolated velocity and temperature fields and Lagrangian particle tracing experiments. Redrawn from Ishizaka (1990 a).

Eulerian model and CZCS chlorophyll distributions to estimate the processes that contribute to the local time change of chlorophyll fields on the southeastern U.S. continental shelf. This paper gives a review of these studies published previously, and makes comments concerning future approaches for coupling physical-biological models with ocean color data.

2. Comparison to Temperature and Flow Fields

A starting point in the analysis of ocean color data is to compare the patterns observed in temperature and flow fields with those seen in the satellite-derived pigment distributions. Such comparisons illustrate the relationship between pigment distributions and physical forcing and gives insight into possible interactions. Yentsch (1983, 1984) first pointed out that CZCS-derived phytoplankton distributions for the Gulf of Maine resembled sea surface temperature patterns obtained by infra-red channel of CZCS and also resembled distribution of tidal mixing predicted by a model (Garrett *et al.* 1978). Yentsch (1984) also explained the high chlorophyll water observed north of the Gulf Stream and in Gulf Stream cold-core eddies by conceptual models of the density distribution of the Gulf Stream front and the cold-core eddy. From these comparisons, he concluded that chlorophyll distribution patterns were dependent on the mixing of water masses which brings nutrients to the euphotic zone. Furthermore, he suggested the importance of pattern recognition in comparisons of numerical model distributions with those obtained from satellite data. Particles were released in the high-chlorophyll region observed in the southern model domain on April 10.Trajectories indicate that eddy event 3 (see figure 1) formed the high chlorophyll plume. Dotted lines are contours of chlorophyll concentrations of 0.5 µg l-1 and 1.0 µg l-1. Triangles represent the position of the particles, and solid lines are the trajectory followed by the particles during 1 day. Redrawn from Ishizaka (1990 a).

McClain *et al.* (1986) combined CZCS images, hydrographic data, and a numerical model of wind-induced upwelling for the east coast of Spain. The CZCS data and field observations indicated that the observed chlorophyll distribution was the result of upwelling and that the coastal circulation was caused by two wind events during a 10-day period. Correspondence of patterns in the CZCS images and simulated circulation distributions, which reproduced the upward movement of the isopycnals and the circulation along the coast in response to the wind events, provided further evidence that wind-driven upwelling is the major mechanism producing the changes seen in the CZCS pigment fields.

Walsh *et al.* (1987) compared CZCS distributions from the Mid-Atlantic Bight continental shelf with wind-driven circulation patterns obtained from a numerical model. On the basis of these comparisons, they concluded that wind events resuspended phytoplankton as well as producing upwelling which enhances phytoplankton growth. After the wind event has terminated, the phytoplankton sink rapidly and are eventually advected offshore.

Gulf Stream frontal eddies are an important physical forcing along the outer southeastern U.S. continental shelf during spring. These events propagate northward with the Gulf Stream front and the upwelling associated with the frontal eddies supplies nutrient-rich Gulf Stream subsurface water to the outer shelf. The warm-tongue and cold-core structures of Gulf Stream frontal eddies and their northward propagation is seen in hydrographic, current meter, and sea surface temperature data (cf. Lee *et al.* 1981). Moreover, Yoder *et al.* (1981) found that ship-observed chlorophyll patchiness corresponded to the upwelled cold-core of the frontal eddy detected by AVHRR.

McClain *et al.* (1984) used current meter observations from the southeastern U.S. continental shelf to track features in CZCS images. However, it is sometimes difficult to compare current and temperature time series data with the two-dimensional time-separated measurements from satellite. Ishizaka (1990 a) applied an optimal interpolation technique to the same current meter observations. Optimal-interpolation provides a time series of two-demensional flow and temperature fields (figure 2) which are directly comparable to satellite data (figure 1). For the southeastern U.S. continental shelf, comparisons between optimally-derived flow and temperature fields showed good correspondence between features in the CZCS images and frontal eddy events detected by the current meters (Ishizaka 1990 a). In his analysis, Ishizaka (1990 a) used length scales for the optimal interpolation that were similar to the values estimated from CZCS pigment distributions (Yoder *et al.* 1987). However, the exact relationship between correlation length scales for current, temperature, and pigment distributions remains to be determined. The optimally-interpolated fields do not explicitly

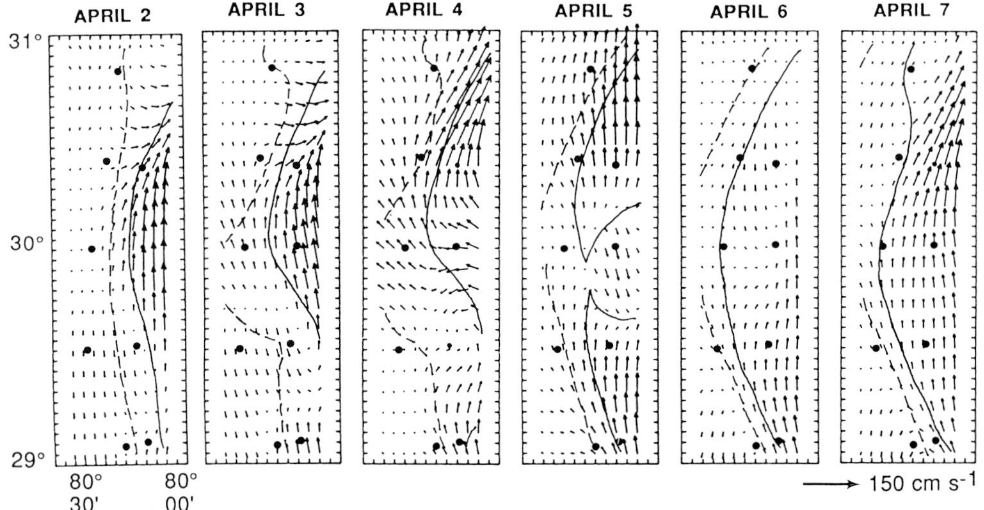

Figure 2. Optimally-interpolated flow and temperature fields during April 1-7, 1980, that show the movement of a Gulf Stream frontal eddy event (event 2 of figure 1). Arrows indicate flow directions and magnitudes. Contour lines indicate the 22°C (solid) and 20°C (broken) isotherms, respectively. The current meters used in the optimal interpolation are indicated by the circles. Redrawn from Ishizaka (1990 a).

provide information on the active physical processes, such as could be obtained from a circulation model. However, this approach has the advantage of making direct comparisons between distribution fields easy.

3. Lagrangian Particle Tracing

Lagrangian particle tracing studies provide a quantitative way to compare patterns in ocean color satellite data with those in flow fields. Once time and space-dependent flow fields are constructed, it is straightforward to trace particles in the flow using:

$$x\left(t+\Delta t\right) = x\left(t\right) + u\left(x, y, t\right)\Delta t \tag{1}$$

$$y\left(t+\Delta t\right) = y\left(t\right) + v\left(x, y, t\right)\Delta t \tag{2}$$

where $x\,(t)$ and $y\,(t)$ are the current position of the particle at time (t), and u and v are velocities in the x and y directions, respectively. The new position of the particle is obtained by adding to the current location the distance the particle is moved by the flow field in a time interval, Δt.

Ishizaka (1990 a) used Lagrangian particle tracing to follow the movement of features seen in CZCS images for a limited region on the outer southeastern U.S. continental shelf (figure 3). The results of three particle tracing experiments showed that some of the high-chlorophyll filament structures seen in the CZCS data developed in response to the passing of frontal eddy events. Also, particles in the frontal eddies moved northward and remained with the features for at least several days.

Lagrangian calculations are particularly useful for analysis of CZCS data because these measurements are frequently separated by at least one day and usually by more than a few days. Consequently, it is not easy to determine how features, like frontal eddies, affect the evolution of the patterns seen in the chlorophyll distributions. Also, unlike the Eulerian model discussed below, Lagrangian particle tracing models do not require many a priori assumptions and need only minimal computational time. Lagrangian models also have the benefit of providing a check on the accuracy of the flow fields.

Another application of Lagrangian particle tracing is to determine the rate of change of chlorophyll concentration following a water mass. Abbott and Zion (1985) used this approach to obtain the growth rate of phytoplankton in a coastal upwelling plume. They tracked an actual Lagrangian drifter through a sequence of CZCS-derived pigment maps and inferred growth rates from the chlorophyll changes. McClain *et al.* (1990) used this same approach, but with numerical Lugrangian particle tracing, to obtain the rate of change of chlorophyll on the outer southeastern U.S. continental shelf. With a numerical model, many particles can be followed and distributions of rate of pigment concentration change, which does not include horizontal advective effects, can be obtained.

The advective portion of the flow field can be derived from CZCS data by pattern matching (Garcia and Robinson, 1989). This method was first used with satellite sea surface temperature data (Emery *et al.* 1986). However, ocean color data has the advantage in that it extends to one optical depth, which may be several meters deep, rather than just the few micrometers that are sensed for sea surface temperature. However, ocean color data should be used carefully because there is no guarantee that biological processes do not change the patterns in the satellite image.

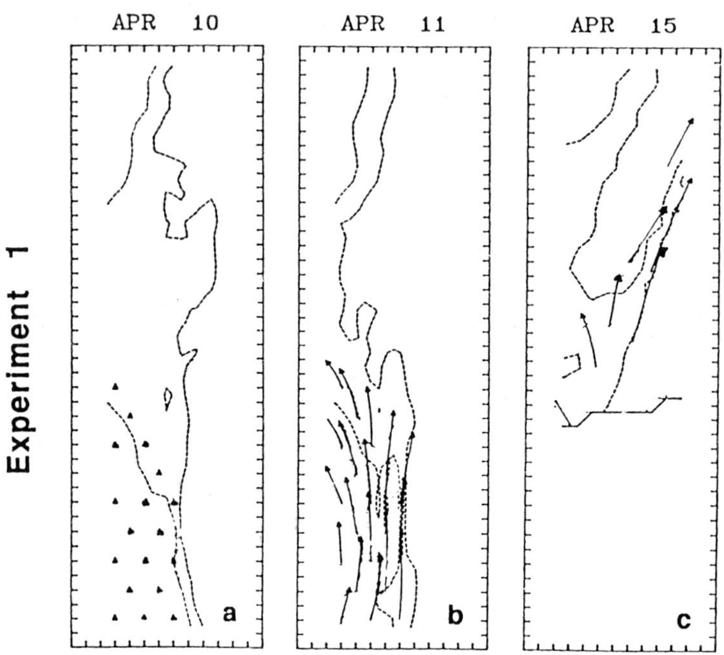

Figure 3. CZCS chlorophyll concentrations and particle trajectories from the Lagrangian particle tracing experiments.

4. Eulerian Model

Lagrangian particle tracing models are not truly physical-biological models since the pigment concentration is treated as a conservative quantity. Recently there have been attempts to include biological behavior in Lagrangian particle tracing models (cf. Woods and Onken, 1982). However, the more matured approach for describing the changes in the distribution of biological material in a moving fluid in an Eulerian model. These models consist of a system of partial differential equations that describe the advection and diffusion of a non-conservative material.

Wroblewski *et al.* (1988) explained the North Atlantic chlorophyll distributions seen in composite CZCS images (Esaias *et al.* 1986) with a simple biological model that depended upon the distribution of the mixed layer depth. This model is essentially a number of time-dependent biological models distributed on a horizontal grid, and does not include any horizontal and vertical interactions except for changes in the depth of the mixed layer. However, even with this simple model, the simulated chlorophyll distributions showed good agreement with the composite CZCS chlorophyll distributions. Wroblewski (1989) extended this analysis to include time-dependent variations. These solutions show better agreement with the developments of the spring bloom in the North Atlantic, which can be seen clearly in the CZCS chlorophyll distributions.

Walsh *et al.* (1988) presented the results of a wind-driven, three-dimensional ecosystem model of the Mid-Atlantic Bight, which was initialized with a CZCS-derived chlorophyll distribution from this area. On the basis of simulated surface chlorophyll distributions, obtained with differing conditions, Walsh *et al.* (1988) concluded that certain levels of growing stress and high particle sinking rates were required to produce the chlorophyll distributions observed in the Mid-Atlantic Bight CZCS images. Further model verification was obtained by comparison with time series from moored fluorometers and with ship observations. The use of CZCS data by Walsh *et al.* (1988) is a step beyond the applications in Wroblewski *et al.* (1988) and Wroblewski (1989) in that the CZCS data were used for model initialization as well as verification. However, all of the studies used visual comparisons of patterns in the CZCS and simulated chlorophyll distributions for model verification.

Eslinger (1990) also compared results of a Mid-Atlantic Bight physical-biological model with CZCS chlorophyll distribution. He combined a biological model with a number of time-dependent vertically-one-dimensional layer model to reproduce the spring phytoplankton bloom condition. For the comparison of model results with CZCS chlorophyll data, he used Empirical Orthogonal Function to extract dominant features. The first mode represented the time average of chlorophyll distribution in the Mid-Atlantic Bight and the second mode represented the spring bloom condition. His model reasonably reproduced both these two modes.

Ishizaka (1990 b) extended the method of comparison between CZCS and simulated chlorophyll distributions, by using simple statistical descriptions. This analysis was based upon distributions obtained with a physical-biological model of the form:

$$\frac{\partial C}{\partial t}+u\frac{\partial C}{\partial x}+v\frac{\partial C}{\partial y}=K_x\frac{\partial^2 C}{\partial x^2}+K_y\frac{\partial^2 C}{\partial y^2}$$

$$+\text{Vertical } Process\ Terms + Biological\ Process\ Terms$$

(3)

where C is a non-conservative quantity such as nutrients, phytoplankton, zooplankton and detritus. The first term on the left represents the local time change of C, and the next two terms are the horizontal advective changes. The horizontal velocities, u and v, are obtained from an optimal interpolation of current meter data, as described in Ishizaka (1990 a). The first two terms on the right represent horizontal diffusive changes, where K_x and K_y are the horizontal eddy diffusion coefficients, which are assumed to be constant. The vertical process terms are estimated from the difference between the simulated and interpolated temperature distributions. The biological process terms are obtained from a biological model. Details of the technique used to estimate the vertical terms and the formulation of the biological model are given in Ishizaka (1990 b).

As a first step, Ishizaka (1990b) took the time-average of 9 CZCS images from the southeastern U.S. continental shelf. The resultant distribution (figure 4) showed an across-shelf gradient in chlorophyll and high variance at the shelf break. This provided the basic pattern that would be expected from a physical-biological model constructed for this region. Simulated chlorophyll distributions, for various model cases, were averaged in the same way as the CZCS data. The resulting time-average and variance fields were compared with the CZCS-derived field using spatial correlation coefficients and root mean square error estimates.

This approach reduces comparisons between model and CZCS-derived chlorophyll distributions to two statistical parameters, which facilitates the comparison of many cases. The correlation coefficient indicates similarity of pattern, and the root mean square error mainly shows differences in magnitude. On the basis of simulations for many different conditions, Ishizaka (1990 b) concluded that horizontal advection is responsible for the chlorophyll patterns observed by the CZCS. Shelf edge upwelling and biological processes produce the magnitude of the observed across-shelf gradient and are necessary to maintain the observed patterns.

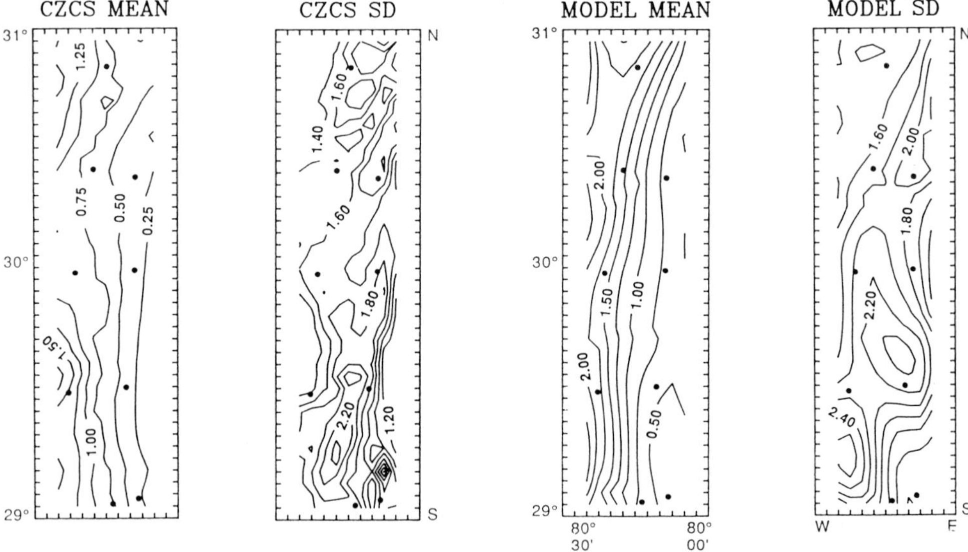

Figure 4. Time mean (left) and standard deviation (right) of CZCS-derived (top) and simulated (bottom) chlorophyll distributions on the southeastern U.S. continental shelf during spring of 1990. Current meter locations are indicated by circles. Redrawn from Ishizaka (1990 b).

It is also possible to run the model with a range of parameter values and evaluate the effect of these on the model by doing a sensitivity analysis on the statistical parameters. For example, figure 5 shows the variation of the correlation coefficient and root mean square error between satellite and model-derived chlorophyll fields for a range of phytoplankton and zooplankton loss rates. These results show that the loss rate parameters chosen for the model are in the range of high correlation and close to the minimum error. This type of analysis suggests that such an approach could be used to bracket parameter ranges for the biological model, especially for poorly known parameters. However, the implicit assumption is that the parameter being adjusted is the source of error in the model. Otherwise, the model results may be worse than before the adjustment.

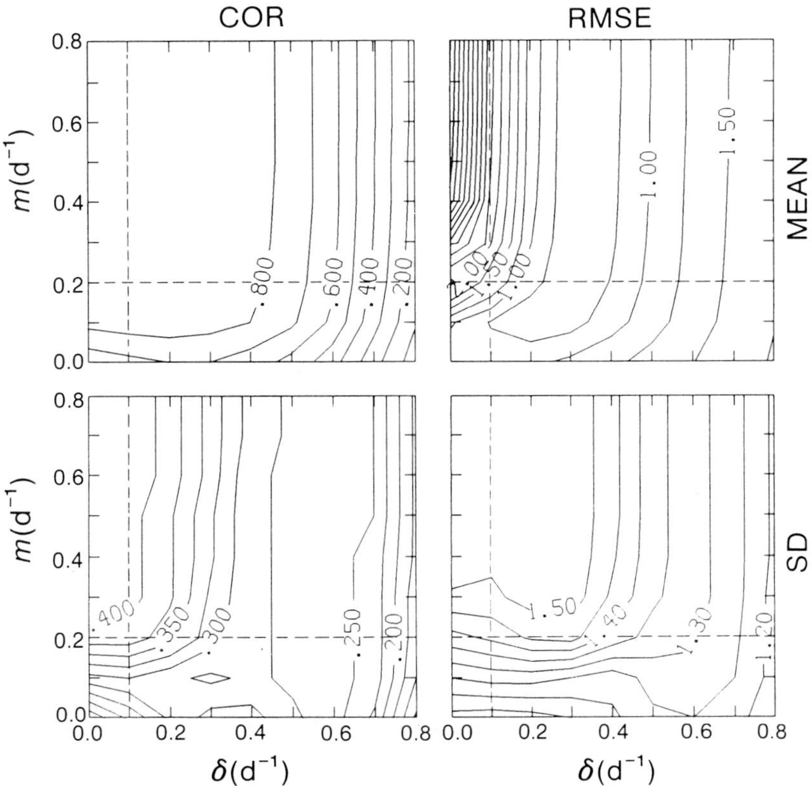

Figure 5. Changes in the correlation coefficient (COR) and root mean square error (RMSE) for changes in the phytoplankton loss rate, δ, and zooplankton loss rate, m. (a) and (b) for the correlation coefficient, and (b) and (d) for the root mean square error. (a) and (b) for the time mean, and (c) and (d) for the standard deviation. Dotted lines indicated the standard values for the parameters used in the physical-biological model. Redrawn from Ishizaka (1990 b).

5. Assimilation of Ocean Color Data into a Physical-Biological Model

Assimilation of ocean color data into a physical-biological model is the logical extension of the above analyses. Assimilation is defined as replacement of simulated chlorophyll fields with the chlorophyll fields derived from the CZCS. The intent is to improve the accuracy of the phytoplankton distributions, which are used for estimating the across-shelf flux of phytoplankton carbon. Ishizaka (1990 c) presents some simple approaches for assimilation of CZCS-chlorophyll fields into the physical-biological model developed for the southeastern U.S. continental shelf.

In the data assimilation procedure (figure 6), the chlorophyll fields are replaced with CZCS-derived chlorophyll concentrations. However the other components of the model ecosystem, *e.g.*, nitrogen, zooplankton and detritus, also need to be updated to be consistent with the new chlorophyll fields. Three methods were tested to obtain new nitrogen, zooplankton and detritus fields

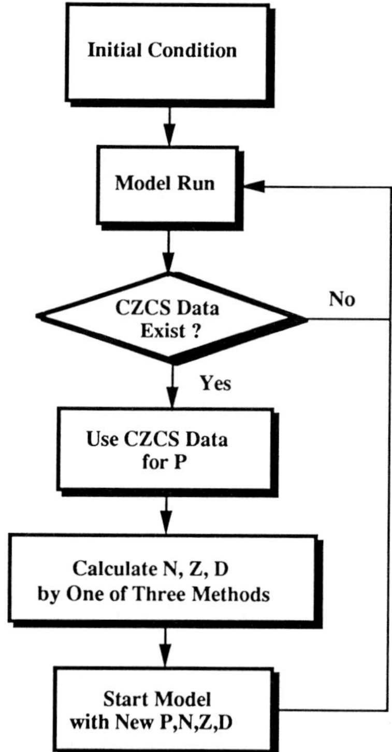

Figure 6. Schematic of the procedure used for assimilation of CZCS data into the physical-biological model of southeastern U.S. continental shelf ecosystem model. Phytoplankton, nutrient, zooplankton and detritus fields are indicated as P, N, Z and D, respectively.

(Ishizaka, 1990 c). The first method assumed that the zooplankton and nitrogen remained constant at the values used for the initial conditions. The detritus component was calculated from the difference between the total nitrogen estimated from temperature and the total of the P-N-Z nitrogen. The second used the simulated nutrient and zooplankton concentrations, while the detrital component was calculated as in the first method. The third assumed that a ratio among nutrient, zooplankton and detritus existed. The new N-Z-D distributions were calculated from the difference between the total nitrogen

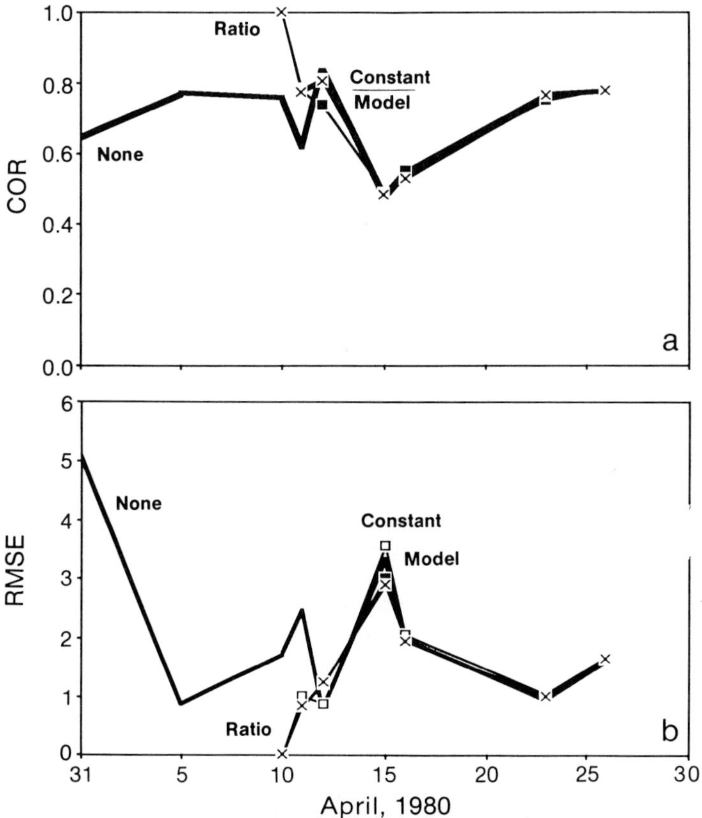

Figure 7. Time variations in the correlation coefficient and root mean square error. The heavy line indicates the non-upgraded model (none). Other lines indicate the results obtained from the model that was upgraded with CZCS data from April 10, 1980. Those results that used constant nutrient and zooplankton distributions are indicated by the white square (constant). Results obtained with the model nutrient and zooplankton distributions are indicated by the solid square (model). Results obtained using a ratio of nutrient, zooplankton and detritus are indicated by the cross (ratio). Redrawn from Ishizaka (1990 c).

estimated from temperature and the phytoplankton concentration obtained from the CZCS.

All methods gave an improvement in model performance in that the simulated fields matched better the CZCS fields. However, there was no significant difference between the results of the three methods. Within a few model days following data assimilation, the error in the simulated fields increased and reached the level that was obtained from the model with no data assimilation (figure 7). Ishizaka (1990 c) also tested the assimilation of partial data fields and found that the improvement in the simulated distribution was intermediate between the full assimilation and no assimilation cases. In general, this study showed that ocean color data can be assimilated into physical-biological models and that assimilation of these data improve the model-derived phytoplankton fields.

6. Calculation of Processes

McClain *et al.* (1990) presented another approach for using ocean color data and physical-biological models to estimate the contributions of physical and biological processes to the time change of chlorophyll distributions. They estimated the space and time derivatives in equation (3) from a sequence of CZCS images from the southeastern U.S. continental shelf. These calculations used current meter measurements to specify the advective velocities and the horizontal diffusion coefficients. The contribution of the biological and vertical processes was inferred by requiring that the terms in equation 3 balance. A similar set of calculations was done using the Eulerian model. Both approaches confirmed the importance of advective processes in determining chlorophyll distribution patterns on the southeastern U.S. continental shelf. Differences in the magnitude and direction of the terms computed from the model and observations showed the sensitivity of these calculations to the errors and assumptions underlying the two methods. An additional calculation consisted of comparing the distribution of the biological-vertical-diffusive term with the same quantity estimated from the Lagrangian calculations described in section 3.

7. Discussion

We have described five approaches for using ocean color data with physical-biological models. The first approach is a comparison of patterns in ocean color data with those in temperature and flow fields obtained by a circulation model or optimal interpolation of current meter measurements. This approach is simple but it provides insight into possible interactions between the physical environment and patterns in ocean color data. Second, numerical Lagrangian particle tracing experiments allow investigation of the evolution of features in ocean color data that are separated by several days and verification of the movement of specific features. These calculations also have the potential for obtaining the rate of change in chlorophyll concentrations caused by the combination of biological, vertical, and diffusive processes. The third approach is a quantitative statistical comparison between ocean color and model-derived chlorophyll distributions. This provides a means to investigate model sensitivity to a large range of parameter values, and is particulary useful for determining ranges of values for poorly known parameters. The fourth approach is assimilation of ocean color data into a physical-biological model. Use of CZCS data in this way significantly improves the predictive capability of physical-biological model. The final approach uses ocean color data to estimate the terms in the governing equation for the advection and diffusion of a non-conservative quantity.

The above approaches, while promising, are relatively simple. More sophisticated methods for data assimilation are being developed and will be available in the near future (cf. Haidvogel and Robinson, 1989). However, there are a number of problems and uncertainties associated with coupling ocean color data with physical-biological models. For example, the satellite ocean color measurements represent the chlorophyll concentration averaged over one optical depth and do not provide any information on the vertical distribution of chlorophyll. Assimilation of a satellite-derived phytoplankton field into a three-dimensional physical-biological field will require that the subsurface phytoplankton field be known. Also, the horizontal and vertical distributions of other model ecosystem components, that are consistent with the upgrade phytoplankton field, will need to be known. Hurlburt (1986) discussed methods for upgrading an eddy-resolving primitive equation circulation model with altimeter data and suggested procedures for obtaining subsurface pressure distributions from the combination of altimeter data and model results. A similar approach may be feasible for assimilation of ocean color data into three-dimensional physical-biological models.

Sensitivity analyses of the statistical comparisons between CZCS and model distributions showed that unknown parameters in the biological model can be adjusted so that correlations are maximum and errors are minimum. However, while this is useful, many combinations of parameter values may be chosen from the statistical analysis. For example, figure 5 shows that zooplankton loss rate (m) can be any value greater than 0.2 day 1^{-1} as long as phytoplankton loss rate (δ) is 0.3 day^{-1}. This makes it difficult to adjust the unknown biological parameters by simply comparing satellite- and model derived chlorophyll distributions. Additional information, such as that from moored fluorometers (*e.g.* Walsh 1988) and nutrient measurements (Friederich *et al.* 1986) and acoustically-derived zooplankton concentration maps or time series (GLOBEC, 1991) is needed to adjust the biological parameters.

A further complication is that the physical-biological model is relatively sensitive to the physical forcing, which itself may not be perfect. The chlorophyll fields obtained from the Eulerian model described in Ishizaka (1990 b) typically overestimated the CZCS chlorophyll fields. One potential reason for this overestimation is that the simple approach used to estimate the vertical processes overestimated the nutrient input from upwelling. Thus, one possibility is to adjust the physical model first with some other satellite data, such as sea-surface temperature, sea surface evaluation from an infrared satellite and sea-surface elevation from altimetry, and then adjust the biological model with the phytoplankton fields derived from ocean color.

It is difficult to construct a model that covers a wide range of time and space scales. It is important that the models and the data being assimilated into them have compatible length and time scales. For example, McClain *et al.* (1990) showed that estimation of biological-diffusive-vertical processes is

284

sensitive to the length scales used for spatial averaging of the CZCS and model-derived chlorophyll distributions. Careful analysis of the coupling of ocean color data with different time and space scales are interrelated (cf. Powell, 1989). There have been a few attempts to model basin scale phytoplankton distributions (Wroblewski *et al.* 1988, Wroblewski, 1989, Sarmiento *et al.* 1989). However, many unknown factors remain to be resolved before regional models, such as the ones discussed here, can be extended to basin scales (Hofmann, 1991).

Currently extensive studies of the optical characteristics of the upper ocean are underway (cf. Spinrad, 1989). The sophisticated spectral optical models (*e.g.* Gregg and Carder, 1990, Sathyendranath and Platt, 1988) that are now being developed can be incorporated into physical-biological models that use ocean color data. Recent hypotheses about light absorption by phytoplankton affecting upper layer physical dynamics (Lewis *et al.* 1983, Sathyendranath *et al.*, 1991) can be investigated with truly coupled physical-biological models that use ocean color data.

The OCTS, which is scheduled to be launched on the Japanese ADEOS satellite in 1995, has ocean color and temperature sensors. Furthermore, NSCAT, which will give wind speed information, is scheduled for the same satellite. Solar radiation can be estimated from the cloud cover measured by the GMS satellite (Harashima, 1991). Thus, this mission should provide measurements of physical and biological responses to wind events. This will provide an opportunity to use physical and biological models with multiple satellite sensor data (figure 8) to obtain estimates of primary production.

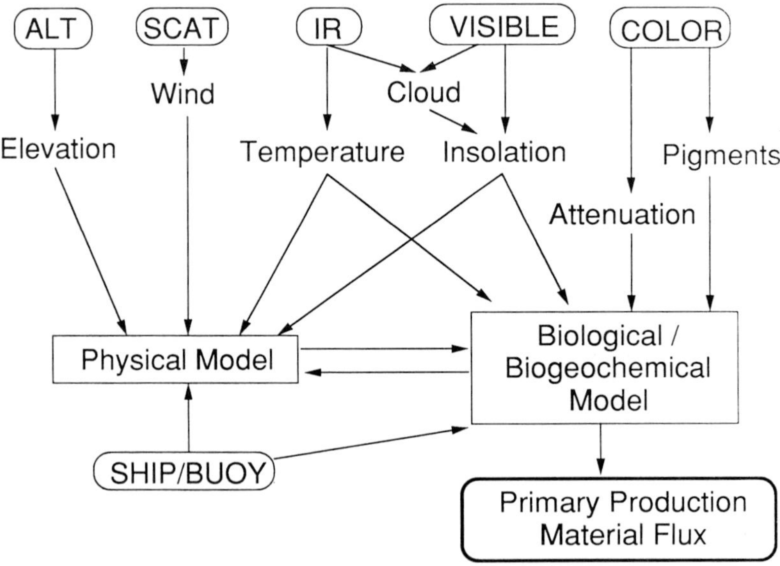

Figure 8. Possible future coupling of multiple satellite sensor data with physical and biological models.

Acknowledgements

This study was mainly funded by National Astronaut and Space Agency of the Unite States. It was partially supported by New Energy Developmental Organization and Environmental Agency of Japan.

References

Abott, M. R. and Chelton, D. B. (1992) 'Advances in passive remote sensing of the ocean', Revieuws of Geophysics, 571-589.

Abbott, M. R. and Zion, P.M. (1985) 'Satellite observations of phytoplankton variability during an upwelling event', Continental Shelf Research 4, 661-680.

Emery, W. J., Thomas, A. C., Collins, M. J., Crawford, W. R., and Mackas, D. L. (1986) 'An objective method for computing advective surface velocities from sequential infrared satellite images', Journal of Geophysical Research 91, 12865-12878.

Esaias, W.E., Feldman, G. C., McClain, C. R., and Elrod, J. A. (1986) 'Monthly satellite-derived phytoplankton pigment distribution for the North Atlantic ocean basin', EOS Transactions AGU 67, 835.

Eslinger, D. L. (1990) 'The effects of convective and wind-driven mixing on springtime phytoplankton dynamics as simulated by a mixed-layer model', 127 pp., Ph. D Dissertation, Florida State University.

Friederich, G. E., Kelly, P. J., and Codispoti, L. A. (1986) 'An inexpensive moored water sampler for investigating chemical variability', in M. J. Bowman, C. M. Yentsch and W. T. Peterson (eds.), Tidal Mixing and Plankton Dynamics, Springer-Verlag, New York, pp. 463-482.

Garcia, C. A. and Robinson, I. S. (1989) 'Sea surface velocities in shallow seas extracted from sequential Coastal Zone Color Scanner satellite data', Journal of Geophysical Research 94, 12681-12691.

Garrett, C. J. R., Keeley, J. R., and Greenberg, D. A. (1978) 'Tidal mixing versus thermal stratification in the Bay of Fundy and the Gulf of Maine', Atmos. Ocean. 16,403-423.

GLOBEC (1991) 'Initial Science Plan', Joint Oceanographic Institutions, 93 pp.

Gregg, W. W. and Carder, K. L. (1990) 'A simple spectral solar irradiance model for cloudless maritime atmospheres', Limnology and Oceanography 35, 1657-1675.

Haidvogel, D. B. and Robinson, A. (1989) 'Special issue : data assimilation', Dynamics of Atmospheres and Oceans 13, 171-513.

Harashima, A. (1991) 'Remote sensing for modelling of variation in primary production field', in K. Takano (ed.), Oceanography of Asian Marginal Seas, Elsevier Science Publishers, Amsterdam, pp. 75-84.

Hofmann, E. E. (1991) 'How do we generalize coastal models to global scale?', in R. F. C. Mantoura, J. M. Martin and R. Wollast (eds.), Ocean Margin Processes in Global Change, John Wiley & Sons Ltd, pp. 401-417.

Hurlburt, H. E. (1986) 'Dynamic transfer of simulated altimeter data into subsurface information by a numerical ocean model', Journal of Geophysical Research 91, 2372-2400.

Ishizaka, J. (1990 a) 'Coupling of Coastal Zone Color Scanner data to a physical-biological model of the southeastern U.S. continental shelf ecosystem 1. CZCS data description and Lagrangian particle tracing experiments', Journal of Geophysical Research 95, 20167-20181.

Ishizaka, J. (1990 b) 'Coupling of Coastal Zone Color Scanner data to a physical-biological model of the southeastern U.S. continental shelf ecosystem 2. An Eulerian model', Journal of Geophysical Research 95, 20183-20199.

Ishizaka, J. (1990 c) 'Coupling of Coastal Zone Color Scanner data to a physical-biological modle of the southeastern U.S. continental shelf ecosystem 3. Nutrient and phytoplankton fluxes and CZCS data assimilation', Journal of Geophysical Research 95, 20167-20181.

Lee, T. N., Atkinson, L. P., and Legeckis, R. (1981) 'Observations of a Gulf Stream frontal eddy on the Georgia continental shelf, April 1977', Deep-Sea Research 28, 347-378.

Lewis, M. R., Cullen, J. J., and Platt, T. (1983) 'Phytoplankton and thermal structure in the upper ocean: consequences of nonuniformity in chlorophyll profile', Journal of Geophysical Research 88, 2565-2570.

McClain, C. R., Chao, S.-Y., Atkinson, L. P., Blanton, J. O., and Castillejo, F. d. (1986) 'Wind-driven upwelling in the vicinity of Cape Finisterre, Spain', Journal of Geophysical Research 91, 8470-8486.

McClain, C. R., Ishizaka, J., and Hofmann, E. E. (1990) 'Estimation of the processes controlling variability in phytoplankton pigment distributions on the southeastern U. S. continental shelf', Journal of Geophysical Research 95, 20213-20235.

McClain, C. R., Pietrafesa, L. J., and Yoder, J. A. (1984) 'Observation of Gulf Stream-induced and wind-driven upwelling in the Georgia Bight using ocean color and infrared imagery', Journal of Geophysical Research 89, 3705-3723.

Nihoul, J. C. J. (1984) 'Contribution of remote sensing to modelling', in J. C. J. Nihoul (ed.), Remote Sensing of Shelf Sea Hydrodynamics, Elsevier, New York, pp. 25-36.

Powell, T. M. (1989) 'Physical and biological scales of variability in lakes, estauries, and the coastal ocean', in J. Roughgarden, R. M. May and S. A. Levin (eds.), Perspectives in Theoretical Ecology,Princeton Univ. Press, Princeton, N. J., pp. 157-176.

Sarmiento, J., Fasham, M. J. R., Siegenthaler, U., Najjar, R., and Toggweiler, J.R. (1989) 'Models of chemical cycling in the ocean: Progress Report II. Ocean Tracers Laboratory Technical Report No. 6', Princeton University, Princeton, 46 pp.

Sathyendranath, S., Gouveia, A. D., Shetye, S. R., Ravindran, P., and Platt, T. (1991) 'Biological control of surface temperature in the Arabian Sea', Nature 349, 54-56.

Sathyendranath, S. and Platt, T. (1988) 'The spectral irradiance filed at the surface and in the interior of the ocean: A model for applications in oceanography and remote sensing', Journal of Geophysical Research 93, 9270-9280.

Spinrad, R. W. (1989) 'Special issue. Hydrologic Optics', Limnology and Oceanography 34, 1389-1761.

Walsh, J. J. (1988) 'On the nature of Continental Shelves', Academic, San Diego, Calif., 520pp.

Walsh, J. J., Dieterle, D. A., and Esaias, W. E. (1987) 'Satellite detection of phytoplankton export from the Mid-Atlantic Bight during the 1979 spring bloom', Deep-Sea Research 34, 675-703.

Walsh, J. J., Dieterle, D. A., and Meyers, M. A. (1988) 'A simulation analysis of the fate of phytoplankton within the Mid-Atlantic Bight', Continental Shelf Research 8, 757-787.

Woods, J. D. and Onken, R. (1982) 'Diurnal variation and primary production in the ocean-preliminary results of a Lagrangian ensemble model', Journal of Plankton Research 4, 735-756.

Wroblewski, J., Sarmiento, J. L., and Flierl, G. R. (1988) 'An ocean basin scale model of plankton dynamics in the North Atlantic. 1. Solutions of the climatological oceanographic conditions in May', Global Biogeochemical Cycle 2, 199-218.

Wroblewski, J. S. (1989) 'A model of the spring bloom in the North Atlantic and its impact on ocean optics', Limnology and Oceanography 34, 1563-1571.

Yentsch, C. S. (1983) 'Remote sensing of biological substances', in A. P. Cracknell and D. Reidel (eds.), Remote Sensing Applications in Marine Science and Technology, pp. 263-297.

Yentsch, C. S. (1984) 'Satellite representation of features of ocean circulation indicated by CZCS colorimetery', in J. C. J. Nihoul (ed.), Remote Sensing of Shelf Sea Hydrodynamics, Elsevier, New York, pp. 337-354.

Yoder, J. A., Atkinson, L. P., Lee, T. N., Kim, H. H., and McClain, C. R. (1981) 'Role of Gulf Stream frontal eddies in forming phytoplankton patches on the outer southeastern shelf', Limnology and Oceanography 26, 1103-1110.

Yoder, J. A., McClain, C. R., Blanton, J. O., and Oey, J.-Y. (1987) 'Spatial scales in CZCS-chlorophyll imagery of the southeastern U.S. continental shelf', Limnology and Oceanography 32, 929-941.

OCEAN COLOUR IN RELATION TO BIOLOGICAL PATTERNS AND PROCESSES IN A BIOGEOGRAPHICAL PROVINCE

J. PELAEZ-HUDLET
Instituto de Oceanografía Satelital
ICML - UNAM
A.P. 811
Mazatlán, Sinaloa 82000
México

ABSTRACT. Satellite images of ocean colour (CZCS) are used to investigate biological patterns, and associated processes, in the California Current at two different scales. At the mesoscale (hundreds of kilometers, months to years), the major biological patterns, and associated processes, in summer are the following. A meandering jet, entraining high pigment filaments that occur on the inshore side of the jet, and interacting strongly with anticyclonic, low pigment eddies on the offshore side of the jet. The jet does a sharp shoreward turn far offshore Ensenada (about 32°N), and flows onshore generating a zonal front (the Ensenada front). A recurrent, high pigment region occurs in the location of the Southern California Eddy. There is a semipermanent, oligotrophic intrusion of southern and offshore water in the Southern California Bight. In late fall and winter (after the fall transition), "spoked" cyclonic rings of high pigment content can occur far offshore. At a smaller scale (tens of kilometers, days to weeks), biological patterns and processes are investigated in connection with a "red tide" episode that occurred in a coastal area, off southern California. Vertical water motion (coastal upwelling) can introduce nutrients into the euphotic zone. Horizontal advection of high pigment water converging with the coastal boundary (coupled with a buoyant behaviour of cells), can result in accumulation, or concentration, processes that can generate patterns with abnormally high pigment content (*e.g.*, red tides). Conversely, onshore advection of warm (nutrient depleted) water can result in rapid dissipation of biological pattern. Thus, horizontal and vertical water motion (and the associated nutrient supply) appear to be directly responsible for the generation, maintenance, and dissipation of biological patterns observed in the satellite images of ocean colour.

1. Introduction

Recent technological development has given the possibility of observing the oceans from artificial satellites in orbit around the earth. Observations performed from satellites, though often less accurate than *in situ* measurements, have the advantage of synopticity and repetitive coverage. Because of synopticity, satellite observations are very useful to determine spatial patterns at various scales. Because of repetitive coverage, it is possible to assess the temporal changes, or persistence, of the observed patterns.

V. Barale and P.M. Schlittenhardt (eds.),
Ocean Colour: Theory and Applications in a Decade of CZCS Experience, 289–318.
© 1993 *ECSC, EEC, EAEC, Brussels and Luxembourg. Printed in the Netherlands.*

The Coastal Zone Colour Scanner (CZCS) on board the Nimbus-7 satellite was launched in October 1978, with the objective of estimating ocean colour and phytoplankton abundance (Hovis *et al.*, 1980). The CZCS data has greatly increased the possibilities of analysing the patterns of distribution of phytoplankton pigments, both in space and in time. The assessment of phytoplankton heterogeneity, and the identification of its spatial structure constitute important information in biological oceanography, because of the nonuniform distribution of marine organisms, *i.e.*, the so-called patchiness problem.

The present case study, is an example on the use of remote observations (coupled with *in situ* data) for the assessment of biological patterns and processes in a biogeographical province. This study will focus on the California Current region. When analysing patterns, the notion of scale is important. Here, spatial patterns, and their associated temporal variations, will be considered at two different scales. First, the mesoscale patterns (hundreds of kilometers) with temporal variations of months to years will be considered. Then, smaller-scale patterns (tens of kilometers) with temporal variations of days to weeks will be analysed, in connection with a "red tide" episode that occurred in a coastal area, off southern California. Possible mechanisms and processes that may underlie pattern formation, and pattern modification will be examined.

2. Mesoscale Biological Patterns of the California Current

Mesoscale biological patterns of the California Current were observed using three years of CZCS data (Peláez and Guan, 1982, Peláez, 1984, Peláez and McGowan, 1986). In this section, the major patterns observed in those studies will be briefly recalled. Then, the mesoscale patterns will be examined in relation to patterns of sea surface temperature, and quasi-concurrent ocean circulation patterns. Finally, the biological patterns will be considered in relation to interdisciplinary studies of the California Current system, recently appeared in the literature.

The patterns observed in the satellite images of phyhtoplankton pigments (1,260 km on a side) show a high degree of heterogeneity, at scales that range from few kilometers to hundreds of kilometers (figure 1). However, recurrent, or persistent mesoscale patterns have been identified. The major ones are shown in figure 1 (where lighter gray tones correspond to higher phytoplankton pigment concentrations).

1. A zonally oriented, sharp front several hundreds of kilometers long, occurs offshore the city of Ensenada, in northern Baja California. Because of its usual location, it is often referred to as the "Ensenada front". This boundary separates high pigment waters to the north of the front, from oligotrophic waters to the south of it (concentrations can change from more than 1.0 mg chl/m3 to less than 0.1 mg/m3 in a few kilometers distance).

2. A longitudinally oriented, meandering boundary occurs far offshore (100-500 km). The meandering boundary is several hundreds of kilometers long. Its edge looks scalloped, possibly due to shear between the neighbouring water masses. Its scalloped, meandering character indicates that this boundary is located in a highly dynamic region.

3. The inshore side of the longitudinal boundary is characterized by tongues or filaments of high pigment concentration (more than 1.0 mg/m3) that can extend several hundreds of kilometers offshore.

4. The offshore side of the longitudinal boundary is charactertized by large anticyclonic eddies (100-200 km in diameter) of low pigment concentration (0.1- 0.5 mg/m3) lined up in the longitudinal direction. Figure 1 suggests that there is a strong interaction among the inshore filaments, the meandering boundary, and the offshore eddies.

5. A large region (about 300 by 100 km) of low phytoplankton pigment concentration (0.1-0.5 mg/m3) characterizes the Southern California Bight. This area is usually continuous with the oligotrophic region to the south of the Ensenada front.

6. A large region (about 200 by 100 km) of high pigment content (higher than 1.0 mg/m3) occurs immediately offshore of the Southern California Bight, in the location of the Southern California Eddy (about 150 km offshore San Diego, in figure 1).

These features are recurrent biological patterns of the California Current system. In addition, the mesoscale patterns of phytoplankton pigments display a remarkable continuity throughout a year. The patterns are strong and distinct in spring and summer, weaken through the fall (except for a slight intensification in October), and become weakest and poorly defined in late fall-early winter (Peláez, 1984, Peláez and McGowan, 1986). Furthermore, the patterns for a given season tend to reappear from one year to another (but see Fiedler 1984, and Strub *et al.* 1990 for important changes during El Niño years). Such recurrency is significant because the patterns change, and some disappear, within a given year.

Summarizing, (1) there are recurrent, mesoscale, biological patterns in the California Current system, (2) these patterns may change and eventually disappear, but (3) very similar patterns will reform later in time. The existence of recurrent patterns implies the existence of driving forces responsible for their generation. In what follows, possible mechanisms and processes related to the generation and maintenance of the observed patterns will be examined.

3. Processes in the California Current Region

In summer, the California Current is not light limited and the mixed layer is shallower than the critical depth allowing for net growth of phytopklankton populations. Therefore, in the region covered by figure 1, in summer, the limiting factor for phytoplankton growth is nutrient concentration (Eppley, 1979).

Remotely sensed ocean colour represents only a shallow surface layer of the ocean. Thus, the question arises: How representative is figure 1 (obtained on 15 June 1981) in terms of the phytoplankton pigment content of the entire water column? *In situ* chlorophyll measurements of this region were obtained during the California Cooperative Oceanic Fisheries Investigations (CalCOFI) cruises of May-June 1981. These measurements showed a highly significant correlation ($r = 0.86$, $p < 0.01$) between chlorophyll concentrations at the surface, and concentrations integrated throughout the euphotic zone (Hayward and Venrick, 1982, Peláez and Guan, 1982). It should be stressed,

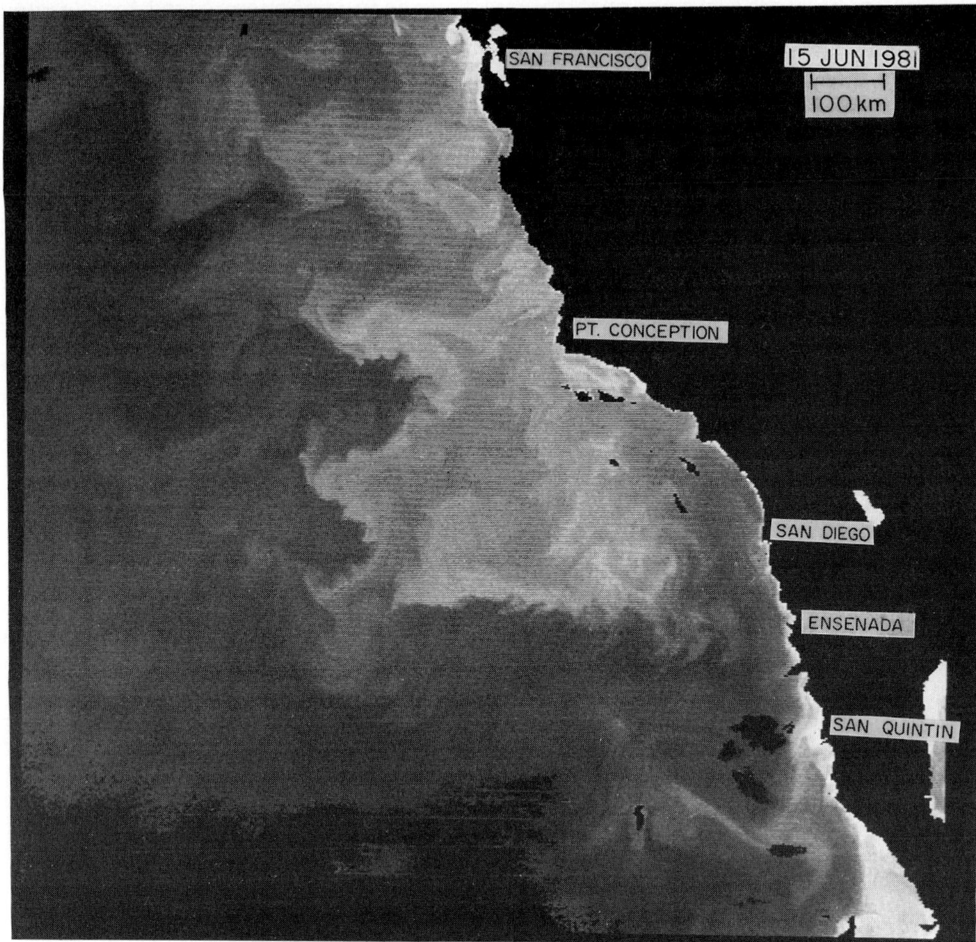

Figure 1. Phytoplankton pigment image off the Californias, near noon 15 June 1981, derived from the Nimbus-7 CZCS imagery. Lighter gray tones correspond to higher pigment concentrations. The range of values for shipboard measured, extracted chlorophyll at 10 m, as sampled over this area 18 May-12 June 1981 was 0.06 to 8 mg/m^3 (Scripps Institution of Oceanography Data Report 85-12). Note the zonally oriented, sharp boundary off Ensenada, and the large degree of heterogeneity in the phytoplankton distribution patterns. The image is about 1,260 km on a side (from Peláez and McGowan, 1986).

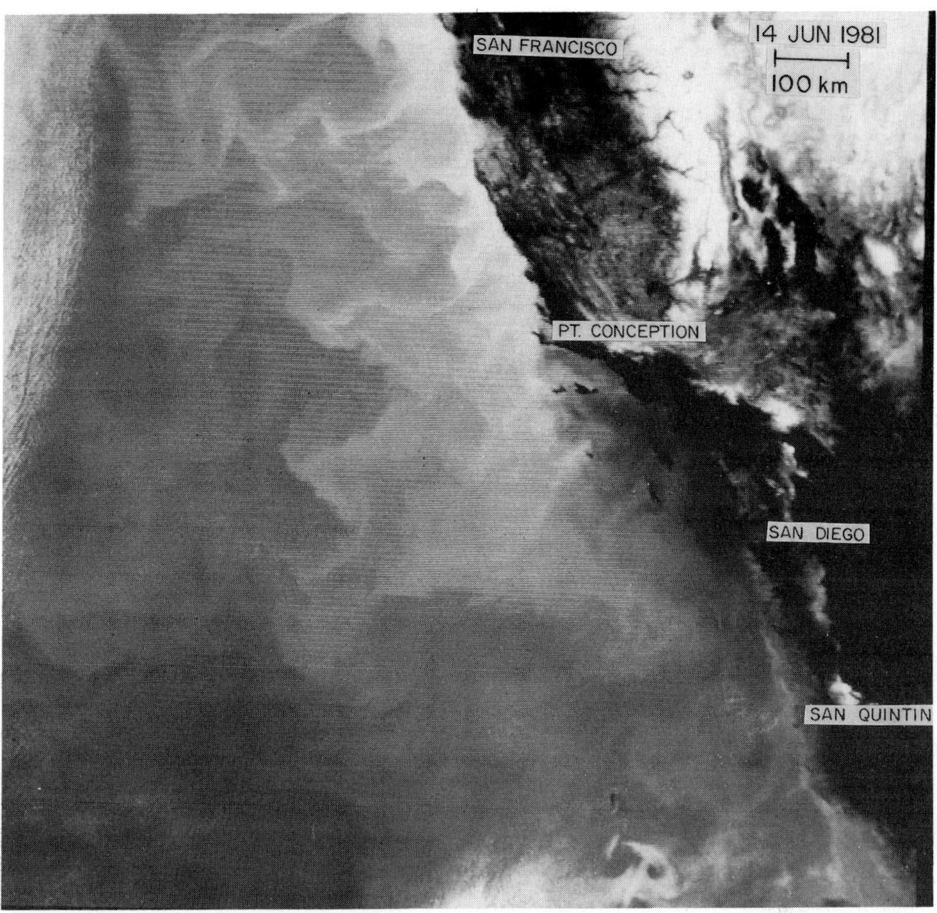

Figure 2. Sea surface temperature image off the Californias, obtained from data collected by the Advanced Very High Resolution Radiometer (AVHRR) of the NOAA-6 satellite, on the evening of 14 June 1981. Lighter gray tones represent cooler sea surface temperatures (puffy white features in the bottom, and upper left of the image are clouds). Temperature resolution is about 0.12°C. Figures 1 and 2 have been registered to the same earth location, but there are about 16 h difference in the data collection times. Note the striking similarity between the phytoplankton pigment patterns and the sea surface temperature patterns in figures 1 and 2, respectively (from Peláez and McGowan, 1986).

however, that this is valid for the California Current in late spring 1981, but it is not a result of general validity (see Mueller and Lange, 1989, for a discussion of ocean colour in relation to vertical profiles of bio-optical properties in this region). On the other hand, such a result gives more meaning to any relation between the biological patterns in figure 1, and possible mechanisms or processes related to their generation and maintenance.

3.1 SEA SURFACE TEMPERATURE

Figure 2 shows a satellite image of sea surface temperature obtained by the Advanced Very High Resolution Radiometer (AVHRR) of the NOAA-6 satellite (lighter gray tones correspond to cooler sea surface temperatures). Figures 1 and 2 have been registered to the same earth location, but they are separated in time by approximately 16 hours (figure 2 was obtained prior to figure 1). It is remarkable that many of the patterns in figures 1 and 2 are almost identical (including some minor details), considering that figure 1 shows biological patterns (phytoplankton pigment concentrations) and figure 2 shows physical patterns (sea surface temperatures). The similarity is better for the larger- scale patterns in offshore waters, than for the smaller patterns in coastal waters.

The clear correspondence between the mesoscale patterns of phytoplankton pigments and those of sea surface temperature, implies a strong cause-and-effect relationship, with the larger-scale population biology being driven by the physical processes responsible for the sea surface temperature patterns. This could come about in at least two related ways.

First, various studies have shown a relationship between temperature and nutrient content of the waters (Zentara and Kamykowski, 1977, Silva, 1982, Traganza *et al.*, 1983, Abbott and Zion, 1985, Kamykowski and Zentara, 1986, Thomas and Emery, 1988). Thus, the sea surface temperature patterns would indicate nutrient input, and the mesoscale biological patterns in figure 1 would be driven by the physical processes responsible for the supply of nutrients.

Second, several studies have shown that the sea surface temperature patterns observed in the satellite images are closely related to ocean circulation (Bernstein *et al.*, 1977, Van Woert, 1982, Vastano and Reid, 1985, Emery *et al.*, 1986, Svejkovsky, 1988, Kelly, 1989, Wahl and Simpson, 1990).

3.2 OCEAN CIRCULATION

Hydrographic measurements of the California Current system obtained during the CalCOFI cruises of May-June 1981, allowed a direct comparison of ocean circulation with the patterns of phytoplankton pigments. Figure 3 shows the isopleths of dynamic height (0-500 db) superimposed on the patterns of phytoplankton pigments of figure 1. For a geostrophic flow (which is a good approximation in this region), water motion occurs along lines of equal dynamic height, with stronger flow associated with more closely spaced isopleths.

Figure 3 shows that there is a strong correspondence between the mesoscale patterns of phytoplankton pigments and the flow of the current. The Ensenada front is related to a shoreward sweep of the California Current. The

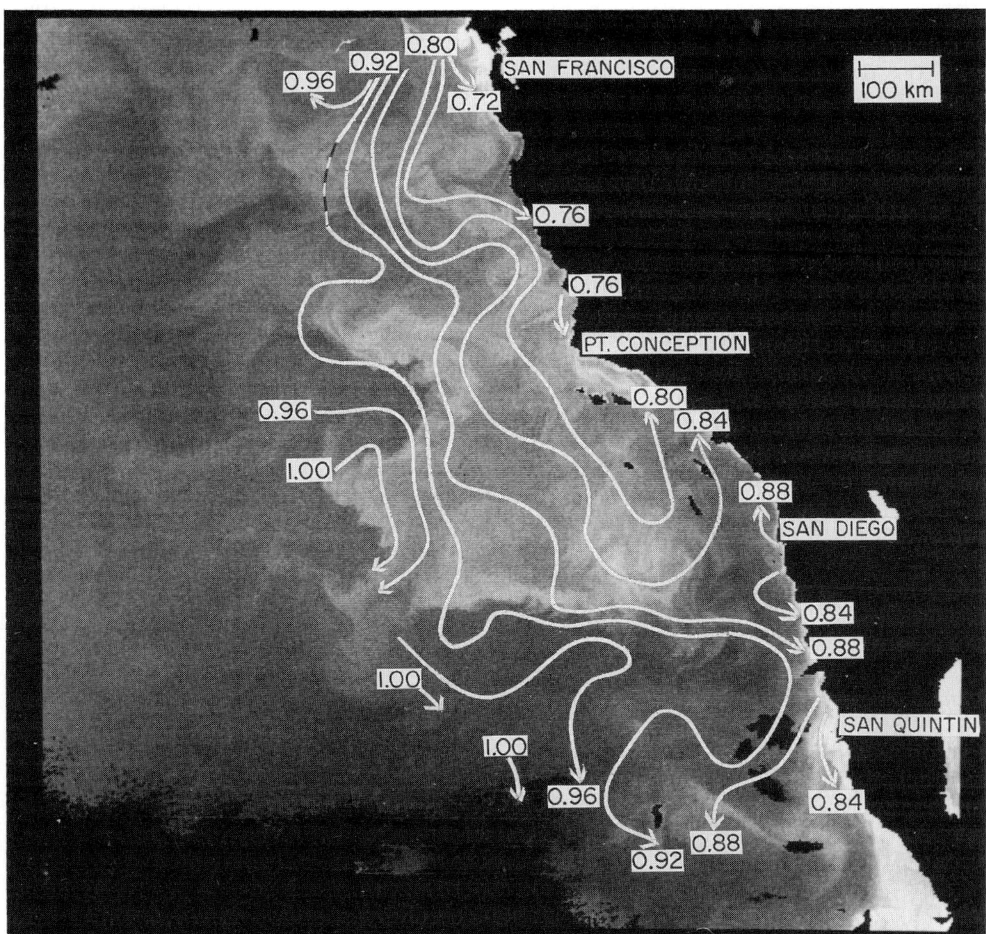

Figure 3. Surface dynamic height in dynamic meters (contour is 0.04 dyn m) relative to 500 dbar obtained from shipboard data (18 May-12 June: Scripps Institution of Oceanography Data Report 85-12), superimposed on the phytoplankton pigment image of figure 1. Water motion occurs along streamlines. Stronger flow is associated with closer streamlines. Despite the very different data collection times and spatial resolutions of satellite and shipboard measurements (see text), there is very good correspondence between the mesoscale patterns of phytoplankton pigments and the flow of the current (modified from Peláez and McGowan, 1986).

longitudinal boundary is associated with meanders of the region of strong flow (the jet) of the California Current. The oligotrophic intrusion in the Southern California Bight (just off San Diego) is related to oligotrophic, offshore water flowing towards the coast, off northern Baja California. The high pigment region about 150 km off San Diego, occurs in a region of eddy-like circulation.

Shifts between the two distributions may be due to different data collection times (25 days for the ship data, less than four minutes for the satellite data). Very different spatial resolutions (point stations tens of kilometers apart for the ship data, contiguous squares about 1 km^2 for the satellite data) can also account for some of the differences, especially in the smaller scale patterns. But the overall resemblance of the mesoscale biological patterns with the flow of the California Current is obvious.

Thus, figures 2 and 3 show that ocean circulation and nutrient supply are very tightly coupled with the mesoscale biological patterns observed in the satellite images.

In what follows, the various mesoscale patterns in figure 1 will be considered in relation to recent investigations of the California Current system performed using *in situ* data.

3.3 THE ZONAL FRONT (OR ENSENADA FRONT)

The three-dimensional circulation of this pattern, as occurred in July 1985, was studied in detail by Niiler *et al.* (1989). The front was present in 1985 demonstrating that it is indeed recurrent. However, the satellite image in Niiler *et al.* shows that the Ensenada front was weaker in 1985 than in 1981. Niiler *et al.* performed hydrographic (CTD and XBT casts), lagrangian particle drift (satellite-tracked mixed layer drifters), and wind measurements. Their major conclusions are the following.

A 50 km wide onshore-flowing "filament" of cold and low salinity water was observed. The flow in the filament is surface intensified. Patterns in the mesoscale horizontal circulation have a vertical length scale of at least the sampling depth (300 m). Near surface divergence estimations show that some areas are upwelling (like the core of the filament) and that others are downwelling. At 50-100 m depth, tracers indicate a general sinking of subarctic water. The authors suggest that such sinking, along with horizontal mixing and surface heating, may be responsible for the abrupt surface disappearance of the cold and pigment rich water. No significant signature of a wind driven circulation was apparent. Thomas and Strub (1990) also find little relation between wind forcing and strength, or position, of the Ensenada front.

The shoreward component of the flow of the California Current, at about 32°N, is also described by Lynn and Simpson (1987), who used 28 years (1950-1978) of CalCOFI data to study the large-scale patterns of the California Current system. Pares-Sierra and O'Brien (1989) used a reduced gravity model to study the dynamics of the California Current sytem. Their model consistenly reproduces the shoreward bend, at about 32°N, of the southeastward-flowing waters of the California Current.

Therefore, the Ensenada front in figure 1 is associated with the onshore sweep of the California Current, at about 32°N. This feature of the current is now clearly established (Lynn and Simpson, 1987, Niiler *et al.*, 1989, Pares-Sierra and O'Brien, 1989). Ocean circulation is, therefore, directly responsible for the generation and maintenance of the sharp biological front off Ensenada.

3.4 LONGITUDINAL BOUNDARY

Figure 1 suggests that there is a strong interaction among the longitudinal boundary, the high pigment filaments inshore of the boundary, and the low pigment anticyclonic eddies offshore of the boundary. Therefore, these patterns should be considered together.

Several multi-institutional efforts have been done to study some of these patterns. The main problem is that the region is too vast: The western boundary of the California Current system occurs about 900 km offshore (Lynn and Simpson, 1987), and dense sampling (both in space and in time) is needed to characterize these patterns and their changes.

Major programmes that have recently studied the California Current system are CalCOFI (Lynn and Simpson, 1987, Chelton, 1984, Hickey, 1979, Wyllie, 1966); CODE (Coastal Ocean Dynamics Experiment, 1987); OPTOMA (Ocean Prediction Through Observations, Modeling, and Analysis; Rienecker and Mooers, 1989, Mooers and Robinson, 1984); and CTZ (Coastal Transition Zone programme, 1991).

Results obtained by the CTZ group show that the longitudinal boundary in figure 1 coincides with the meandering jet of the California Current (Strub *et al.* 1991). The results of Kosro *et al.* (1991) and Huyer *et al.* (1991) show that the longitudinal boundary occurs in the area of rapid change in dynamic height and strong equatorward flow, as in figure 3. Futhermore, Hood *et al.* (1990 and 1991), Mackas *et al.* (1991), and Chavez *et al.* (1991) show that the meandering jet is a physical and also a biological boundary, with different groups of species on either side of the jet. The meandering jet itself carries low salinity, low nutrient water of northern origin (Hayward and Mantyla, 1990, Kosro *et al.* 1991, Huyer *et al.*, 1991, Chavez *et al.*, 1991) and species of northern affinity (Mackas, 1991). Thus, the longitudinal, meandering boundary in figure 1 corresponds to the region of strongest flow (*i.e.*, the center of the alongshore jet) of the California Current.

3.5 INSHORE FILAMENTS

Because of large spatial and temporal variability of filaments, full awareness of these patterns came only with the advent of satellite imagery (Bernstein *et al.* 1977). The CODE programme showed that shelf waters could become entrained into a filament whose core penetrated to a depth of about 100 m, far beyond the surface mixed layer (see Journal of Geophysical Research Vol. 92, No. C2, February 15, 1987 for a full set of reports on CODE). Subduction of near surface waters in the filament was shown by Flament *et al.* (1985). Kelly (1985), and Abbott and Zion (1987) analysed satellite data of sea surface temperature and phytoplanktkon pigment concentrations, respectively, in filaments and found evidence for wind forcing effects.

The study of filaments represents the main focus of the Coastal Transition Zone programme (CTZ Group, 1988, Brink and Cowles, 1991). A full set of papers on this subject appeared recently in the Journal of Geophysical Research (Vol. 96, No. C8, August 15, 1991). The results of Hood *et al.* (1990 and 1991), and Chavez *et al.* (1991) show that dynamic processes asssociated with the meandering jet (a geostrophic adjustment of the flow) result in significant biological enrichment (high surface nitrate, high chlorophyll, and high rates of primary production) on the low dynamic height (cold or inshore) side of the jet. The coldest, most chlorophyll-rich water in these tongues and filaments is located in relatively slow moving water, but entrainment of coastal water into the jet is apparent. Subduction of surface waters of the filament during offshore transport is also reported by Kadko *et al.* (1991), and by Washburn *et al.* (1991).

An instrumented drifter was deployed in a filament by Abbott *et al.* (1990). The authors report gradual depletion of nutrients by a growing phytoplankton population during two days of offshore advection. Then, the drifter apparently lost track of the original water parcel (possibly due to subduction of surface waters). Abbott and Barksdale (1991) used a 4-year CZCS data set to study phytoplankton pigment patterns and wind forcing in relation to filaments. They suggest that the wind stress curl may force the variability of phytoplankton pigment filaments.

The conceptual model of filaments advocated by the Coastal Transition Zone Group (CTZ Group, 1988, Strub *et al.*, 1991) consists of a continuous southward jet, meandering offshore and onshore. During its onshore excursions, the jet may entrain coastally upwelled water and create filaments of cold, rich water which extend offshore in the next meander. Here, the jet is the primary structure and the source of energy. Alternate conceptual models are "squirts", and mesoscale eddies. Squirts are jets transporting coastally upwelled water to the deep ocean, with broader and weaker return flow between squirts, or in subsurface flow. Squirts are generated by nearshore convergences (Stern, 1986, Huyer and Kosro, 1987, Send *et al.*, 1987). The third conceptual model consists of offshore mesoscale eddies drawing recently upwelled water away from the coast, thus creating offshore-flowing filaments of cold, rich water (Mooers and Robinson, 1984, Rienecker and Mooers, 1989, see also Simpson *et al.*, 1984). Here, the source of energy is the offshore eddy field.

The results published by the CTZ programme (Journal of Geophysical Research, Vol. 96 No. C8, 1991) support the model of a meandering jet as primary structure and source of energy. The interpretation that the large-scale filaments observed in the satellite images are squirts, is apparently not suppported by most of the evidence (Strub *et al.*, 1991). The offshore eddy model needs further consideration, because neither the CTZ nor the CODE hydrographic data extend far enough offshore, to determine the role of anticyclonic eddies that may occur offshore of the jet. The offshore OPTOMA surveys have shown a mesoscale eddy field off northern California (Rienecker and Mooers 1989). Off central, southern, and Baja California, recurrent eddies have been documented by various studies (Bernstein *et al.*, 1977, Simpson *et al.*, 1984 and 1986, Haury *et al.*, 1986, Lynn and Simpson, 1987, Simpson and Lynn, 1990).

3.6 OFFSHORE EDDIES

Mesoscale variability has long been recognized as an important component of the total variability of eastern boundary currents (Wooster and Reid, 1963). Bernstein *et al.* (1977) using satellite infrared data concurrently with more traditional measurements showed, for the first time, the complex patterns of filaments, fronts, and eddies that occur in the California Current system. This study focused on a persistent, anticyclonic, mesoscale eddy that occurs southwest of Point Conception, California. This eddy was studied again in

1981, using infrared satellite data in conjuction with concurrent shipboard data. The eddy, about 150 km in diameter, consists of a 75 m thick surface layer, a cold-core region to about 200 m, and a warm-core eddy extending to at least 1,450 m. Significant interaction with the surrounding waters occurs through frontal boundary processes, and lateral entrainment. The surface manifestation, interior dynamics, chemistry, and plankton distributions are described in detail in Simpson *et al.* (1984).

Simpson *et al.* (1986) and Haury *et al.* (1986) extended the analysis of this eddy to include CZCS data, and additional *in situ* measurements. The satellite images of phytoplankton pigments can detect the near-surface chlorophyll structure of the eddy. But the CZCS data do not reflect features deeper than about 25 m of this oligotrophic eddy, including the contribution of the deep chlorophyll maximum to the integrated chlorophyll values. Physical and biological properties show entrainment of waters of nonlocal origin into the upper layers of the eddy, and possibly at depth. Offshore transport of coastal species occurs in the form of large entrained filaments. The eddy appears to have a significant effect on the distribution of both oceanic and nearshore organisms. Analyses of 28 years of CalCOFI data, and of remotely sensed data, show that the Point Conception eddy is a recurrent feature in the offshore California Current system. In addition, mesoscale eddies occur (and recur at preferred locations) within the transition zone of the California Current.

Lynn and Simpson (1987) find three domains in the California Current system: coastal, oceanic, and an intervening transition zone. The transition zone, a broad band centered approximately 200-300 km offshore and parallel to the coast, is coincident with the core of flow of the California Current (Lynn and Simpson, 1987). Single mesoscale eddies, and dipole pairs (also referred to as "mushrooms," "hammerheads," or "T" structures) occur offshore, within the transition zone. The cores of these offshore eddies contain California Undercurrent water (Simspon and Lynn, 1990). These authors suggest that a bathymetrically induced instability of the California Undercurrent is a likely generation mechanism for the offshore eddies. This would explain the recurrence of offshore mesoscale eddies at preferred locations within the transition zone.

A strong interaction between the core of the California Current and the mesoscale eddy field is evident (Lynn and Simpson, 1987). Such interaction could be seen as a large-scale "eddy diffusion," *i.e.*, a quasi-random motion of parcels of water exchanged among a large number of offshore mesoscale eddies populating the transition zone, and parameterized by a large-scale, anisotropic eddy diffusivity (see Mooers and Robinson, 1984, Strub *et al.*, 1991). Or, alternatively, the interaction can be seen as a more orderly process; where a few, large, anticyclonic eddies on the offshore side of the jet, interact with jet and filaments (see figure 1), entraining coastal waters into the upper layers of the eddies.

But the main points in relation to offshore eddies are the following. First, the satellite images of phytoplankton pigments can detect the near-surface structure of an oligotrophic eddy, but do not provide a good estimate of the integrated chlorophyll field. Second, the biological patterns observed on the offshore side of the jet correspond to anticyclonic eddies that can extend to 1,450 m depth. Moreover, that entrainment and frontal-boundary processes result in biological patterns associated with the eddy. Thus, the observed biological patterns appear to be tightly coupled with water motion.

3.7 MODELING OF MEANDERS, FILAMENTS, AND EDDIES

Ikeda and Emery (1984) used a quasi-geostrophic model (simplified bathymetry and vertical resolution of structures) to show that an initially longshore, baroclinic flow (equatorward surface jet over a poleward undercurrent, typical of summer and fall in the California Current) could be unstable, leading to meanders, eddies, and features associated with filaments. They show the importance of coastline irregularity (coastal capes) in perturbing the flow. The mechanism involved is not directly dependent on the wind stress field, although it may depend indirectly, to the extent that the basic California Current-Undercurrent dynamics are dependent on the seasonal wind fields.

Two-layer laboratory experiments in a tank that included rotation, surface stresses (with and without curl), idealized capes and bathymetric ridges are presented by Narimousa and Maxworthy (1989). Capes and ridges appear to be important for the development of meanders and filaments. Furthermore, the results suggest that a positive offshore maximum of wind stress curl produces offshore eddies which draw the coastal water offshore.

Haidvogel *et al.* (1991) using a primitive-equation model confirm that irregularities in the shelf-slope topography are important for enhancing the growth of instabilities. Such topographic irregularities represent points where the filaments are loosely "anchored." The model reproduces the observed downwelling within filaments, and suggests that winds are not directly important in the basic dynamics of filaments.

The model of McCreary *et al.* (1991), however, forced by an annually oscillating, spatially uniform (*i.e.*, without curl) wind stress field, develops a jet which becomes unstable in spring-summer, producing meanders, eddies, and filaments, without coastal capes or bathymetric features. In winter, the system decays to weaker, larger-scale eddies. Thus, the role of the wind in the generation of mesoscale patterns is apparently not yet completely understood. It seems likely that the main role of the wind is in driving the large-scale California Current, which then becomes unstable.

Hence, depending on parameters and forcings, various models can produce meanders, filaments, and eddies similar to the observations. But each model considers only a few of the processes that influence ocean circulation and, therefore, cannot evaluate the relative importance of all the different processes. Thus, the models are still unable to demonstrate conclusively which physical processes are responsible for generating and maintaining the physical patterns observed in the California Current.

3.8 THE OLIGOTROPHIC INTRUSION IN THE BIGHT

The oligotrophic character of the Southern California Bight (Eppley *et al.*, 1978, Eppley *et al.*, 1985) is the result of a periodic intrusion of warm, oligotrophic water, coming from the south and offshore. The intrusion occurs every year, starting to develop far south and offshore in June and intruding into the Southern California Bight in July, on the average. Apparently, a slackening of the southern portion of the California Current (off Baja and southern California) by late spring or early summer progressively allows southern offshore waters to intrude into nearshore southern California, on a yearly basis. This is possibly related to the late spring-early summer lessening of the winds, as they progressively weaken from lower to higher latitudes (see Peláez, 1984, Peláez and McGowan, 1986).

Strub *et al.* (1990), and Thomas and Strub (1990) in their comprehensive analyses of the surface pigment variability of the California Current also characterize the Southern California Bight as an oligotrophic region of very low seasonality, as compared with the rest of the California Current. Furthermore, they point out that off Baja California, there are higher pigment concentrations in summer, which are similar in magnitude and timing to those of central California. This clarifies the fact that the low concentrations in the eastward-flowing current entering the bight are more of an intrusion over the California Current, rather than the southern end of the productive region that occurs to the north of the Ensenada front (Strub *et al.*, 1990). Therefore, the oligotrophic region in the Southern California Bight is strongly related to ocean circulation.

3.9 THE HIGH PIGMENT PATTERN OFF THE BIGHT

The high pigment region located about 150 km off San Diego (figures 1 and 3) occurs in the location of the Southern California Eddy (Sverdrup and Fleming, 1941, Reid *et al.*, 1958). This high pigment pattern (Sverdrup and Allen, 1939, Sargent and Walker, 1948) is geographically fixed over the bathymetric ridge to the southeast of Point Conception (Emery 1960). The western boundary of the bight is characterized by shallow banks and islands, and the centre of the eddy encircles San Nicolas island (figure 3).

The Southern California Eddy develops in June-July (Lynn and Simpson, 1987), concurrently with the appearance of the oligotrophic intrusion flowing north into the Southern California Bight (Peláez and McGowan, 1986). The juxtaposition of a strong northern flow into the bight, and a strong California Current flowing southeast on the offshore side of the bathymtric ridge, results in a strong cyclonic eddy during the summer months (Lynn and Simpson, 1987). Cool water on the offshore side of the ridge (figure 2) suggests that a bathymetrically induced adjustement of the flow occurs in this area, as the sea floor rises rapidly from more than 3,000 m to less than 500 m (Emery, 1960). Note that the correspondence of circulation patterns with pigment patterns is better, than with sea surface temperature patterns (figures 2 and 3). Thus, biological patterns are again strongly related to ocean circulation.

3.10 RINGS

An additional mesoscale biological pattern that may occur in late fall and winter is shown in figure 4. Three cyclonic rings (100-200 km in diameter), with relatively high pigment content (0.3-0.4 mg/m3) occur far offshore (500-600 km off the coast). Three additional rings appear to be forming far offshore Monterey (south of San Francisco), Point Conception, and northern Baja California. These rings have small protuberances or "spokes" on their edges, and a cyclonic sense of rotation. These features are different from the cyclonic cut-off eddies that occur in spring and summer (Strub *et al.*, 1991), which do not have this "spoked" appearance. To my knowledge, California Current

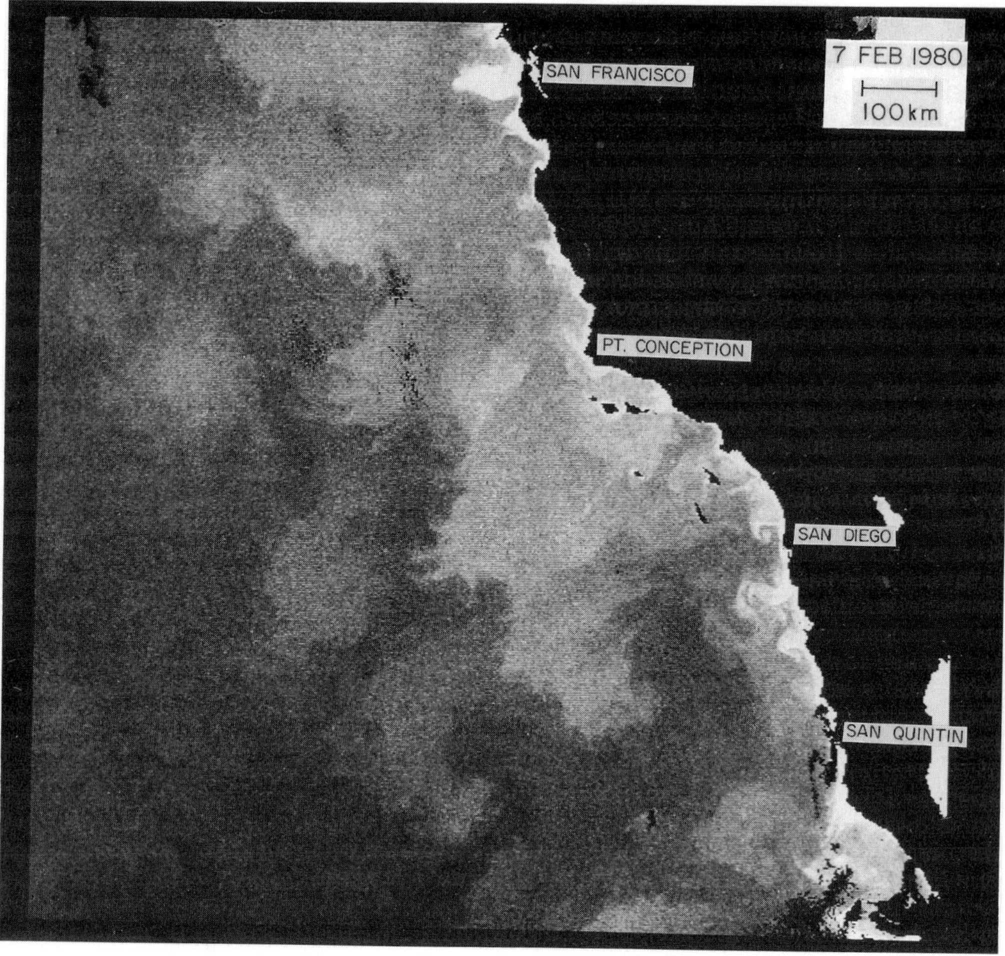

Figure 4. Similar to figure 1, but in winter (7 February 1980). Note three California Current rings far offshore, and three additional rings in the process of formation offshore Monterey (south of San Francisco), Point Conception, and south of San Diego (from Peláez and McGowan, 1986).

rings have not yet been studied, possibly because they occur quite far offshore, and because of harder study conditions in winter for ships (sea conditions) and satellites (sun angle).

Rings have been observed in late fall and winter, after the fall transition (Strub and James, 1988). During this period, the meandering jet becomes less convoluted, and the interaction between inshore filaments and offshore anticyclonic eddies relaxes (Peláez, 1984, Peláez and McGowan, 1986). Concurrently with these changes, the pigment patterns become very weak, and the size of the pigment structures increases. A winter decay to weaker, larger-scale eddies is also found by McCreary *et al.* (1991), when they force their model with an annually oscillating, spatially uniform wind stress. The Coastal Transition Zone group has also observed that the California Current is significantly different during the winter period. Their winter surveys showed no coherent flow pattern and little eddy activity, nor were there any obvious filaments (Kosro *et al.*, 1991, Brink and Cowles, 1991).

Before closing this section, I would like to summarize the previous discussion using the schematic representation of figure 5 (adapted from Peláez, 1984). Major mesoscale biological patterns, and associated processes, in the California Current during summer are: a meandering jet, entraining

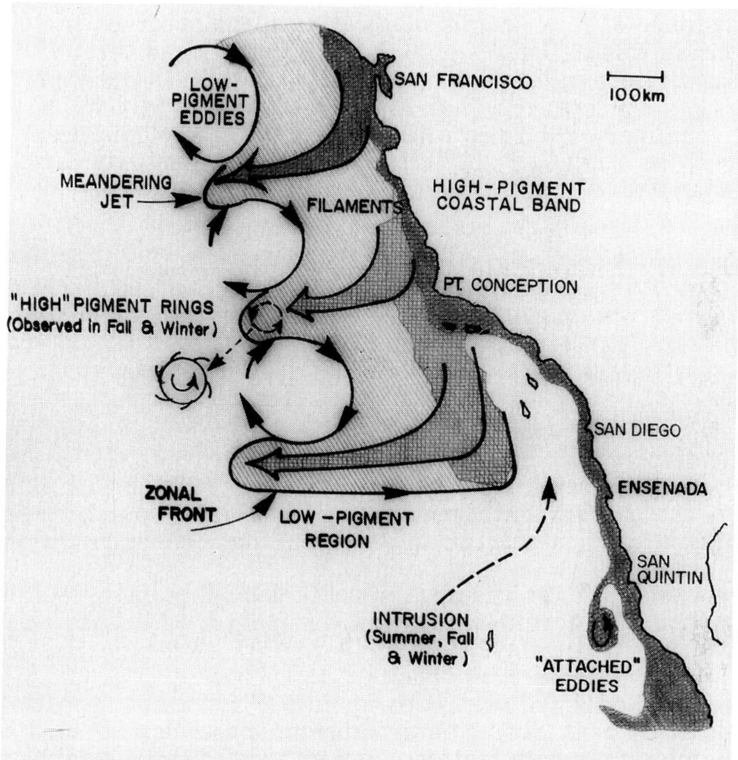

Figure 5. Schematic representation of the major mesoscale biological patterns, and associated processes, in the California Current during summer, considered in the text: A meandering jet, entraining high pigment filaments that occur on the inshore side of the jet, and interacting strongly with anticyclonic, low pigment eddies on the offshore side of the jet. The jet does a sharp shoreward turn far offshore Ensenada (about 32°N), and flows onshore generating a zonal front (the "Ensenada front"). A recurrent, high pigment region occurs in the location of the Southern California Eddy. There is a semipermanent, oligotrophic intrusion of southern and offshore water in the Southern California Bight. In late fall and winter (after the fall transition), "spoked," cyclonic rings of "high" pigment content can occur far offshore (adapted from Peláez, 1984).

high pigment filaments that occur on the inshore side of the jet, and interacting strongly with anticyclonic, low pigment eddies on the offshore side of the jet. The jet does a sharp shoreward turn far offshore Ensenada (about 32°N), and flows onshore generating a zonal front (the "Ensenada front"). A recurrent, high pigment region occurs in the location of the Southern California Eddy. There is a semipermanent, oligotrophic intrusion of southern and offshore water in the Southern California Bight. In late fall and winter (after the fall transition), "spoked," cyclonic rings of "high" pigment content can occur far offshore.

Thus, water motion appears to be directly responsible for the generation, and maintenance of mesoscale biological patterns observed in the satellite images of ocean colour. This is apparently accomplished through turning on and shutting off the availability of nutrient-rich water in the euphotic zone. It should be stressed, however, that these conclusions apply to the California Current in summer, when the limiting factor for phytoplankton growth is nutrient concentration (*i.e.*, light is not limiting, and the mixed layer is shallower than the critical depth).

4. Smaller-Scale Biological Patterns and Processes

Small-scale (few kilometers to tens of kilometers), recurrent patterns with high pigment content (1-10 mg/m3) do exist, especially in association with the coastal region. These patterns change more rapidly than those considered previously.

A narrow coastal band (few kilometers wide), and shallow regions consistently display higher pigment content (Eppley *et al.*, 1978, Eppley *et al.* 1985, Barale and Wittenberg-Fay, 1986, Mullin, 1986, Smith *et al.*, 1988). Apparently, the shallow bathymetry and the coastal boundary force water motion, inducing vertical mixing and the shoaling of the nutricline (Eppley *et al.*, 1979).

In this section, however, the analysis of biological patterns will be done in connection with the development, and dissipation of a "red tide" episode (Cullen *et al.*, 1982, Huntley, 1982, Peláez, 1987) that occurred in a coastal region, off southern California.

I will try to make the following three points. First, that the spatial patterns of a red tide at its peak can be observed from space. Second, that horizontal advection of a high pigment feature converging with the coastal boundary (in conjunction with the swimming behaviour of dinoflagellates), can result in accumulation processes that generate patterns with abnormally high pigment content, *i.e.*, red tides. Third, that advection of oligotrophic (nutrient depleted) water can result in the dissipation of a red tide. Thus, the goal of this section is to show that water motion (horizontal and vertical) plays a very important role in the generation, maintenance, and dissipation of a red tide, and associated biological patterns.

4.1 THE RED TIDE AT ITS PEAK

The first question to ask is, obviously, can the red tide be observed from space? The answer is yes, and figure 6 shows the spatial patterns associated with the red tide episode off southern California, in July 1980. The region affected by

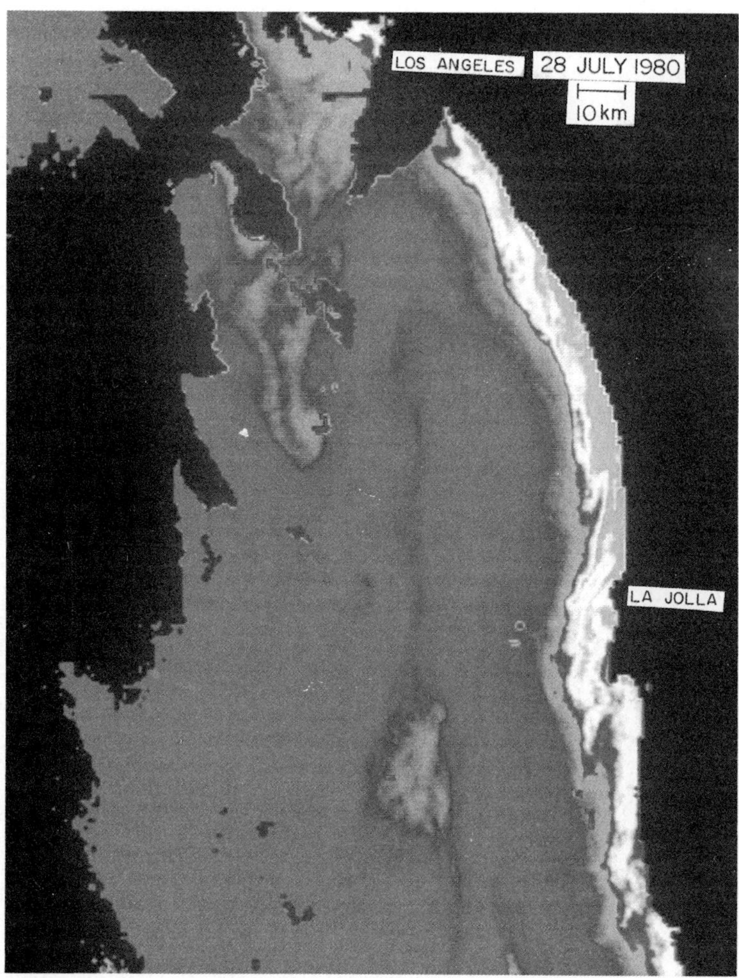

Figure 6. Arbitrary-colour image of phytplankton pigments showing the extent and configuration of the "red tide" when the episode was at its peak (land and clouds are masked in black). The data were obtained by the Nimbus-7 CZCS (ground resolution is about 800 m), near local noon on 28 July 1980. Vertical size of image corresponds to approximately 210 km. Phytoplankton pigments measured *in situ* three hours after the satellite overpass at a station 2-3 km off La Jolla yielded a concentration of 19.9 mg/m³ (Cullen *et al.*, 1982); the satellite estimate at the same location is 15.2 mg/m³. Satellite estimates within the red area reach values up to 30 mg/m³; but a few kilometers offshore, pigment concentrations drop drastically to less than 0.5 mg/m³ in the deep blue region (from Peláez, 1987).

Colorphotograph on p. 367

the red tide consists of a narrow (about 10 km wide), long (about 150 km) coastal band of very high pigment concentration (up to about 30 mg/m3). The dominant organism (80 to 99% of the phytoplankton) was *Gymnodinium flavum* a small (about 35 µm) naked dinoflagellate (Cullen *et al.* 1982). The coastal band of high pigment content has a considerable amount of finer structure, usually in the form of streaks, or sub-bands oriented almost parallel to the coastline. A possible mechanism for the generation of this pattern, through the interaction between phytoplankton and semidiurnal internal tides, was proposed by Kamykowski (1974).

4.2 DEVELOPMENT OF THE RED TIDE

The next question would be, then, is it possible to observe the development of the red tide from space? But before trying to answer this question, it is important to clarify what were the oceanographic conditions prior to the initiation of the red tide episode. Analyses of concurrent infrared and ocean colour satellite data showed that the oligotrophic intrusion of warm waters from the south and offshore into the Southern California Bight was established by late June 1980, prior to the initiation of the red tide (Peláez, 1987). The oligotrophic intrusion is an important component of the red tide episode, because such a regime favours the development of phytoplankton species that can swim vertically searching for nutrients, *e.g.*, dinoflagellates (Eppley and Harrison, 1975, Cullen and Horrigan, 1981).

About two or three weeks after the oligotrophic intrusion was established in the bight, an unusually strong and persistent coastal upwelling event occurred off southern California, and northern Baja California. Sea surface temperatures and salinities at the Scripps Pier (figure 7) showed that the July

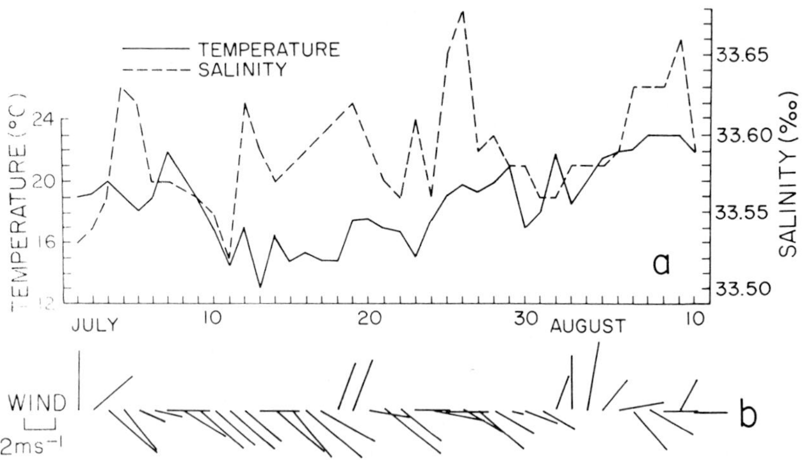

Figure 7. a) Scripps Pier sea surface temperatures and salinities; b) coastal winds measured at Lindbergh Field, about 17 km south of Scripps (Local climatological data, Lindbergh Field, San Diego, California, records on file at the National Climatic Center, Asheville, N.C., 28801). (From Peláez, 1987).

1980 cooling episode had been exceptionally intense (13°C on July 13 is the second coldest day in July, in the 64-year Scripps Pier record); exceptionally rapid (9°C in six days is the strongest cooling event on record); and unusually persistent (three weeks continuously below the July 64-year mean of 19.9°C). Lower temperatures associated with higher salinities (figure 7) indicate that the nutrients needed for the development of the red tide were supplied by the coastal upwelling event. Local winds (figure 7) from the WNW (290° on average) are consistent with a coastal upwelling event in this area (Dorman and Palmer, 1981).

In situ sampling started only on 26 July (Cullen *et al.*, 1982), when the red tide was already fully developed. In the absence of previous *in situ* data, which is often the case for such unexpected events, it was tempting to see if the satellite data of ocean colour could provide any information related to the early stages of the dinoflagellate bloom. Dinoflagellates cannot be identified using only CZCS data. Thus, the analyses had to be restricted to a search of high pigment patterns, and to see how they evolved. One could ask, for example, if early phytoplankton growth, and the associated high pigment patterns, occurred in the region where the red tide was observed at its peak (on 28 July), or elsewhere. That is, if the red tide had mostly developed *in situ*, or if transport processes were significant.

Figure 8 shows that the strongest phytoplankton growth occurred to the south of the region where the red tide was observed at its peak. A large (about 50 km in diameter) "attached," or "anchored" cyclonic eddy of high pigment content (1-2 mg/m^3) developed over the bathymetric promontory associated with the Coronado islands and broader shelf, about 50 km southwest of La Jolla (figure 8a). Figure 8b shows that there is also phytoplankton growth in the region between Los Angeles and La Jolla, where the red tide was observed on 28 July. However, growth in this area is considerably less important than that occurring over the Coronado shelf.

"Attached," or "anchored" eddies can also be seen in the sea surface temperature satellite data (figure 2, for example, shows a weak sea surface temperature signature of a broken "attached" eddy off San Quintín, in figure 1). Apparently, the entrainment of cold, recently upwelled coastal water into the cyclonic eddy results in a spiral patch, due to algal growth in the nutrient-rich filament (figure 8). The "attached," or "anchored" eddies drift with the current as they rotate and develop, remaining connected to the shelf of origin only by a stem, or filament. The results of Haidvogel *et al.* (1991), using a primitive-equation model, also show that irregularities in the coastline and shelf-slope topography represent points where filaments are loosely "anchored."

Figures 8a and 8b show northern transport of the Coronado shelf eddy. In five days, the eddy has displaced north by about 35 km, while being deflected towards the coastline in a clockwise advective motion. This northern transport is consistent with the large-scale oligotrophic intrusion, and typical northern flow in the Southern California Bight during this time of the year (Wyllie, 1966, Tsuchiya, 1980, Lynn and Simpson, 1987).

Holmes *et al.* (1967), and Eppley and Harrison (1975) have pointed out that some mechanism of physically concentrating dinoflagellates at the surface must operate in discoloured water blooms, because the nutrient content of the entire water column is usually insufficient to account for the nutrient content in the cells of the red tide population. Figures 8a, 8b, and 6 suggest that

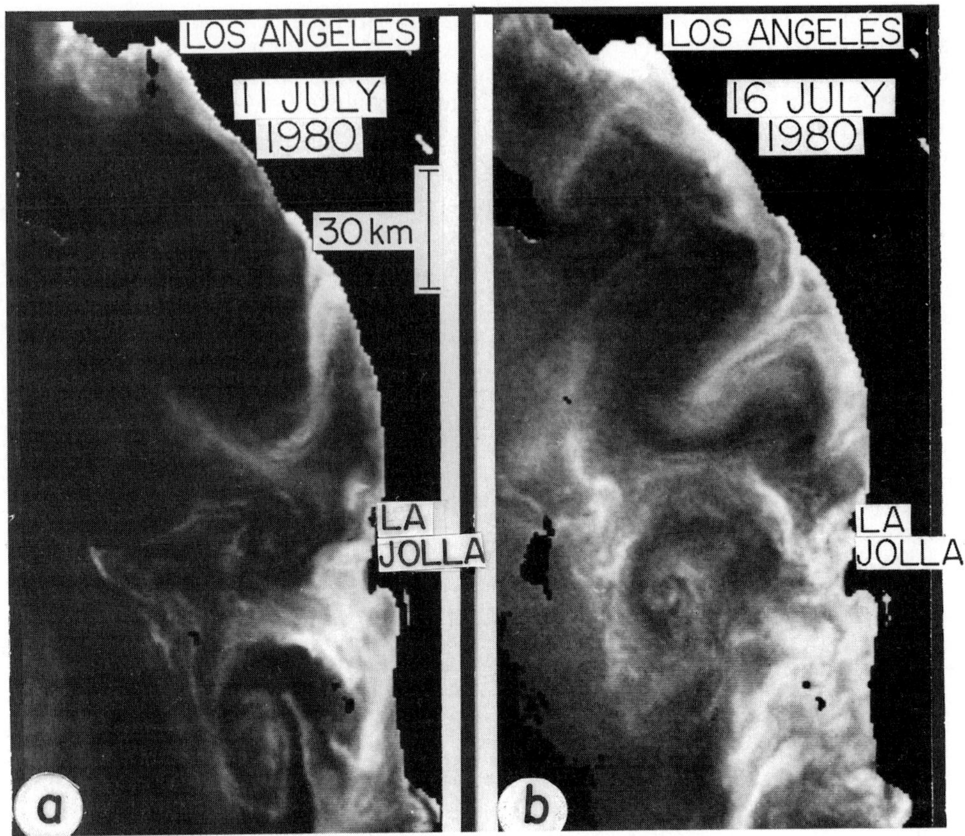

Figure 8. Phytoplankton pigment images off southern California derived from Nimbus-7 CZCS data. Vertical size of images corresponds to approximately 210 km. Lighter gray shades represent higher pigment concentrations (land and clouds are masked in black). Pigment concentrations range from about 0.1 to about 6 mg/m^3. Successive aspects of the "red tide" episode: a) early algal development; b) intense algal growth. See text for further details (modified from Peláez, 1987).

progressive northern transport and clockwise advection of the Coronado shelf eddy converging with the coast, can provide the physical component of a mechanism which concentrates algal cells in a narrow band against the coastline.

An additional component is needed to keep the cells concentrating in a surface layer against the coastline, as the surface water moves north and some of it is forced into deeper layers by a downwelling water mass converging with the coastal boundary. This component is biological. It is provided by the ability of dinoflagellates to swim vertically (Eppley and

Harrison, 1975, Cullen and Horrigan, 1981), *i.e.*, a buoyant behaviour of dinoflagellates opposing entrainment into deeper, darker waters.

The combined effect of physical northern transport of the eddy, and the behavioural ability of dinoflagellates to swim vertically opposing entrainment into darker layers can result in the accumulation, or concentration, of dinoflagellates into a narrow surface layer adjacent to the coastline.

Thus, the very high pigment concentrations (up to 30 mg/m^3) apparently result from a combination of *in situ* growth, and transport of an "attached" eddy by ocean circulation into the coastal area where the red tide was observed.

4.3 DISSIPATION OF THE RED TIDE

The red tide "disappeared" rapidly by the end of July-early August. However, biological processes did not appear to be responsible for the termination of the red tide. Chemical composition of *Gymnodinium flavum* did not indicate progressive nutrient stress (Cullen *et al.*, 1982). Grazing experiments and gut content analyses did not show significant grazing on *Gymnodinium flavum* (Huntley, 1982). Water analyses, however, showed a drastic reduction of nitrate availability by the end of July (Cullen *et al.*, 1982). These results suggested that physical processes were responsible for the dissipation of the red tide.

Satellite infrared data corresponding to the last week of July show an intensification, or surge, of the warm water intrusion upon a weakening coastal upwelling (figure 8a). Higher temperatures associated with higher salinities at the Scripps Pier in late July and early August (figure 7), also indicate the presence of more southern and offshore waters by the coast. Note, however, that warming is not the reverse of cooling (figure 7). Cooling lasted only six days. Slow relaxation of the unusually cold temperatures, and more or less progressive warming lasted at least four weeks. Furthermore, the processes responsible for cooling and warming, as well as their scales, are quite different. Cooling was associated with a coastal upwelling event with spatial scales of tens of kilometers, and time scales of days or weeks. Warming was associated with the warm water, oligotrophic intrusion with spatial scales of hundreds of kilometers, and time scales of months.

The satellite data of sea surface temperature show the convergence towards the coast of the warm water mass (darker area in figure 8a) in late July. Initially, the convergence process appeared to concentrate the motile dinoflagellates in a surface layer against the coastline (figure 6), confirming the existence of an accumulation mechanism. Further onshore advection of this large warm water mass, however, changed the coastal regime to oligotrophic, making nutrients suddenly unavailable to the dinoflagellates (a drastic change to warm, and nutrient-depleted waters was observed *in situ* near La Jolla on July 30, 1980, by Cullen *et al.*, 1982).

By early August, additional onshore movement of the warm water mass results in its progressive collapse against the southern California coastline (figure 8b) under apparent coastal downwelling (Winant, 1980). Onshore movement of this warm water mass washed ashore part of the dinoflagellate population, and dissipated most of the remaining cells (Peláez, 1987; see also Cullen *et al.*, 1982). Therefore, rapid onshore advection of a warm, nutrient-

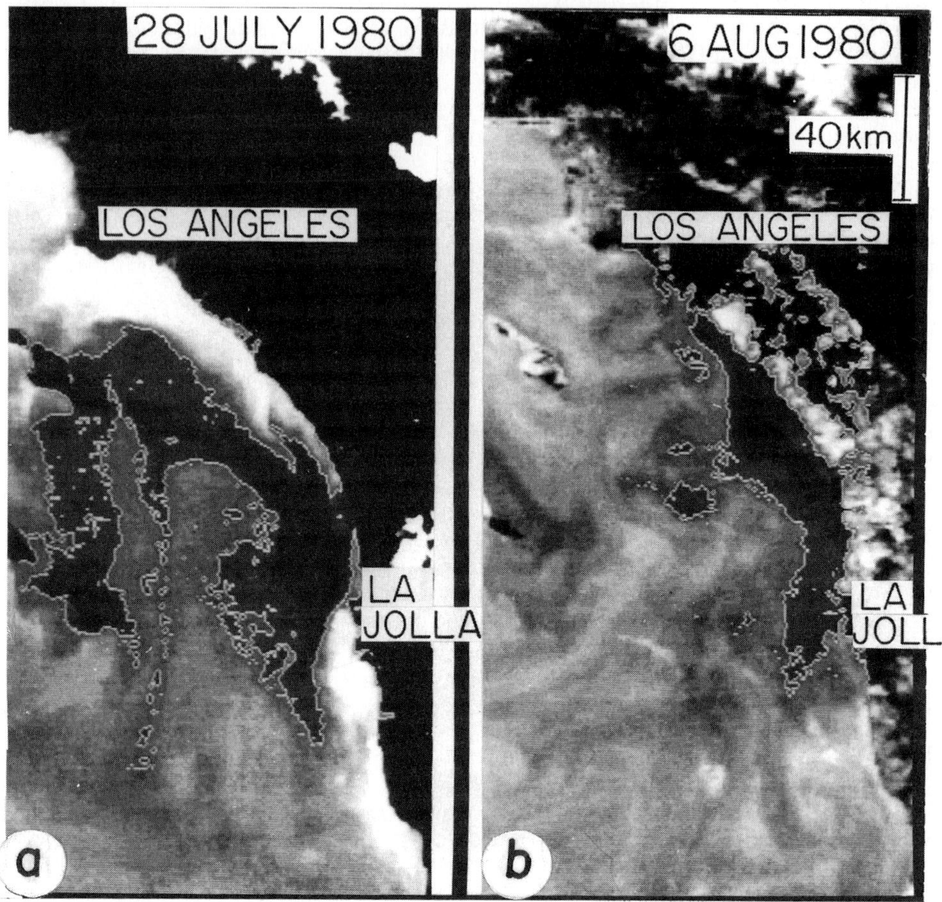

Figure 9. Sea surface temperature images obtained from NOAA-6 AVHRR data (ground resolution is about 1.1 km). The vertical size of images corresponds to approximately 280 km. Water structures related to the dissipation of the "red tide" episode: a) warm water mass (contoured at about 20°C) immediately offshore of the weakening coastal upwelling (lighter features near the coast); b) onshore collapse of the warm water mass (contoured at about 22°C) as it converges with the coastal boundary. See text for further details (modified from Peláez, 1987).

poor water mass appears to be the major factor terminating the red tide episode, and dissipating its associated biological patterns.

Thus, vertical water motion (coastal upwelling) can introduce nutrients into the euphotic zone. Horizontal advection of pigment-rich features (coupled with the swimming behaviour of dinoflagellates), can result in accumulation processes that generate patterns with very high pigment content (*e.g.*, red tides). Conversely, onshore advection of warm, nutrient-depleted water can result in the rapid dissipation of biological pattern.

Summarizing, a complex physical situation was associated with the occurrence of this red tide.

1. The oligotrophic intrusion was established in the Southern California Bight by late June, as it often does every year, favouring the development of phytoplankton species that can swim vertically searching for nutrients (*e.g.*, dinoflagellates).
2. About two or three weeks after the establishment of the intrusion, an unusually strong and persistent coastal upwelling event supplied nutrients to the euphotic zone of this otherwise oligotrophic environment.
3. Strongest phytoplankton growth did not occur in the area where the red tide was observed at its peak, but south of it.
4. Northern transport of pigment rich waters into the bight, in conjunction with the vertical swimming ability of dinoflagellates, resulted in the accumulation of cells in a narrow surface layer against a segment of the coastline. Thus, accumulation processes, besides *in situ* growth, were significant for the development of this red tide.

The occurrence and timing of somewhat antagonistic physical processes occurring at quite different scales (mesoscale, oligotrophic intrusion into the bight followed by a smaller-scale, but very intense coastal upwelling event) may account for the occurrence, or absence, of red tides in the Southern California Bight, in different years (Allen, 1941, Holmes *et al.*, 1967, Sweeney, 1975). Thereafter, which species will dominate (Anderson *et al.*, 1985, Taylor and Seliger, 1979, LoCicero, 1975) probably depends on biological aspects (Holligan 1985), *e.g.*, seed population (local or transported from elsewhere, presence of benthic cysts), behaviour (vertical migration), and physiology (control of carbon and nitrogen assimilation).

Acknowledgments

I thank the American Society of Limnology and Oceanography, and Gauthier-Villars for their permissions to make use of the material in Peláez and McGowan (1986), and Peláez (1987) published by Limnology and Oceanography and Oceanologica Acta, respectively. I thank the editors of the present volume, V. Barale and P. Schlittenhardt for their patience. I acknowledge gratefully the support and encouragement of the Joint Research Centre, Commission of the European Communities.

References

Abbott, M.R. and Barksdale, B. (1991) 'Phytoplankton pigment patterns and wind forcing off central California', Journal of Geophysical Research 96, 14649-14667.

Abbott, M.R. and Zion, P.M. (1985) 'Satellite observations of phytoplankton variability during an upwelling event', Continental Shelf Research 4, 661-680.

Abbott, M.R. and Zion, P.M. (1987) 'Spatial and temporal variability of phytoplankton pigment off northern California during Coastal Ocean Dynamics Experiment 1', Journal of Geophysical Research 92, 1745-1755.

Abbott, M.R., Brink, K.H., Booth, C.R., Blasco, D., Codispoti, L.A., Niiler, P.P., and Ramp, S.R (1990) 'Observations of phytoplankton and nutrients from a Lagrangian drifter off northern California', Journal of Geophysical Research 95, 9393-9409.

Allen, W.E. (1941) 'Twenty years' statistical studies of marine plankton dinoflagellates of southern California', American Midland Naturalist 26, 603-635.

Anderson, D.M., White, A.W., and Baden, D.G., (Eds.) (1985) 'Toxic Dinoflagellates', Elsevier, New York.

Barale, V., and Wittenberg-Fay, R. (1986) 'Variability of the ocean colour field in central California near-coastal waters as observed in a sesonal analysis of CZCS imagery', Journal of Marine Research 44, 291-316.

Bernstein, R.L., Breaker, L., and Whritner, R. (1977) 'California Current eddy formation: ship, air, and satellite results', Science 195, 353-359.

Brink, K.H. and, Cowles, T.J. (1991) 'The coastal transition zone program', Journal of Geophysical Research 96, 14,637-14,647.

Chavez, F.P., Barber, R.T., Kosro, P.M., Huyer, A., Ramp, S.R., Stanton, T.P., and Rojas de Mendiola, B. (1991) 'Horizontal transport and the distribution of nutrients in the coastal transition zone off northern California: effects on primary production, phytoplankton biomass, and species composition', Journal of Geophysical Research 96, 14833-14848.

Chelton, D.B. (1984) 'Seasonal variability of alongshore geostrophic velocity off central California', Journal of Geophysical Research 89, 3473-3486.

Coastal Ocean Dynamics Experiment (1987) Journal of Geophysical Research special issue 92 (C2), 1455-1884.

Coastal Transition Zone Group (1988) The coastal transition zone programme. EOS Transactions of the American Geophysical Union 69 (27), 697-707.

Coastal Transition Zone Programme (1991) Journal of Geophysical Research special issue 96 (C8), 14637-15052.

Cullen, J.J., and Horrigan, S.G. (1981) 'Effects of nitrate on the diurnal vertical migration, carbon to nitrogen ratio, and photosynthetic capacity of the dinoflagellate *Gymnodinium splendens*', Marine Biology 62, 81-89.

Cullen, J.J., Horrigan, S.G., Huntley, M.E., and Reid, F.M.H. (1982) 'Yellow water in La Jolla bay, California, July 1980. I. A bloom of the dinoflagellate, *Gymnodinium flavum* Kofoid and Swezy', Journal of Experimental Marine Biology and Ecology 63, 67-80.

Dorman, C.E., and Palmer, D.P. (1981) 'Southern California summer coastal upwelling', in F.A. Richards (ed.), Coastal Upwelling, EOS Transactions of the American Geophysical Union, Washington D.C., pp. 44-56.

Emery, K.O. (1960) 'The Sea off Southern California', John Wiley and Sons, Inc., New York.

Emery, W.J., Thomas, A.C., Collins, M.J., Crawford, W.R., and Mackas, D.L. (1986) 'An objective method for computing advective surface velocities from sequential infrared satellite images', Journal of Geophysical Research 91, 12865-12878.

Eppley, R.W. and Harrison, W.G. (1975) 'Physiological ecology of *Gonyaulax polyedra*, a red water dinoflagellate off southern California', in V.R. LoCicero (ed.), Proceedings First International Conference on Toxic Dinoflagellate Blooms, Massachusetts Science and Technology Foundation, Wakefield, Massachusetts, 11-22.

Eppley, R.W., Renger, E.H., and Harrison, W.G. (1979) 'Nitrate and phytoplankton production in southern California coastal waters', Limnology and Oceanography 24, 483-494.

Eppley, R.W., Sapienza, C., and Renger, E.H. (1978) 'Gradients in phytoplankton stocks and nutrients off southern California', in 1974-76. Estuarine and Coastal Marine Science 7, 291-301.

Eppley, R.W., Stewart, E., Abbott, M.R., and Heyman, U. (1985) 'Estimating ocean primary production from satellite chlorophyll. Introduction to regional differences and statistics for the southern California bight', Journal of Plankton Research 7, 57-70.

Fiedler, P.C. (1984) 'Satellite observations of the 1982-1983 El Niño along the U.S. Pacific coast', Science 224, 1251-1254.

Flament, P., Armi, L., and Washburn, L. (1985) 'The evolving structure of an upwelling filament', Journal of Geophysical Research 90, 11765-11778.

Haidvogel, D.B., Beckmann, A., and Hedström, K.S. (1991) 'Dynamical simulations of filament formation and evolution in the coastal transition zone', Journal of Geophysical Research 96, 15017-15040.

Haury, L.R., Simpson, J.J., Peláez, J., Koblinsky, C.J., and Wiesenhahn, D. (1986) 'Biological consequences of a recurrent eddy off Point Conception, California', Journal of Geophysical Research 91, 12937-12956.

Hayward, T.L. and Mantyla, A. W. (1990) 'Physical, chemical, and biological structure of a coastal eddy near Cape Mendocino', Journal of Marine Research 48, 825-850.

Hayward, T.L. and Venrick, E.L. (1982) 'Relation between surface chlorophyll, integrated chlorophyll, and integrated primary production', Marine Biology 69, 247- 252.

Hickey, B.M. (1979) 'The California Current system-Hypothesis and facts', Progress in Oceanography 8, 191-279.

Holligan, P.M. (1985) 'Marine dinoflagellate blooms - growth strategies and environmental exploitation', in D.M. Anderson, A.W. White, and D.G. Baden (eds.) Toxic Dinoflagellates, Elsevier, New York, pp. 133-139.

Holmes, R.W., Williams, P.M., and Eppley, R.W. (1967) 'Red water in La Jolla bay, 1964-1966', Limnology and Oceanography 12, 503-512.

Hood, R.R., Abbott, M.R., and Huyer, A. (1991) 'Phytoplankton and photosynthetic light response in the coastal transition zone off northern California in June 1987', Journal of Geophysical Research 96, 14769-14780.

Hood, R.R., Abbott, M.R., Huyer, A., and Kosro, P.M. (1990) 'Surface patterns in temperature, flow, phytoplankton biomass, and species composition in the coastal transition zone off northern California', Journal of Geophysical Research 95, 18081-18094.

Hovis, W.A., Clark, D.K., Anderson, F., Austin, R.W., Wilson, W.H., Baker, E.T., Ball, D., Gordon, H.R., Mueller, J.L., El-Sayed, S.Z., Sturm, B., Wrigley, R.C., and Yentsch, C.S. (1980) 'Nimbus-7 coastal zone colour scanner: system description and initial imagery', Science 210, 60-63.

Huntley, M.E. (1982) 'Yellow water in La Jolla bay, California, July 1980. II. Suppression of zooplankton grazing', Journal of Experimental Marine Biology and Ecology 63, 81-91.

Huyer, A.E. and Kosro, P.M. (1987) 'Mesoscale surveys over the shelf and slope in the upwelling region near Point Arena, California', Journal of Geophysical Research 92, 1655- 1682.

Huyer, A., Kosro, P.M., Fleischbein, J., Ramp, S.R., Stanton, T., Washburn, L., Chavez, F.P., Cowles, T.J., Pierce, S.D., and Smith, R.L. (1991) 'Currents and water masses of the coastal transition zone off northern California, June to August 1988', Journal of Geophysical Research 96, 14809-14831.

Ikeda, M. and Emery, W.J. (1984) 'Satellite observations and modeling of meanders in the California Current system off Oregon and northern California', Journal of Physical Oceanography 14, 1434-1450.

Kadko, D.C., Washburn, L., Jones, B. (1991) 'Evidence of subduction within cold filaments of the northern California coastal transition zone', Journal of Geophysical Research 96, 14909-14926.

Kamykowski, D. (1974) 'Possible interactions between phytoplankton and semidiurnal internal tides', Journal of Marine Research 32, 67-89.

Kamykowski, D. and Zentara, S.J. (1986) 'Predicting plant nutrient concentrations from temperature and sigma-t in the upper kilometer of the world ocean', Deep-Sea Research 33, 89-105.

Kelly, K.A. (1985) 'The influence of winds and topography on the sea surface temperature patterns over the northern California slope', Journal of Geophysical Research 90, 11783-11798.

Kelly, K.A. (1989) 'An inverse model for near-surface velocity from infrared images', Journal of Physical Oceanography 19, 1845-1864.

Kosro, P.M., Huyer, A., Ramp, S.R., Smith, R.L., Chavez, F.P., Cowles, T.J., Abbott, M.R., Strub, P.T., Barber, R.T., Jessen, P., and Small, L.F. (1991) 'The structure of the transition zone between coastal waters and the open ocean off northern California, winter and spring 1987', Journal of Geophysical Research 96, 14707-14730.

LoCicero, V.R., editor (1975) Proceedings First International Conference on Toxic Dinoflagellate Blooms, Massachusetts Science and Technology Foundation, Wakefield, Massachusetts.

Lynn, R.J. and Simpson. J.J. (1987) 'The California Current system: the seasonal variability of its physical characteristics', Journal of Geophysical Research 92, 12947-12966.

Mackas, D.L., Washburn, L., Smith. S.L. (1991) 'Zooplankton community pattern associated with a California Current cold filament', Journal of Geophysical Research 96, 14781- 14797.

McCreary, J.P., Fukamachi, Y., and Kundu, P.K. (1991) 'A numerical investigation of jets and eddies near an eastern ocean boundary', Journal of Geophysical Research 96, 2515- 2534.

Mooers, C.N.K. and Robinson, A.R. (1984) 'Turbulent jets and eddies in the California Current and inferred cross-shore transports', Science 223, 51-53.

Mueller, J.L. and Lange, R.E. (1989) 'Bio-optical provinces of the northeast Pacific ocean: A provisional analysis', Limnology and Oceanography 34, 1572-1586.

Mullin, M.M. (1986) 'Spatial and temporal scales and patterns', in R.W. Eppley (ed.), Plankton Dynamics of the Southern California Bight, Lecture Notes on Coastal and Estuarine Studies, Vol. 15, Springer-Verlag, New York, pp. 216-273.

Narimousa, S. and Maxworthy, T. (1989) ' Application of a laboratory model to the interpretation of satellite and field observations of coastal upwelling', Dynamics of Atmospheres and Oceans 13, 1-46.

Niiler, P.P., Poulain, P.M., and Haury, L.R. (1989) 'Synoptic three-dimensional circulation in an onshore-flowing filament of the California Current', Deep-Sea Research 36, 385-405.

Pares-Sierra, A. and O'Brien, J.J. (1989) 'The seasonal and interannual variablity of the California Current system: a numerical model', Journal of Geophysical Research 94, 3159- 3180.

Peláez, (1984) 'Phytoplankton pigment concentrations and patterns in the California Current as determined by satellite', Ph.D. thesis, University of California, San Diego.

Peláez, J. (1987) 'Satellite images of a "red tide" episode off southern California', Oceanologica Acta 10, 403-410.

Peláez, J. and Guan, F. (1982) 'California Current chlorophyll measurements from satellite data', California Cooperative Oceanic Fisheries Investigations Reports 23, 212-225.

Peláez, J. and McGowan, J.A. (1986) 'Phytoplankton pigment patterns in the California Current as determined by satellite', Limnology and Oceanography 31, 927-950.

Reid, J.L., Roden, G.I., and Wyllie, J.G. (1958) 'Studies of the California Current system', California Cooperative Oceanic Fisheries Investigations Reports 6, 27-57.

Rienecker, M.M. and Mooers, C.N.K. (1989) 'Mesoscale eddies, jets, and fronts off Point Arena, California, July 1986', Journal of Geophysical Research 94, 12555-12569.

Sargent, M.C. and Walker, T.J. (1948) 'Diatom populations associated with eddies off southern California', Journal of Marine Research 7, 490-505.

Send, U., Beardsley, R.C., and Winant, C.D. (1987) 'Relaxation from upwelling in the Coastal Ocean Dynamics Experiment', Journal of Geophysical Research 92, 1683-1699.

Silva, V.M. (1982) 'Thermal calibration of satellite infrared images and correlation with sea surface nutrient distribution', M.S. thesis, Naval Postgraduate School, Monterey, California.

Simpson, J.J., Koblinsky, C.J., Haury, L.R., and Dickey, T.D. (1984) 'An offshore eddy in the California Current system', Progress in Oceanography 13, 1-111.

Simpson, J.J., Koblinsky, C.J., Peláez, J., Haury, L.R., and Wiesenhahn, D. (1986) 'Temperature-plant pigment-optical relations in a recurrent offshore mesoscale eddy near Point Conception California', Journal of Geophysical Research 91, 12919-12936.

Simpson, J.J. and Lynn, R.J. (1990) 'A mesocale eddy dipole in the offshore California Current', Journal of Geophysical Research 95, 13009-13022.

Smith, R.C., Zhang, X., and Michaelsen, J. (1988) 'Variability of pigment biomass in the California Current system as determined by satellite imagery. 1. Spatial variability', Journal of Geophysical Research 93, 10863-10882.

Stern, M.E. (1986) 'On the amplification of convergences in coastal currents and the formation of "squirts", Journal of Marine Research 44, 403-421.

Strub, P.T. and James, C. (1988) 'Atmospheric conditions during the spring and fall transitions in the coastal ocean off western United States', Journal of Geophysical Research 93, 15561-15584.

Strub, P.T., James, C., Thomas, A.C., and Abbott, M.R. (1990) 'Seasonal and nonseasonal variability of satellite-derived surface pigment concentrations in the California Current', Journal of Geophysical Research 95, 11501-11530.

Strub, P.T., Kosro, P.M., Huyer, A., and CTZ collaborators (1991) 'The nature of the cold filaments in the California Current system', Journal of Geophysical Research 96, 14734- 14768.

Svejkovsky, J (1988) 'Sea surface flow estimation from advanced very high resolution radiometer and coastal zone colour scanner imagery: a verification study', Journal of Geophysical Research 93, 6735-6743.

Sverdrup, H.V., and Allen, W.E. (1939) 'Distribution of diatoms in relation to the character of water masses and currents off southern California in 1938', Journal of Marine Research 2, 131-144.

Sverdrup, H.V., Fleming, and R.H. (1941) 'The waters off the coast of southern California March to July 1937', Bulletin of the Scripps Institution of Oceanography 4, 261-378.

Sweeney, B.M. (1975) 'Red tides I have known', in V.R. LoCicero (ed.), Proceedings First International Conference on Toxic Dinoflagellate Blooms, Massachusetts Science and Technology Foundation, Wakefield, Massachusetts, pp. 225-234.

Taylor, D.L. and Seliger, H.H. (eds.) (1979) 'Toxic Dinoflagellate Blooms', Elsevier, New York.

Thomas, A.C. and Emery, W.J. (1988) 'Relationships between near-surface plankton concentrations, hydrography, and satellite-measured sea surface temperature', Journal of Geophysical Research 93, 15733-15748.

Thomas, A.C. and Strub, P.T. (1990) 'Seasonal and interannual variability of pigment concentrations across a California Current frontal zone', Journal of Geophysical Research 95, 13023-13042.

Traganza, E.D., Silva, V.M., Austin, D.M., Hanson, W.L., and Bronsink, S.H. (1983) 'Nutrient mapping and recurrence of coastal upwelling centers by satellite remote sensing: its implication to primary production and the sediment record', in Coastal upwelling: its sediment record. North Atlantic Treaty Organization Conference Series 4, Vol. 10A, Plenum, pp. 61- 83.

Tsuchiya, M. (1980) 'Inshore circulation in the southern California bight, 1974-1977', Deep-Sea Research 27 (A), 99-118.

Vastano, A.C. and Reid, R.O. (1985) 'Sea surface topography estimation with satellite infrared imagery', Journal of Atmospheric Technology 2, 393-400.

Winant, C.D. (1980) 'Downwelling over the southern California shelf', Journal of Physical Oceanography 10, 791-799.

Van Woert, M.L. (1982) 'The subtropical front: satellite observations during FRONTS 80', Journal of Geophysical Research 87, 9523-9536.

Wahl, D.D. and Simpson, J.J. (1990) 'Physical processes affecting the objective determination of near-surface velocity from satellite data', Journal of Geophysical Research 95, 13511-13528.

Washburn, L., Kadko, D.C., Jones, B., Hayward, H. T., Kosro, P.M., Stanton, T.P., Ramp, S., and Cowles, T. (1991) 'Water mass subduction and the transport of phytoplankton in a coastal upwelling system', Journal of Geophysical Research 96, 14927-14945.

Wooster, W.S. and Reid, J.L. (1963) 'Eastern boundary currents', in M.N. Hill (ed.), The Sea Vol. 2, Interscience, New York, pp. 253-260.

Wyllie, J.G. (1966) 'Geostrophic flow of the California Current at the surface and at 200 meters', California Cooperative Oceanic Fisheries Investigations Atlas 4, 288 p.

Zentara, S.J. and Kamykowski, D. (1977) 'Latitudinal relationships among temperature and selected plant nutrients along the west coast of North and South America', Journal of Marine Research 35, 321-337.

SEASONAL AND MERIDIONAL VARIABILITY OF THE REMOTELY SENSED FRACTION OF EUPHOTIC ZONE CHLOROPHYLL PREDICTED BY A LAGRANGIAN PLANKTON MODEL

K.U. WOLF
Institut für Meereskunde an der Universität Kiel
Düsternbrooker Weg 20
2300 Kiel, Germany

V. STRASS
Alfred Wegener Institut für Polar- und Meeresforschung
Columbusstrasse P.O. Box 120161
2850 Bremerhaven, Germany

ABSTRACT. A Lagrangian plankton model is used to calculate the seasonal and meridional variations in the satellite-sensed fraction of the euphotic zone chlorophyll (EZC), which are caused by changes in the vertical chlorophyll distribution related to the succession of growth phases during the phytoplankton seasonal cycle in mid-latitudes. In order to assess how accurately satellite color signals represent the total euphotic zone phytoplankton biomass, the satellite-sensed fraction of the EZC content as well as the ratio between the mean vertical EZC concentration estimated from satellite data (*e.g.* CZCS) and *in-situ* model data are calculated. According to the model the ratio of the remotely sensed to the *in-situ* EZC concentration is close to 1 early in the year before the spring bloom. It takes values up to 2 when the bloom is at full spate and decreases to about 0.5 after nutrient depletion in the mixed layer when most phytoplankton is in the deep maximum. Subject to the same range of variation, the satellite-sensed fraction is about 10% of the EZC content on average. Hence the seasonal variations of the *in-situ* phytoplankton are overestimated by a factor of 4 if projected only from the remotely sensed color signal. These model predictions closely coincide with an analysis of measured chlorophyll data.

1. Introduction

Ocean color sensing devices flown on satellites (*e.g.* the Coastal Zone Color Scanner, CZCS, on Nimbus 7; see Hovis *et al.*,1980) have provided the scientific community with a novel and stimulating view of the horizontal distribution of the near-surface phytoplankton concentration and of how it changes in time. Because of the important role of phytoplankton in the global carbon cycle and hence of its potential influence on climate change it is tempting to use the remotely sensed color data, converted to chlorophyll concentration, also for estimating the carbon fixation by phytoplankton primary production.

The problem with estimating ocean primary production from satellite color images is that the remotely sensed signal is emitted only from the near-surface layer, whereas primary production takes place in the whole deeper-

319

V. Barale and P.M. Schlittenhardt (eds.),
Ocean Colour: Theory and Applications in a Decade of CZCS Experience, 319–329.
© 1993 *ECSC, EEC, EAEC, Brussels and Luxembourg. Printed in the Netherlands.*

320

extending euphotic zone. The key to estimating ocean primary production from satellite color images is to project accurately the total EZC content (as a measure of phytoplankton biomass) from the remotely sensed chlorophyll signal. Sensitivity studies (Platt *et al.*, 1988, 1991), based on generalized chlorophyll profiles, and using algorithms for estimating primary production from satellite images proposed by Platt (1986) and Platt *et al.* (1988), concluded that the error in the estimate of production is dominated by the uncertainty of the euphotic zone chlorophyll content projected from the remotely sensed signal. Since the chlorophyll distribution varies with depth, the utilization of satellite color images for estimating ocean primary production is controversial (*e.g.* Smith *et al.*, 1982; Platt and Herman, 1983; Eppley *et al.*, 1985; Platt *et al.*, 1988; Sathyendranath *et al.*, 1989).

An earlier simulation of satellite measurements based on *in-situ* data collected at different times of the phytoplankton growth season in the open North Atlantic (Strass, 1990) has shown that the satellite-sensed fraction of the euphotic zone chlorophyll content accounts for only 7.5 % of the total, on average, and that it varies meridionally and seasonally by several times of its mean value. The major variations are attributed to changes in the vertical chlorophyll distribution, related to the succession of growth phases during the phytoplankton seasonal cycle.

In this study we will make a different approach, namely by use of an advanced technique in plankton modelling - the Lagrangian ensemble method (Woods and Onken, 1982; Wolf and Woods; 1988; Wolf, 1991). This also makes possible to improve the earlier investigation in two respects:

(1) to allow the attenuation of light to vary also with depth;

(2) to achieve a better temporal resolution as well as a complete seasonal cycle of the ratios of the remotely sensed EZC.

2. Methods

2.1. THE PLANKTON MODEL

In order to model plankton bloom dynamics usually the Eulerian continuum method has been used (*e.g.* Steele, 1962; Jamart *et al.*, 1979 or Evans and Parslow, 1985). The Eulerian method has the advantage of relatively low needs for computing time and CPU memory. This is achieved by parametrisations which simplify the diversity in the properties of cells as well as the non-linear interaction between physics, biolology and chemistry. To take this interaction into account Woods and Onken (1982) introduced the Lagrangian ensemble method where the whole phytoplankton population is represented by a finite number of individual Lagrangian particles. In comparative model simulations McGillicuddy (1989) found differences between Eulerian and Lagrangian formulations amounting to *e.g.* 50% in biomass after a 50 day model run. A further developed version of the Lagrangian ensemble method is used in an extended, one-dimensional plankton model (Wolf, 1991) for our simulations. In this method all phytoplankton- "particles" suffer their own individual fate (mixing rate, nutrient-concentration, light and zooplankton grazing) depending on their actual depth. The model equations are integrated separately for each of those particles. Every Lagrangian particle consists of a number of cells varying due to cell division, zooplankton grazing and mortality. Besides the number of cells and the actual depth, the other attributes of an ensemble particle are light adaptation and the contents of the individual nutrient and energy pools which are responsible for cell divisions. The depth of a particle is determined differently if within or below the mixed layer. Within the mixed layer the vertical movement of a particle is dominated by turbulent motion, in the model represented by random shuffling; below the mixed layer its vertical displacement is determined by the sinking velocity of the cells. The depth of the mixed layer is provided by a one-dimensional (depth) Kraus-Turner type mixed layer model (Woods and Barkmann, 1986) which interacts with the biological part. The model comprises one phyto- and one zooplankton species. In order to represent spring bloom (generally diatoms) and summer (generally flagellates) phytoplankton populations, the sinking velocity of the particles switches between 1 m/d during eutrophic and 0.3 m/d during oligotrophic conditions.

Similar to the phytoplankton part, the zooplankton biomass is divided into individual Lagrangian cohorts which move independently in the vertical. Every zooplankton cohort is characterized by five attributes: number of individuals, depth, weight, age and satiation. These variables influence vertical movement, ingestion and reproduction of the zooplankton and therefore the grazing pressure on the phytoplankton biomass. For further description of the plankton model used see Wolf and Woods (1988) and Wolf (1991). For our study we choose the temperate open North Atlantic for calculations so that we can compare with the observations of Strass (1990), and so that we can use the climatological Bunker-data (Isemer and Hasse, 1987) as regionally and seasonally varying boundary conditions for the model simulations.

2.2. SIMULATION OF SATELLITE CHLOROPHYLL IMAGERY

The backscattered chlorophyll signal accessible to passive remote sensing depends on the attenuation of light on its way down to the phytoplankton and back to the surface. Using profiles of chlorophyll concentration and solar irradiance it is possible to calculate which chlorophyll concentration would be measured by satellite remote sensing, and of which depth range the remotely sensed signal is representative. Neglecting such error sources as atmospheric influences or species-dependent backscatterring differences, the remotely sensed signal could be converted to the mean EZC concentration easily if there is no variability of attenuation with depth due to the structure of chlorophyll profiles. The calculations carried out here are guided by the technical performance of the CZCS, and largely follow the procedure employed by Strass (1990), in order to allow for a direct comparison of the present calculations based on model profiles with the earlier evaluations of Strass (1990) using *in-situ* measurements.

With Chl(z) denoting the in-water chlorophyll concentration,

$$C_e = \int_0^{ze} Chl(z)\, dz \tag{1}$$

gives the chlorophyll content of the euphotic zone, when the integration extends from the sea surface to z_e, the depth of the euphotic zone. Then,

$$Chl_e = \frac{C_e}{z_e} \tag{2}$$

gives the mean chlorophyll concentration in the euphotic zone. The chlorophyll concentration which would be detected by satellite remote sensing can be estimated (Gordon *et al.*, 1980) from

$$Chl_s = \frac{C_s}{z_s} \tag{3}$$

where

$$C_s = \int_0^{z_e} Chl(z) \cdot W(z)\, dz \tag{4}$$

and

$$z_s = \int_0^{z_e} W(z)\, dz \tag{5}$$

C_s, if scaled by C_e, gives the fraction of the EZC content that is accessible to satellite imagery. The quantity z_s, being in units of depth and amounting to nearly half the attenuation length, is called the effective satellite visibility depth. It gives the depth range from which the satellite would receive its total signal (as given by eq.3) if its sensitivity were distributed homogeneously over the water column. In case of a vertically constant attenuation, the weighting function

$$W(z) = e^{-zk_d \cdot z} \tag{6}$$

is given analytically by a single exponential, taking into account the attenuation of the downwelling solar irradiance as well as that of the backscattered light.

However, for our calculations, based on modelled profiles of chlorophyll and irradiance, where we take into account the vertical variation of attenuation (due the feedback of the phytoplankton on the irradiance profile), eqs. (4) and (5), are expressed in form of the sums

$$C_s = \sum_{i=1}^{n} Chl(z_i) \cdot \left(\frac{I(z_i)}{I_0} \right)^2 \Delta z$$

and

$$z_s = \sum_{i=1}^{n} \cdot \left(\frac{I(z_i)}{I_0} \right)^2 \Delta z$$

where I_0 denotes the downwelling irradiance at the surface, $Chl(z_i)$ is the chlorophyll concentration and $I(z_i)$ the irradiance at the discrete depth levels of the model with $\Delta z = 1m$ and $n = z_e/\Delta z$. From the 19 spectral bands used in the model those representing the wavelengths employed by the CZCS for chlorophyll remote sensing (443, 520, and 550 nm) are selected to give I_0 and $I(z_i)$.

3. Results

Every spring, roundabout equinox the mixed layer shallows, when the ocean begins to gain heat (see figure 1a,b). At the same time the critical depth deepens so that when both meet, the Sverdrup Criterion (Sverdrup, 1953) is fulfilled. This determines the start of the phytoplankton spring bloom in the model (Wolf, 1991) and the phytoplankton biomass increases rapidly (figure 1c,d).

At the end of the spring bloom the nutrients in the mixed layer are consumed. The exhaustion of nutrients inhibits further New Production above the nutricline, while the increased zooplankton population is able to sweep this layer in a few days, improving light conditions below. The depletion of the mixed layer nutrients marks an important change in ecosystem boundary conditions resulting in the transition to oligotrophy. In figure 1 a,b this transition is represented by the first occurence of nutricline which in the model is defined by a concentration criterion of 0.5 µg N/l.

In the depth range where nutrients are still available, the light conditions are best at the nutricline itself which, hence, marks the most favorable depth for New Production. This leads to the formation of a subsurface phytoplankton maximum, at a depth where still enough light and nutrients are available.

The nutrient consumption caused by this production deepens slowly both the nutricline and the subsurface maximum (see figure 2). At the end of summer, when the insolation is reduced, the deepening stops. But when in autumn the mixed layer deepens again, mixed layer depth and nutricline meet, causing entrainment of nutrients into the mixed layer from below and possibly triggering an autumn bloom.

Although the model takes the diurnal variability in mixed layer dynamics (Woods and Onken, 1982) into account figure 1a and 1b show only, for graphical clarification, the seasonal variation of the diurnal maximum mixed layer depth as well as the plankton- induced nutricline simulated by our model. The relatively high phytoplankton abundance in spring is concentrated in the mixed layer while in summer most of the phytoplankton biomass is present in the deep maximum (see figure 2).

The depth and duration of the subsurface maximum vary with latitude and cause the most significant regional differences in the remotley sensed EZC. The annual variation of the effective satellite visibility depth (figure 3a,b) shows an expected minimum during the spring bloom of about 3m as well as a deepening to more than double that value in summer. The depth of the subsurface phytoplankton maximum is always well below z_s, i.e. below the depth range to which satellites are sensitive. The effect can be seen in figure 3 where the ratios between remotely sensed and modelled chlorophyll content

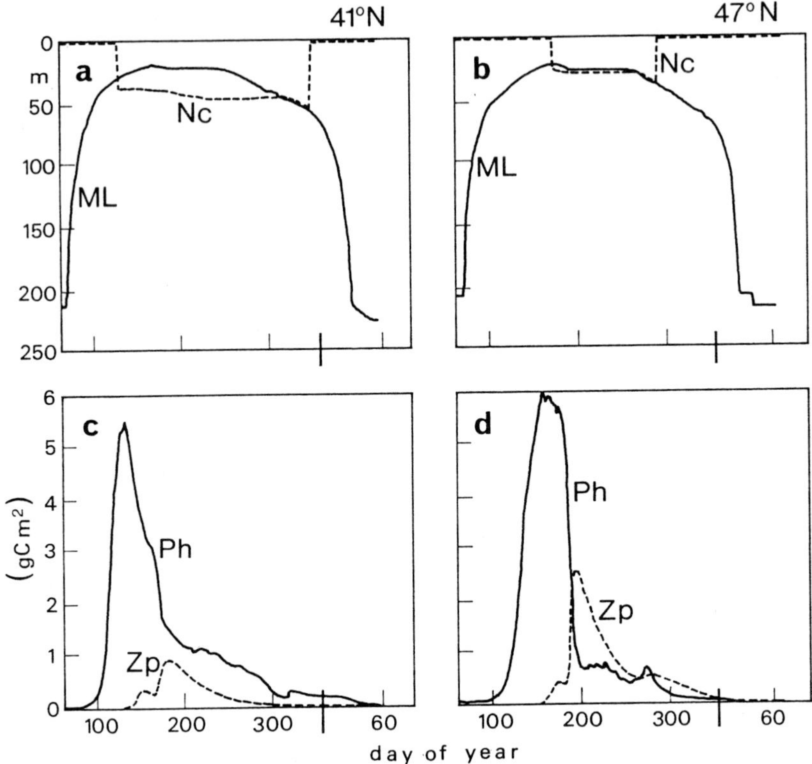

Figure 1.   (a,b) Modelled seasonal variation of mixed layer- (ML) and nutricline (Nc) depth at 41°N and 47°N; (c,d) corresponding total phytoplankton (solid line) and zooplankton (dashed line) biomass in the water column.

as well as concentration are shown. A comparison of the results from 41°N and 47°N shows that the deeper the subsurface maximum the less are the remotely sensed fractions of EZC, while during spring blooms the ratio of Chl_s/Chl_e generally takes values up to 2 and in autumn up to 1.5.

Although it is quite easy to evaluate with the model the seasonal and regional variations of z_e caused by the variability of attenuation in depth and time, we choose the euphotic zone depth $z_e = const.$

This is reasonable because information on ze generally is not availiable for remotely sensed data. On the other hand the satellite ratio of the chlorophyll content C_s/C_e as well as of the chlorophyll concentration Chl_s/Chl_e depend on the choice of z_e.

The modelled phytoplankton profiles (see figure 2) show that it is sufficient to set $z_e = 70m$ in order to include almost all the chlorophyll in the integrals. Deeper z_e will enhance overestimation in spring and decrease differences in summer; shallower z_e will have the opposite effect.

The seasonal depth of the chlorophyll maximum decreases with latitude (Strass and Woods, 1991; Wolf, 1991), so that the latitudinal constant z_e causes at 47°N (see figure 3) the effect described above.

We also investigated the direct relation of surface chlorophyll concentration to total production of the euphotic zone using the model results at different latitudes, but were not able to detect significant correlations.

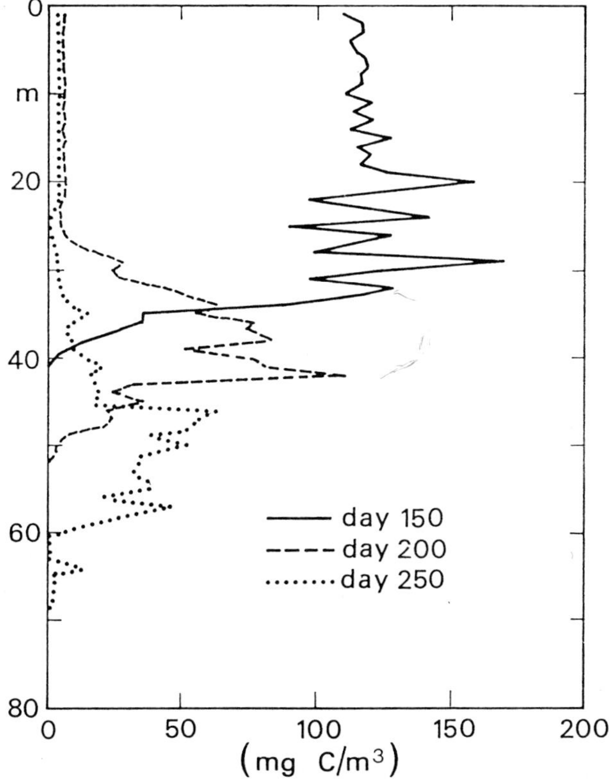

Figure 2. Three selected phytoplankton profiles representing a spring bloom (day 150) and two oligotrophic summer situations (days 200 and 250) at 41°N.

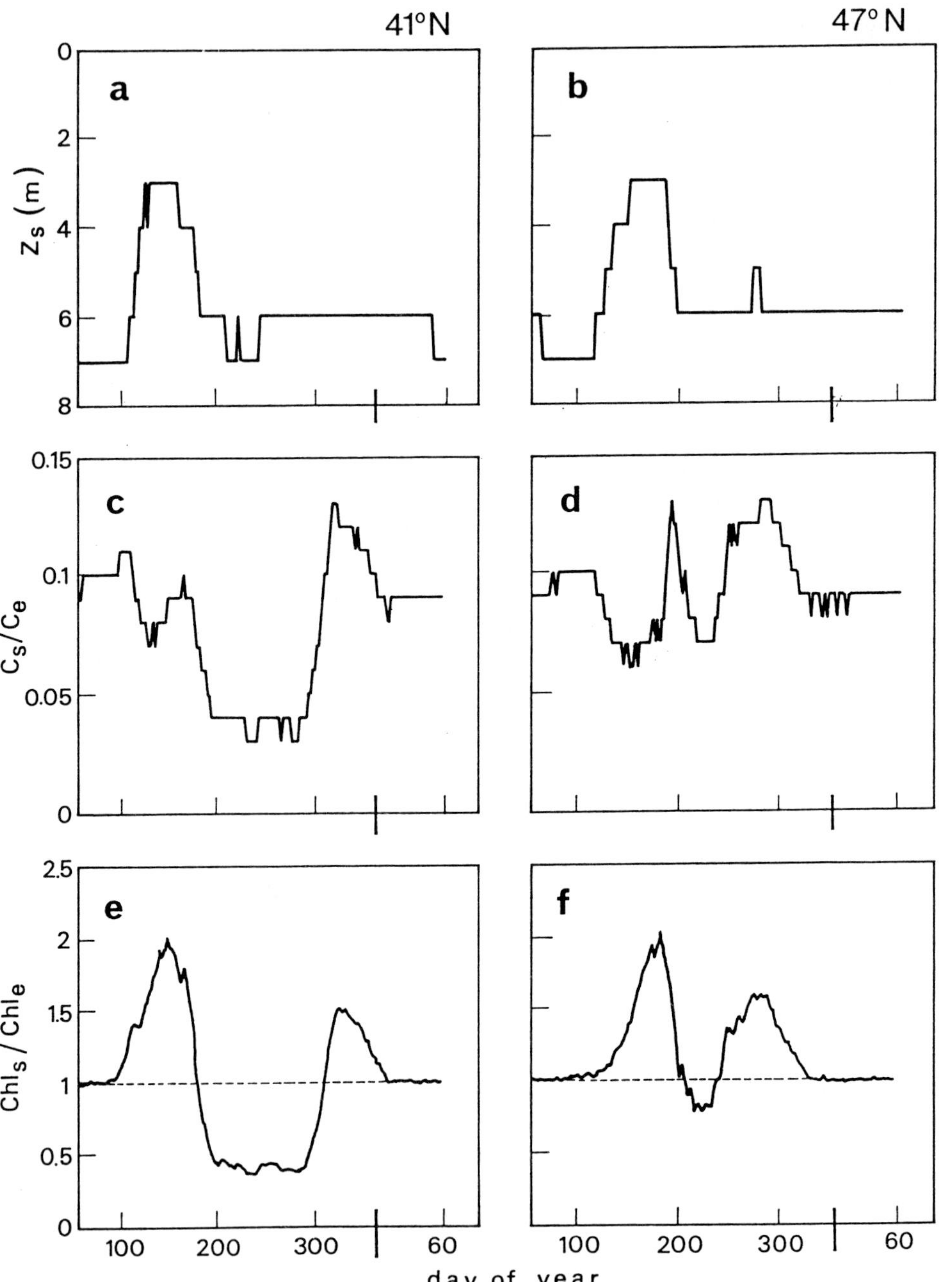

Figure 3. Seasonal variation of the satellite visibility depth Z_s (a,b), the fraction C_s/C_e of EZC content accessible to remote sensing (c,d) and the mean EZC concentration sensed by satellite

4. Discussion

The comparision of our model results with the *in-situ* data presented by Strass (1990) suggests that the model predicts at least the seasonal and meridional variations of the biomass profile reasonably well. It was shown that the remotely sensed fraction of phytoplankton content as well as phytoplankton concentration are strongly affected by the shape of the phytoplankton profile. These profiles vary seasonally and regionally. So seasonal oligotrophy and the related occurrence of a deep chlorophyll maximum are common features in mid-latitudes. In particular, it has been shown that the model simulates well the strong relation of the nutricline to the chlorophyll maximum which was suggested by Venrick *et al.* (1973). In this study we used an attenuation coefficient $k_d(z)$ varying with depth, whereas a vertically constant coefficient was used in the analysis of the *in-situ* data by Strass (1990).

The results achieved by both methods are virtually identical though the seasonal variation of Ze is larger in the model results. Our simulations show that determination of EZC by CZCS measurements without considering the variability of chlorophyll profiles, overestimates the seasonal variation by a factor of 4. In order to improve satellite analysis for determination of the EZC content and oceanic primary production Platt *et al.* (1991) introduced a method which uses standard phytoplankton profiles varying between 9 regions and 4 seasons for the North Atlantic. From our model simulations with relatively high temporal resolution (1h) we have seen that at most latitudes in the open ocean the total phytoplankton growth cycle is completed within less than tree months. So the promising attempt by Platt *et al.* still lacks temporal and most probably regional resolution. Therefore we particularly recommend to consider other remotely sensed parameters like temperature (Strass, 1990; Wolf, 1991) or sea surface roughness to get more information on the timing of the spring bloom, its transition to oligotrophy, and the actual wind induced mixed layer depth, which all influence the shape of phytoplankton profiles.

Acknowledgements

The authors acknowledge valuable comments from Dr. G. Evans and help with the figures by A. Eisele.

References

Eppley, R.W., Steward, E., Abbott, M., and Heyman, U. (1985) 'Estimating ocean primary production from satellite chlorophyll. Introduction to regional differences and statistics for southern California Bight', Journal of Plankton Research 7, 57-70.

Evans, G.T. and Parslow, J.S. (1985) 'A model of annual plankton cycles', Biological Oceanography 3, 713-715.

Gordon, H.R., Clark, D., Mueller, J., and Hovis, W. (1980) 'Phytoplankton pigments from Nimbus-7 CZCS: Comparison with surface measurements', Science 210, 63-66.

Hovis, W.A., Clark, D., Anderson, F., Austin, R., Wilson, W., Baker, E., Ball, D., Gordon, H., Mueller, J., El-Sayed, S., Sturm, B., Wrigley, R., and Yensch, J. (1980) 'Nimbus-7 Coastal Zone Color Scanner: system description and initial imagery', Science 210, 60-63.

Isemer, H.J. and Hasse, L. (1987) 'The Bunker climate atlas of the North Atlantic Ocean', vol. 2, Air-sea interactions, Springer Verlag.

Jamart, B.M., Winter, D., and Banse, K. (1979) 'Sensitivity analysis of a mathematical model of phytoplankton growth and nutrient distribution in the Pacific Ocean off the northwestern U.S.- coast', Journal of Plankton Research 1, 267-290.

McGillicuddy, D. (1989) 'One dimensional numerical simulation of new primary production; Lagrangian and Eulerian formulations', Havard University, unpublished manuscript.

Platt, T. and Herman, A. (1983) 'Remote sensing of phytoplankton in the sea: surface layer chlorophyll as an estimate of water column chlorophyll and primary production', International Journal of Remote Sensing 4, 343-351.

Platt, T. (1986) 'Primary production of the ocean water column as a function of surface light intensity: Algorithms for remote sensing', Deep-Sea Research 33, 149-163.

Platt, T. and Sathyendranath, S. (1988) 'Ocean primary production: Estimation by remote sensing at local and regional scales', Science 241, 1613-1620.

Platt, T., Sathyendranath, S., Caverhill, C., and Lewis, M. (1988) 'Ocean primary production and available light: further algorithms for remote sensing', Deep-Sea Research 35, 855-879.

Platt, T., Cavernhill, C., and Sathyendranath, S. (1991) 'Basin scale estimates of oceanic primary production by remote sensing: The North Atlantic', Journal of Geophysical Research 96-C8, 15,147-15,159.

Sathyendranath, S., Platt, T., Caverhill, C., Warnock, R., and Lewis, M. (1989) 'Remote sensing of oceanic primary production: Computations using a spectral model', Deep-sea Research 36, 431-453.

Smith, R.L., Eppley, R., and Baker, K. (1982) 'Correlation of primary production as measured aboard a ship in southern California coastal waters and as estimated from satellite chlorophyll images', Marine Biology 66, 269-279.

Steele, J.H. (1962) 'Environmental control of photosynthesis in the sea', Limnology and Oceanography 7, 137-150.

Strass, V. (1990) 'Meridional and seasonal variations in the satellite sensed fraction of the euphotic zone chlorophyll', Journal of Geophysical Research 95 (C10), 18,289-18,301.

Strass, V. H. and Woods, J. D. (1991) 'New production in the summer revealed by the meridional slope of the deep chlorophyll maximum', Deep-Sea Research 38, 35-56.

Sverdrup, H.U. (1953) 'On conditions for vernal blooming of phyto- plankton', Journal du Conseil International pour l'Exploration de la Mer 18, 287-297.

Venrick, E.L., McGowan, J., and Mantyla, A. (1973) 'Deep maxima of photosynthetic chlorophyll in the Pacific Ocean', Fisheries Bullettin 71, 41-52.

Wolf, K.U. (1991) 'Meridionale Variabilitaet des physikalischen und planktologischen Jahreszyklus - Lagrange'sche Modellstudien im Nordatlantik.', Reports of the Inst. of Marine Sciences - Kiel, No. 203.

Wolf, K.U., and Woods, J.D. (1988) 'Lagrangian simulation of primary production in the physical environment - the deep chlorophyll maximum and nutricline', in B.J. Rothschild (ed.), Toward a Theory on Biological-Physical Interactions in the World Ocean, Kluwer Academic Publishers, Dortrecht, pp. 51-70.

Woods, J.D. and Barkmann, W. (1986) 'The response of the upper ocean to solar heating. I. The mixed layer', Quarterly Journal of the Royal Meteorological Society 112, 1-27.

Woods, J.D. and Onken, R. (1982) 'Diurnal variation and primary production in the ocean - preliminary results of a Lagrangian ensemble model', Journal of Plankton Research 4, 735-756.

FUTURE SYSTEMS FOR GLOBAL MONITORING OF OCEAN COLOUR

H.VAN DER PIEPEN
Institute of Optoelectronics, DLR,
Oberpfaffenhofen
8031 Wessling Germany

R. DOERFFER
Institute of Physics
GKSS Forschungszentrum Geesthacht
Postfach 1160
2054 Geesthacht, Germany

ABSTRACT. Future earth observation satellites will carry advanced sensors incorporating multispectral scanners and imaging spectrometers. Due to their flexibility especially with regard to spectral channel selection, the latter types permit a variety of applications in regional or global monitoring of the marine and terrestrial biosphere. Planned and approved missions are discussed, such as *e.g.* NASA's Earth Observation System, ESA's Earth Observation Programme and other missions like FY-1 (People's Rep. of China), PRIRODA (USSR), ADEOS (Japan), ATMOS (FRG) and GLOBESAT (France). A brief survey of relevant sensors like MOS, OCTS, POLDER, SeaWiFS, ROSIS, MODIS, MERIS and their performance is presented.

1. Introduction

1.1. HISTORY OF OCEAN COLOUR RESEARCH

The miracle of the ever changing sea has puzzled the human mind as long as reports exist from sailors, fishermen, coastal inhabitants - and even poets. It is thus not surprising that even in early history apparent properties such as the *colour* of the water were used to describe certain parts of the oceans, because empirically it was recognized that there existed a correlation between the water colour and properties such as the abundance of fish, and the water turbidity, etc. Typical examples are the very names of the Red Sea, the Yellow Sea and the Black Sea.

First attempts to use water colour as a means of classifying different lakes, rivers or ocean areas were undertaken by Wasmund 1930, who used colour cards and a visual inspection of different waters from a low-flying aircraft. However, it took modern science and technology so as to quantitatively correlate the *optical* properties of water with relevant biological, biochemical and physical water parameters.

331

V. Barale and P.M. Schlittenhardt (eds.),
Ocean Colour: Theory and Applications in a Decade of CZCS Experience, 331–344.

This research field of *optical oceanography* at first was restricted to tank and ship measurements and was performed at many leading oceanographic research institutes even before the middle of this century. Relevant tasks were investigated *e.g.* by Knudsen 1922, Shuleikin 1923, Atkins and Poole 1933, Hulburt 1934, Pettersson 1934, Kalle 1938, Wattenberg 1938, Jerlov and Liljequist 1938 and Whitney 1938. Thus *optical oceanography* (Jerlov 1976) is the basis for remote sensing of ocean colour.

During the years after the second world war the potential of remote sensing was recognized. First attempts were undertaken to interprete aerial photographs with regard to water features (Nishizawa et al 1954 , Gierloff-Emden 1967). Later these were replaced by airborne radiometers and by multispectral scanners (Ramsey 1968, Clarke et al 1970, Arvesen et al 1973, Mueller 1976, Morel and Prieur 1977, Doerffer 1979, Fischer and Doerffer 1987).

The problems with regard to surface effects and the intervening atmosphere were recognized and correction models were developed (Cox and Munk 1955, Raschke 1971, Kattawar and Humphreys 1976, Plass et al 1976, Gordon 1978, Quenzel and Kaestner 1980, Sturm 1980, Viollier 1982). However, the real breakthrough came with the launch of the Coastal Zone Color Scanner (CZCS) on the Nimbus-7 satellite of NASA in 1978 (Hovis et al 1980). After some problems at the beginning, CZCS data were processed and analyzed in many laboratories around the world. This work contributed to the discovery of new effects in upwellings and ocean circulations. Finally, world maps of the ocean's chlorophyll distribution could be produced (Feldmann *et al.* 1989). Today, after the CZCS has been switched off, scientists are looking forward to the future launch of the next generation of satellites.

1.2 MEASUREMENT PRINCIPLE

The reason why the colour of the sea can be analyzed in terms of water substances lies in the fact that visible sun- and skylight in the spectral range from 400 to 700 nm is able to penetrate both the atmosphere and the upper layers of the water column. By the processes of scattering, absorption and fluorescence it interacts with suspended and dissolved substances. Its intensity and spectral composition is thus changed in a way which is more or less characteristic for these substances (figure 1).

The backscattered portion penetrates the surface and can be detected by multispectral radiometers or scanners. A suitable analysis of the relevant spectral reflectance permits a quantitative interpretation in terms of suspended matter, pigments, yellow substance (Joseph 1950, Duntley *et al.* 1974, Austin 1974, Jerlov 1976, Smith and Baker 1978, Prieur and Sathyendranath 1981, Fischer 1983, Gordon and Morel 1983, Siegel 1986) and pigment fluorescence (Neville and Gower 1977, Gower and Borstad 1981, Kim *et al.* 1985, GKSS 1986).Before the measurement of ocean colour can be used for marine applications, the relationships between measured radiances and the biochemical parameters in the water have to be established quantitatively including their validity under certain environmental conditions (interpretation algorithms). This includes a quantitative evaluation and correction of disturbing effects, such as *e.g.* the reflectance at the water surface or the scattering and absorption of the atmosphere (correction methods).

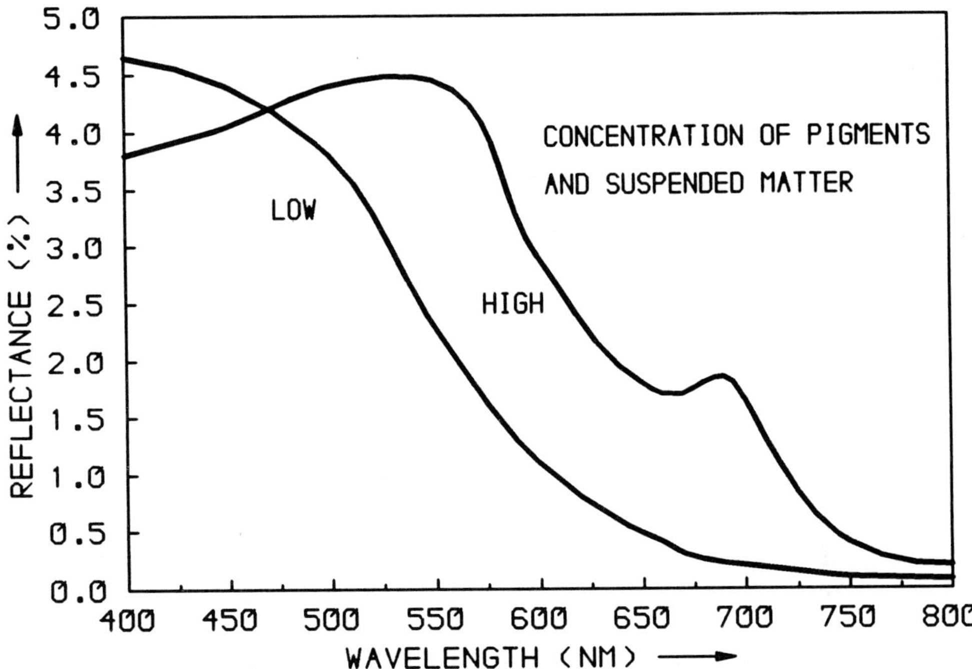

Figure 1. Typical reflectance of water containing different concentrations of phytoplankton and suspended matter.

2. Monitoring Systems

2.1 REQUIREMENTS FOR OCEAN COLOUR SENSORS

Taking into account the successful features, as well as the evident shortcomings of the CZCS experiment, the main requirements for a second-generation sensor for global monitoring of ocean colour are:

- High sensitivity and S/N ratio
- High radiometric resolution
- Large swath width
- Medium geometric resolution
- Narrow spectral bands
- High pixel-to-pixel registration in different channels
- Tilt capability for sunglint avoidance
- Low instrument polarization
- High calibration accuracy (*e.g.*, calibration against the sun)
- Channels in the NIR for atmospheric corrections.

While the geometric resolution is not a design driver as in land applications, a high radiometric resolution and calibration accuracy is required to correct for the atmospheric effects and to apply suitable bio-algorithms (Gordon 1990). The incorporation of the fluorescence measurements for chlorophyll mapping in coastal zones imposes the definition and implementation of very narrow spectral bands for both, the fluorescence channel near 685 nm and baseline reference channels below and above this wavelength. For the monitoring of large scale features at a moderate repetition rate, the swath width has to be large. In addition the sensor must be able to be tilted forward or backward for sunglint avoidance.

2.2 SCANNING TECNOLOGIES

The scanning of a scene by means of a rotating or oscillating 45° mirror (*e.g.*, CZCS, AVHRR) permits sequential imaging of a large FOV on one detector for each spectral channel. S/N can be improved by a larger optics and/or by simultaneously imaging on more than one detector/channel (*e.g.*, OCM). The latter technique also increases the dwell time by reducing the scanning velocity of the rotating mirror.

The incorporation of a linear detector array in the focal plane permits simultaneous imaging of an entire scan line through a suitable interference filter for spectral band selection. Unless beam splitting is used for a separation of spectral bands, a separate optic/detector module has to be provided for each spectral channel. This may cause problems with regard to pixel-to-pixel co-registration amongst different spectral channels.

Imaging spectrometer designs usually incorporate a spectrometer for channel separation and either a set of linear arrays, or a two-dimensional detector array in the focal plane. The latter is used for both spectral selection and for simultaneously imaging along the scan line perpendicular to the flight direction. The main advantage of imaging spectrometers is an easy and adaptable channel selection upon command as spectral information becomes simultaneously available in the entire focal plane. However, a compromise has to be met between the pixel co-registration over a wide spectral range and the size of the entire FOV. Imaging of a wide FOV is possible through the incorporation of a few optic modules, as, in the case of for example the Canadian airborne Fluorescence Line Imager, FLI (Hollinger *et al.* 1987) or of the German Reflective Optics System Imaging Spectrometer, ROSIS (Kunkel *et al.* 1987).

2.3 FUTURE SATELLITE MONITORING SYSTEMS

Many attempts have been undertaken by NASA and ESA during the past to place another ocean color sensor into space as a replacement for the Nimbus-7/CZCS. Figure 2 gives a survey of the various investigations. None of these sensors (except for the OSTA-1/OCE mission) have been launched so far, either because the entire mission was not approved (COMSS/OCM, or the ocean colour instrument was not implemented on a particular platform (NOAA/OCI,ERS-1/OCM, ERS-2/AOCM).

Recently, however, the increasing demands by the oceanographic and climate-oriented user communities (for monitoring ecology, environmental pollution and the carbon fixation in the sea), the involvement of new methods

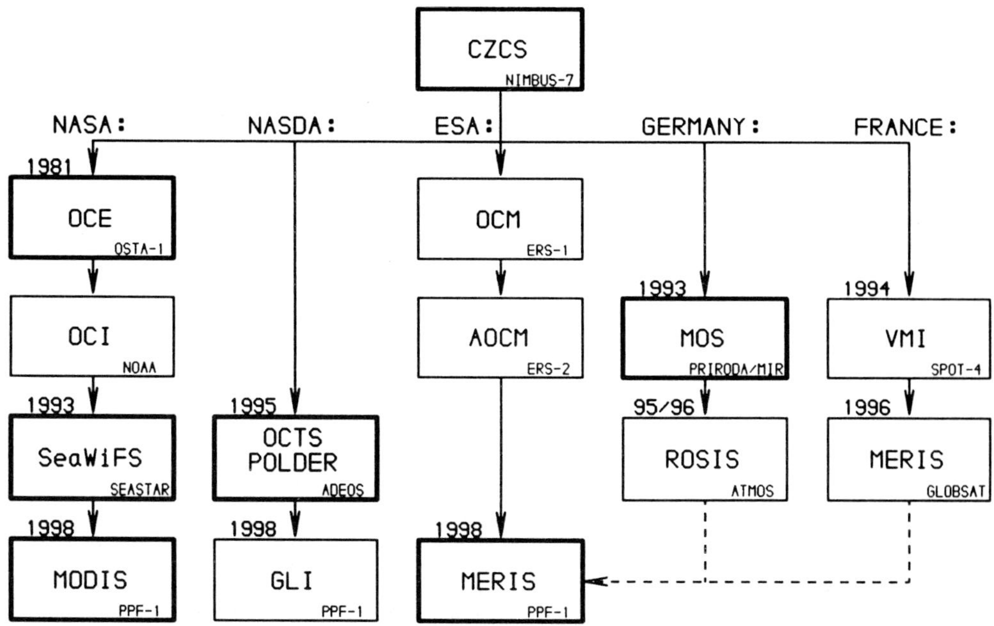

Figure 2. Development of sensors for global remote sensing of water colour within the programmes of NASA, NASDA, ESA, Germany and France. An enhanced frame indicates that the mission either has been flown during the past (CZCS, OCE) or has been approved to be flown (MOS, SeaWiFS, OCTS, POLDER, MODIS, MERIS).

(natural fluorescence as indicator for pigment concentration; (Gower and Borstad 1981; Kim *et al.* 1985; GKSS 1986)) and the incorporation of new sensor/detector technologies (imaging spectrometers using CCD detectors (Hollinger 1987, Kunkel *et al.* 1987)) was the cause for a number of new activities and instrument proposals both, in USA, Japan and in Europe. The background for these activities are national programmes (SEASTAR, ADEOS, ATMOS, GLOBSAT) and the plans for polar orbiting platforms (PPF) as part of the NASA, NASDA, and ESA earth observation programmes.

Meanwhile, in view of the delay and possible changes of the programmes for the polar orbiting platforms, other organizations and other countries have been exploring various alternatives for environmental monitoring from space, which are briefly discussed. The specifications and the foreseen launch date are to be considered as preliminary and reflect the status at the time this paper was written.

- **Wind and Cloud Satellite (1990):** During September 1990 the State Oceanographic Administration of the People's Republic of China launched the second weather satellite Fengyun-1B (FY-1B) with 2 identical sensors combining properties of the CZCS and the AVHRR. (5 channels positioned around 505, 555, 625, 875 nm and 10.5 µm.). The blue and green channels are expected to allow an analysis of global chlorophyll at least in case 1 water, and the thermal channel will permit analysis of sea surface temperature. The format of transmitted data is compatible to that of the NOAA satellites and can be received through S-band from NOAA receiving stations.

- **Modular Optical Scanner on PRIRODA (1993):** The Modular Optical Scanner (MOS) from the former IKF (now: DLR) in Germany is an imaging spectrometer consisting of two sub-assemblies (MOS-A and MOS-B) scheduled to fly on the PRIRODA modul, which will be attached to the former USSR/MIR station in 1993. MOS-B measures in 13 channels (half width: 10nm) in the spectral range between 400 and 1010 nm by means of 128 detector elements in each channel arranged perpendicular to the flight direction. The swath width based on the MIR altitude of 350 km is 82 km (Zimmermann 1990).
  While MOS will not permit *global* monitoring of the biosphere, it will be the first space sensor to perform narrow-band measurements during the forthcoming years and thus will strongly support the preparation of the data analysis for future flights of related instruments.

- **Optical Multispectral Scanner in the former USSR programmes:** Little is known with regard to the specifications, the performance and the approval of ocean colour sensors within the former USSR programme. However, an Optical Multispectral Scanner (MSU-M) is flying on the KOSMOS and the OCEAN satellite. Because of its spectral coverage (0.5 - 1.1 µm measured in 4 channels) it is not clear to what extent this sensor can be used for water colour analysis. An improved sensor, the MSU-SK will fly on the PRIRODA mission and will incorporate 5 channels (within the range 0.5 - 12.6 µm). The next generation, the MSU-SK (M) will foresee 6 channels (within the range 0.41 - 12.4 µm), a swath width of $\pm 40°$ (conical scan) and a pixel size of 200 m (nadir) in the VIS/NIR and 600 m (nadir) in the thermal IR. Future plans also include various imaging spectrometers with narrow bands and more channels.

- **Sea-viewing Wide-Field-of-View Sensor on SEASTAR (1993):** The Sea-viewing Wide-Field-of-View Sensor (Sea-WiFS) is a derivative of the CZCS and was originally planned to be launched on LANDSAT-6. Instead, it is planned now to launch this instrument in 1993 on the platform SEASTAR by means of a new special airborne launch vehicle called PEGASUS. Sea-WiFS measures in 8 visible/NIR channels (half width: 20 to 40 nm) centered around 412, 443, 490, 520, 565, 665, 765, 865 nm, with a pixel size (nadir) of 1.1 km and a swath width of 2800 km.

- **Ocean Colour Temperature Scanner on ADEOS (1995):** Recently the Advance Earth Observation Satellite (ADEOS) from the Japanese National Space Development Agency (NASDA) has been approved to be launched in 1995. Amongst other instruments, ADEOS contains 2 sensors, which are of interest for the measurement of ocean colour, the Ocean Colour Temperature Scanner (OCTS) from NASDA, and an instrument for the measurement of the Polarization and Directionality of the Earth's Radiation (POLDER) as AO instrument from France. ADEOS has a sunsynchronous, descending orbit with an equator crossing time at 10.30 a.m. at an altitude of 796 km. The OCTS is an opto-mechanical scanner with medium resolution (IFOV = 0.85 mrad; nadir pixel size of 700 m), and a large swath width (1400 km). It operates in 8 channels in the VIS/NIR range (halfwidth: 20/40 nm), in 1 channel in the short wave IR and in 3 channels in the thermal IR (0.412, 0.443, 0.490, 0.520, 0.565, 0.665, 0.765, 0.865, 3.7, 8.5, 11.0, 12.0 µm). The sensor can be titled by ±20° for sunglint avoidance.

- **Polarization and Directionality of the Earth's Radiation on ADEOS (1995):** POLDER allows the multi-directional measurement of the backscattered sun light as a function of its degree of polarization (at 0°, 45°, 90°) with a large swath width and a coarse resolution (nadir pixel size: 5 km). The measurements take place in 8 channels in the visible/NIR at wavelengths, which are also of interest for water colour assessment including atmospheric reference channels (435, 490, 520, 565, 670, 760, 820, 880 nm).

- **Vegetation Monitoring Instrument on SPOT-4 (1994):** The vegetation Monitoring Instrument (VMI) on SPOT-4 is a large swath width sensor (FOV: ±50.5°) measuring in 5 channels (0.45 - 1.75 µm),which may be added to the standard payload (HRV) on SPOT-4 with a launch around 1994. Because of the incorporation of the blue spectral range it should be able to measure also water colour in particular chlorophyll-a.

- **Reflectice Optics System Imaging Spectrometer on ATMOS (1995/96):** The Federal Republic of Germany has been studying the Environmental Monitoring Satellite ATMOS, which amongst other sensors will also carry the Reflective Optics System Imaging Spectrometer (ROSIS). ROSIS on ATMOS is a modification of an airborne prototype according to ESA's specifications for the sensor MERIS. The instrument foresees a large swath width (1460 km) at a medium pixel size (250 m) and narrow bandwidth channels (2.5 - 10 nm) within the spectral range from 400-1050 nm. 18 out of 52 spectral channels can be selected upon command and can be recorded and directly transmitted through X-band or (with reduced geometric resolution) through S-band.
ATMOS has a sun-synchronous, descending orbit with an equator crossing time at 11.00 a.m. at an altitude of 793 km.

- **Medium Resolution Imaging Spectrometer on ESA's PPF-1 (1998):** ESA is presently studying the Medium Resolution Imaging Spectrometer (MERIS) to be flown on ESA's first polar platform mission (PPF-1, also called POEM-1). The performance of MERIS is very similar to that of ROSIS, but offers a slightly higher spectral resolution. However, a tilt mechanism for sun glint avoidance is not foreseen on MERIS. The ESA polar orbiting platform will have a sun-synchronous, descending orbit with a equator crossing time at 10.00 a.m. at an altitude of 800 km.

- **Medium Resolution Imaging Spectrometer on GLOBSAT (1996):** ESA's Medium Resolution Imaging Spectrometer (MERIS) is also incorporated in the GLOBSAT mission. which like ATMOS is presently investigated by France as early-launch alternative to ESA's polar platform.

- **Moderate Resolution Imaging Spectrometer on NASA's PPF-1 (1998):**
The Moderate Resolution Imaging Spectrometer (MODIS) is a candidate for
NASA's EOS programme and consists of different instruments (the tiltable
MODIS-T and the nadir-pointed MODIS-N) with a certain overlap of
mission objectives. Because of its capability of sun glint avoidance through
tilt, MODIS-T used to be the main instrument for water colour applications.
However, in view of the budget constraints of EOS and in view of the
approved SeaWiFS mission, the priority for flying MODIS-T has been
lowered recently substantially.
MODIS-N is scheduled to be flown on NASA's first polar platform mission
(PPF-1, designated EOS-A). It makes use of a large FOV ($\pm 45°$) and has a
moderate pixel size (0.25 km at nadir in the VIS/NIR, 1 km in the thermal
IR channels). The spectral range (0.47 - 14.23 µm) is recorded/transmitted
in 36 spectral channels (halfwidth varying between 10 and 300 nm). Special
channels are foreseen so as to measure polarization of the upwelling
radiance. The radiometric quantization is 11 bit (plus 1 bit used as indicator
for the automatic gain control). The orbit of NASA's polar orbiting platform
will be sun-synchronous, descending with an equator crossing time at 13.30
p.m. at an altitude of 705 km.

- **Global Imager on NASDA's PPF-1 (1998):** The Global Imager (GLI) is a
candidate for the first polar platform mission by Japan (JPOP-1, presently
also called ADEOS-2). Similar to the sensors ROSIS, MERIS, MODIS, this
sensor incorporates a wide swath width ($\pm 50°$) for global monitoring in at
least 24 narrow spectral bands (half width 10 nm in the visible) within the
spectral range 0.4 - 12.5 µm at a moderate pixel size of 1 km (at nadir).
The orbit of the NASDA platform will be sun-synchronous, descending with
an equator crossing time at 10.30 a.m. at an altitude of 800 km.

Past, present and future satelliteborne scanners for global monitoring of
water colour are summarized in Table 1.

> TABLE 1. Past and future satellite scanners which permit a
> quantitative analysis of water colour. (A question mark indicates
> that the mission is planned, but has not been approved yet.)

3. List of Abbreviations

ADEOS	Advanced Earth Observation Satellite (NASDA)
AOCM	Advanced Ocean Color Monitor (ESA)
AVHRR	Advanced Very High Resolution Radiometer (NOAA)
CCD	Charge Coupled Device
CNES	Centre National des Etudes Spatiales (France)
CZCS	Coastal Zone Color Scanner (NASA)
COMSS	Coastal Monitoring Satellite System (ESA)
DLR	Deutsche Forschungsanstalt für Luft- und Raumfahrt (FRG)
EOP	Earth Observation Programme (ESA)
EOS	Earth Observation System (NASA)
ERS	ESA Remote Sensing Satellite
EURECA	European Retrievable Carrier (ESA)
FLI	Fluorescence Line Imager (Canada)
FOV	Field-of-View
GKSS	GKSS-Forschungszentrum Geesthacht (FRG)
IKF	Institut für Kosmos-Forschung, now DLR (FRG)
IR	Infrared
JPOP	Japanese Polar Orbiting Platform
MBB	Messerschmitt-Bölkow-Blohm (FRG)
MERIS	Medium Resolution Imaging Spectrometer (ESA)
MODIS	Moderate Resolution Imaging Spectrometer (NASA)
MOS	Modular Optoelectronic Scanner (FRG)
MSS	Multispectral Scanner
NASDA	National Space Development Agency of Japan
NIR	Near-infrared
NOAA	National Atmospheric and Oceanographic Administration (U.S.)
NOSS	Navy Operational Satellite System (U.S.)
OCE	Ocean Color Experiment (NASA)
OCI	Ocean Color Imager (NOAA)
OCM	Ocean Color Monitor (ESA)
OCS	Ocean Color Scanner (NASA)
OCTS	Ocean Color and Temperature Scanner (NASDA)
OCWG	Ocean Colour Working Group (ESA)
OSTA	Office for Space and Terrestrial Applications (NASA)
POEM	Polar Orbiting Earth Mission (ESA)
POLDER	Polarization & Directionality of Earth's Reflectance (CNES)
PPF	Polar Orbiting Platforms (NASA/ESA/NASDA)
ROSIS	Reflective Optics System Imaging Spectrometer
S/N	Signal-to-Noise (ratio)
SOA	State Oceanographic Administration (People's Rep. of China)
SST	Sea Surface Temperature
SeaWiFS	Sea-viewing Wide Field-of-view Sensor (U.S.)
TM	Thematic Mapper (U.S.)

References

Avesen, J.C., Millard, J.P., and Weaver, E.C. (1973) 'Remote Sensing of chlorophyll and temperature in marine and fresh waters', Astronautica Acta 18, 229-239.

Atkins, W.R.G. and Poole, H.H. (1933) 'The photo-electric measurement of penetration of light of various wavelengths into the sea and the physiological bearing of the results', Philos. Trans. R. Soc. London, Ser. B 222, 129.

Austin, R.W. (1974) 'The remote sensing of spectral radiance from below the ocean surface', in N.G. Jerlov and E. Steemann Nielsen (eds.), Optical aspects of oceanography, Academic Press, London New York, pp. 317-344.

Clarke, G.L., Ewing, G.C., and Lorenzen, C.J. (1970) 'Remote measurement of ocean colour as an index of biological productivity', Proc. Sixth Intern. Symp. Remote Sensing of the Environment, Ann Arbor Oct. 13-16 1969, 991-1001.

Cox, C. and Munk, W. (1955) 'Some problems in optical oceanography', Journal of Marine Research 14, 63-78.

Doerffer, R. (1979) 'Untersuchungen über die Verteilung oberflächennaher Substanzen im Elbe-Aestuar mit Hilfe von Fernmessverfahren', Arch. Hydrobiol./Suppl. 43 (2/3), 119-224.

Doerffer, R. (1990) 'How to derive concentrations of chlorophyll, suspended matter and gelbstoff from multispectral radiances of case 2 water', ICES Statutory Meeting, Copenhagen 1990, theme session P, paper C.M. 1990 E:19.

Duntley, S.Q., Austin, R.W., Wilson, W.H., Edgerton, C.F., and Moran, S.E. (1974) 'Ocean color analysis', Scripps, Institution of Oceanography, University of California, San Diego, Technical Report SIO 74-10.

ESA (June 1984) 'Ocean Colour Working Group', Progress report to ESA Earth Observation Advisory Committee ESA BR-20.

ESA (July 1987) 'Ocean Colour', Report by the ESA Ocean Colour Working Group. Villefranche-sur-Mer, November 1986, ESA SP-1083.

Feldmann, G., Kuring, N., Ng, C., Essaias, W., McClain, C.R., Elrod, J., Naynard, N., Endres, D., Evans, R., Brown, J., Walsh, S., Carle, M., and Podesta, G. (1989) 'Ocean Colour: availability of the global data set', EOS 70 (23), 634-641.

Fisher, J. (1983) 'Fernerkundung von Schwebstoffen im Ozean', Hamburger Geophysikalische Einzelschriften, herausgegeben von den Geophysikalischen Instituten der Universität Hamburg, Reihe A Heft 65.

Fischer, J. and Doerffer, R. (1987) 'An inverse technique for remote detection of suspended matter, phytoplankton and yellow substance from CZCS measurements', Advances in Space Research 7 (2), 21-26.

Gierloff-Emden, H.G. (1967) 'Das Luftbild als Hilfsmittel zur Aufklärung der Dynamik von Schwebstoff- und Sinkstofftransport in der Nordsee', Deutsche Hydrographische Zeitschrift 20, 275-278.

Gower, J.F.R. and Borstad, G. (1981) 'Use of the in-vivo fluorescence line at 685 nm for remote sensing surveys of surface chlorophyll-a', in J.F.R. Gower (ed.), Oceanography from space, Marine Science Series Vol. 13, Plemum Press, New York London, pp. 329-338.

GKSS (December 1986) 'The use of chlorophyll fluorescence measurements form space for separating constituents of sea water', ESA contract No. RFQ 3-5059/84/NL/MD, Vol I (Summary Report) and Vol II (Appendices).

Gordon, H.R. (1976) 'Radiative transfer: a technique for simulating the ocean in satellite remote sensing calculations', Applied Optics 15, 1974-1979.

Gordon, H.R. (1978) 'Removal of atmospheric effects from satellite imagery of the oceans', Applied Optics 17 (10), 1631-1636.

Gordon, H.R. (1990) 'Radiometric considerations for ocean color remote sensors', Applied Optics 29 (22), 3228-3236.

Gordon, H.R. and Morel, A.Y. (1983) 'Remote assessment of ocean color for interpretation of satellite visible imagery', Lecture notes on coastal and estuarine studies, Vol. 4 Springer-Verlag, New-York-Berlin-Heidelberg-Tokyo.

Hollinger, A.B., Gray, L.H., Gower, J.F.R., and Edel, H.R. (1987) 'The fluorescence line imager: an imaging spectrometer for ocean and land remote sensing', SPIE Conference on Imaging Spectroscopy II, San Diego 20-21 August 1987, Vol 834, pp. 2-11.

Hovis, W.A., Clark, D.K., Anderson, F., Austin, R.W., Wilson, W.H., Baker, E.T., Ball, D., Gordon, H.R., Mueller, J.L., El Sayed, S.J., Sturm, B., Wringley, R.C., and Yentsch, C.S. (1980) 'Nimbus-7 Coastal Zone Color Scanner: system description and initial imagery', Science 210, 60-63.

Hulburt, E.O. (1934) 'The polarization of light at sea', J. Opt. Soc. Am. 24, 35-42.

Jerlov, N.G. and Liljequist, G. (1983) 'On the angular distribution of submarine daylight and the total submarine illumination', Sven. Hydrogr. - Biol. Komm. Skr., Ny Ser. Hydrogr. 14, pp. 15.

Jerlov, N.G. (1976) 'Marine Optics', Second ed. Elsevier Scientific Publishing Company, Amsterdam Oxford New York.

Jospeh, J. (1950) 'Untersuchungen über Ober- und Unterlichtmessungen im Meere und über ihren Zusammenhang mit Durchsichtigkeitsmessungen', Deutsche Hydrographische Zeitschrift 3, 324-335.

Kalle, K. (1938) 'Zum Problem der Meerwasserfarbe', Ann. Hydrol. Mar. Mitt. 66, 1-13.

Kattawar, G.W. and Humphreys, T.J. (1976) 'Remote sensing of chlorophyll in an atmosphere-ocean environment: a theoretical study, Applied Optics 15, 273-282.

Knudsen, M. (1922) 'On measurement of penetration of light into the sea', Journal du Conseil International pour l'Exploration de la Mer, Publ. Circ. 76, pp. 16.

Kim, H.H., van der Piepen, H., Amann, V., and Doerffer, R. (1985) 'An evaluation of 685 nm fluorescence imagery of coastal waters', ESA Journal 85 (1), 17-27.

Kunkel, B., Blechinger, F., Viehmann, D., van der Piepen, H., and Doerffer, R. (Nov. 1987) 'ROSIS - a candidate instrument for polar platform missions', Proc. SPIE Conference on Optoelectronic Technologies for Remote Sensing, Cannes, Vol. 868.

Morel, A.Y. (1974) 'Optical properties of pure water and pure sea water', in N.G. Jerlov and E. Steemann Nielsen (eds.), Optical aspects of oceanography, Academic Press, London New York, pp. 1-24.

Morel, A.Y., Prieur, L. (1977) 'Analysis of variations in ocean colour', Limnology and Oceanography 22 (4),709-722.

Mueller, J.L. (1976) 'Ocean colour spectra measured off the Oregon coast: characteristic vectors', Applied Optics 15 (2), 394-402.

NASA (1986) Earth Observation System Vol. IIb. MODIS instrument panel report NASA.

NASA (1987) Earth Observation System Vol. IIc. HIRIS instrument panel report NASA.

Neville, R.A. and Gower, J.F.R. (1977) 'Passive remote sensing of phytoplankton via chlorophyll-a fluorescence', Journal of Geophysical Research 82, 3487-3493.

Nishizawa, S., Fukuda, M., Inoue, N. (1954) 'Photographic study of suspended matter and plankton in the sea', Bull. Fac. Fish., Hokkaido Univ. 5, 36-40.

Petterson, H. (1934) 'Scattering and extinction of light in sea water', Medd. Oceanogr. Inst. Goeteborg 9, 1-16.

Plass, G.N., Kattawar, G.W., Guinn, J.A. (1976) 'Radiance distribution over a ruffled sea: contribution from glitter, sky and ocean', Applied Optics 15 (12), 3161-3165.

Prieur, L. and Morel, A.Y. (1975) 'Relation theoretiques entre le facteur de reflexion diffuse de l'eau de mer a diverses profondeurs et les caracteristiques optiques (absorption, diffusion)',IAPSO-IUGGU XVI General Assembly, Grenoble.

Prieur, L., Sathyendranath, S. (1981) 'An optical classification of coastal and oceanic waters based on the specific spectral absorption curves of phytoplankton pigments, dissolved organic matter and other particulate materials', Limnology and Oceanography 26 (4), 671-689.

Quenzel, H., Kaestner, M. (1980) 'Optical properties of the atmosphere: calculated variability and application to satellite remote sensing of phytoplankton', Applied Optics 19 (8), 1338-1344.

Ramsey, R.C. (1968) 'Study of remote measurement of ocean colour', Final Report, TRW, NASW-1658, 1-89.

Raschke, E. (1971) 'Berechnung des durch Mehrfachstreuung entstehenden Feldes solarer Strahlung in einem System Ozean-Atmosphäre', Forschungsbericht, Bundesminister für Bildung und Wissenschaft BMBW-FB W, 71-20.

Shuleikin, V.V. (1923) 'On the colour of the sea', Phys. Rev. 22, 86-100.

Siegel, H. (1987) 'On the relationship between the spectral reflectance and inherent optical properties of oceanic water', Beitr. Meereskd., Berlin 56, 73-80.

Smith, R.C., Baker, K.S. (1978) 'Optical classification of natural waters', Limnology and Oceanography 23, 260-267.

Sturm, B. (1980) 'The atmospheric correction of remotely sensed data and the qualitative determination of suspended matter in marine water surface layers', in A.P. Cracknell (ed.), Remote Sensing in Meteorology, Oceanography and Hydrology, Chichester: Ellis Horwood Ltd., pp. 163-197.

Sturm, B. (1981) 'Ocean colour remote sensing and quantitative retrieval of surface chlorophyll in coastal waters using Nimbus CZCS data', in J.F.R. Gower (ed.), Oceanography from space. Marine Science Series, Vol. 13, Plenum Press, New York, London, pp. 267-279.

Van Der Piepen, H., Doerffer, R., Gierloff-Emden, H.G., Amann, V., Barrot, K.W., and Helbig, H. (1987) 'Kartierung von Substanzen im Meer mit Flugzeugen und Satelliten', Münchener Geographische Abhandlungen, A 37.

Viollier, M. (1982) 'Radiometric calibration of the Coastal Zone Color Scanner on Nimbus-7: a proposed adjustment', Applied Optics 21 (6), 1142-1145.

Wattenberg, H. (1938) 'Untersuchungen über Durchsichtigkeit und Farbe des Seewassers 1., Kieler Meeresforsch 2.

344

Whitney, L. V. (1938) 'Transmission of solar energy and the scattering produced by suspension in lake waters', Trans. Wisc. Acad. Sci. Arts Lett. 31, 201-221.

Zimmermann, G. (1990) 'Spektral hochaufgeloeste Fernerkundung zur Bestimmun oekologischer Veraenderungen', in Gesamtdeutsches Symposium: Raumfahrt und Umwelt, Hannover 15 Mai 1990, DGLR-Bericht 90-03, 44-50.

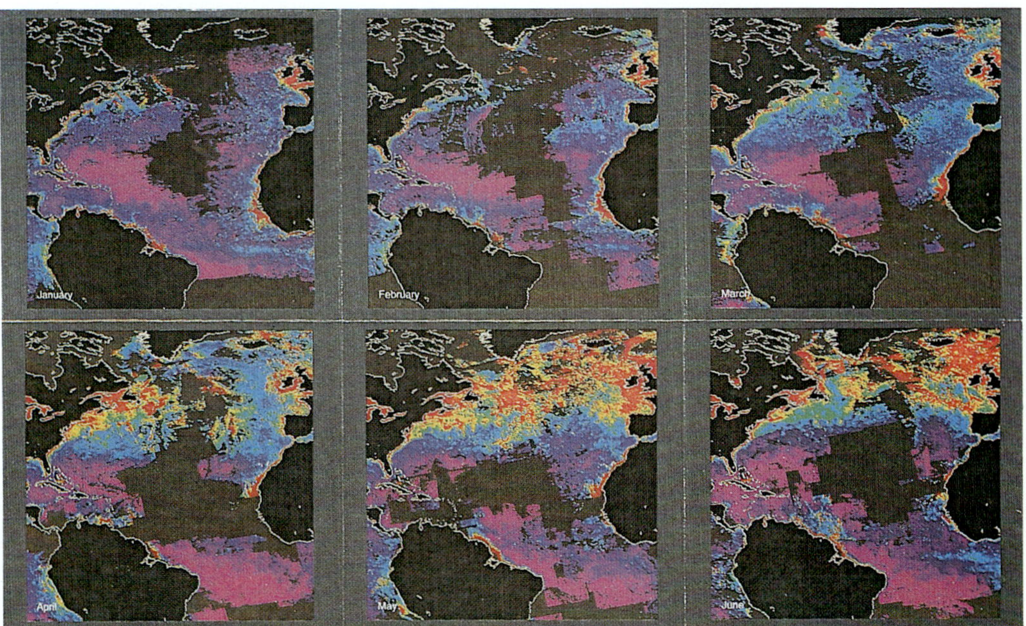

Fig. 6 from p. 23

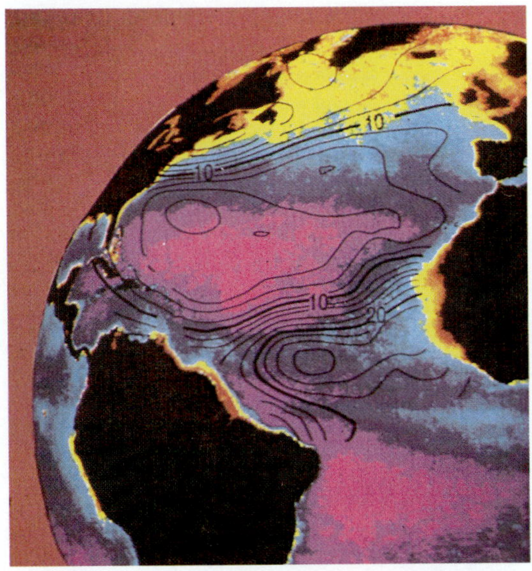

Fig. 12 from p. 28

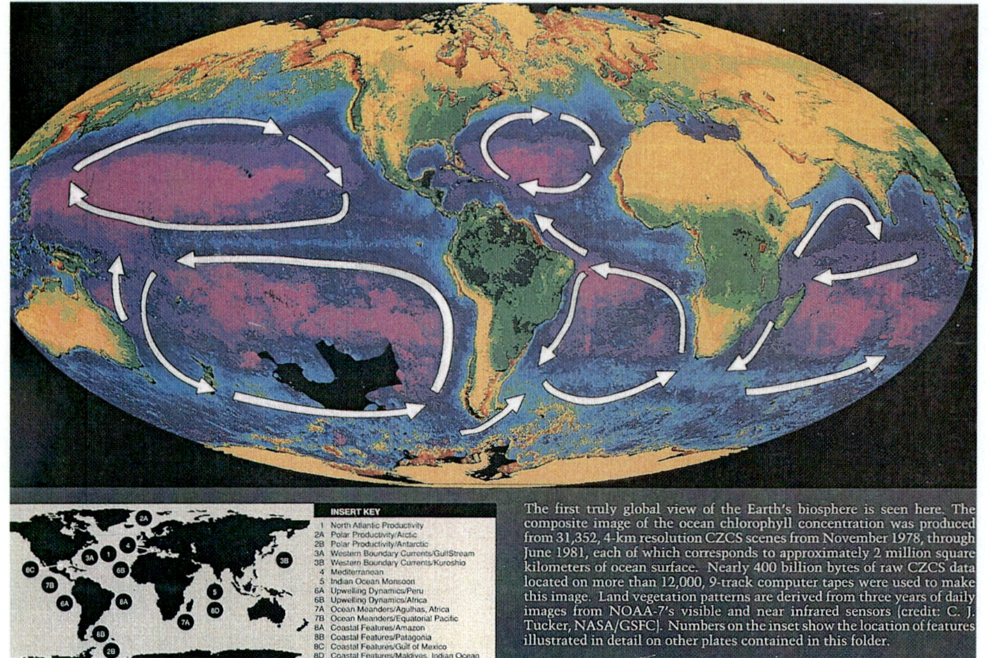

Fig. 14 from p. 30

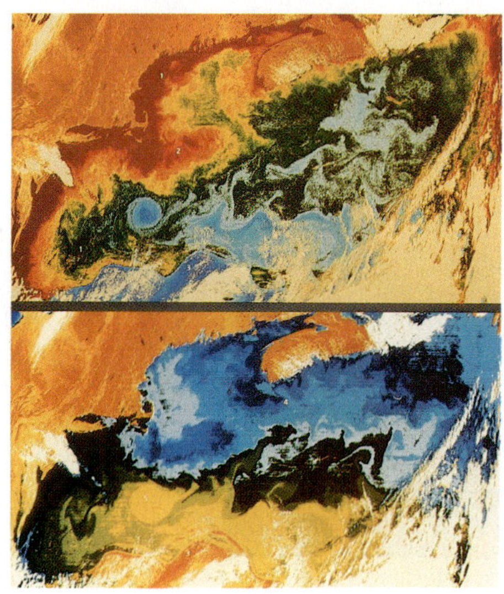

Fig. 2 from p. 171

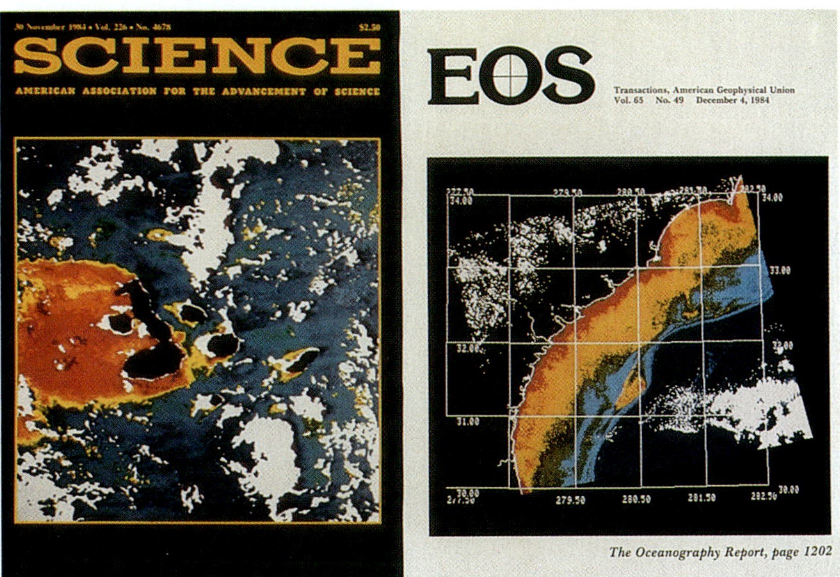

Fig. 3 from p. 171

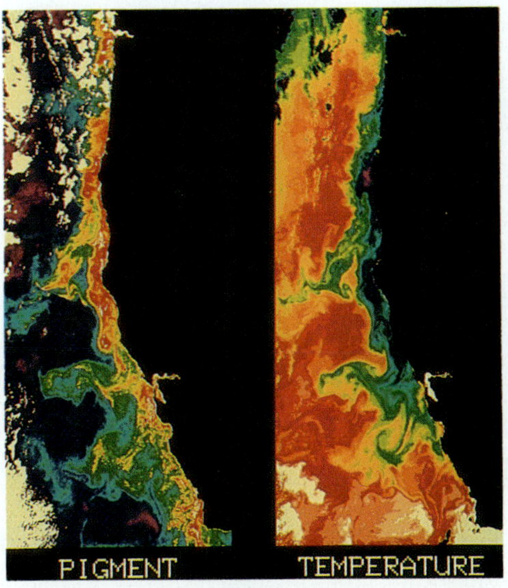

Fig. 4 from p. 172

352

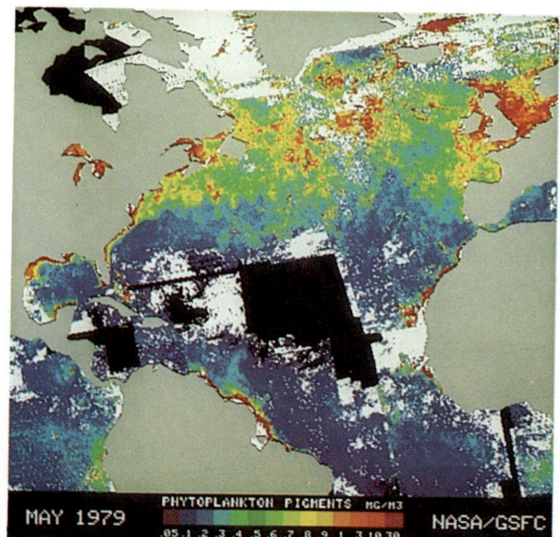

Fig. 5 from p. 172

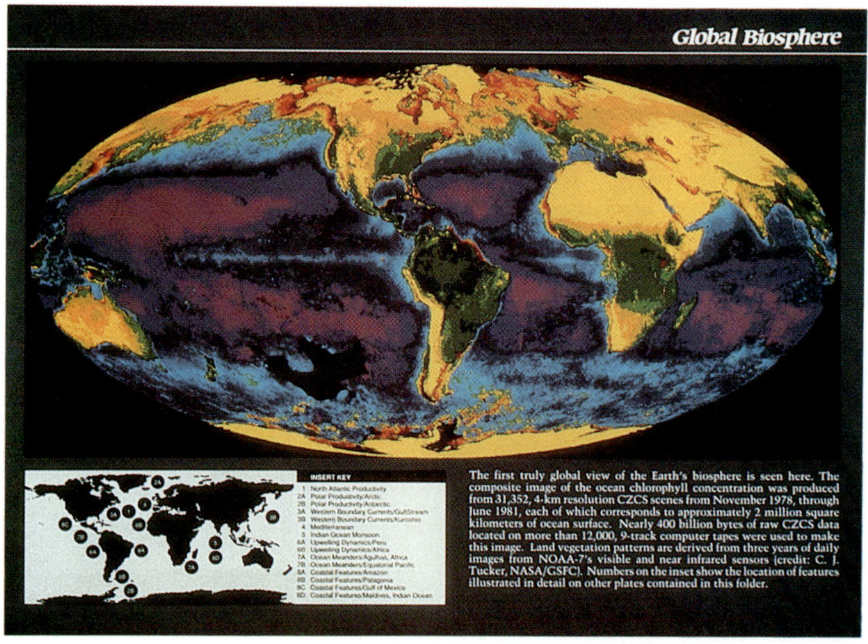

Fig. 6 from p. 173

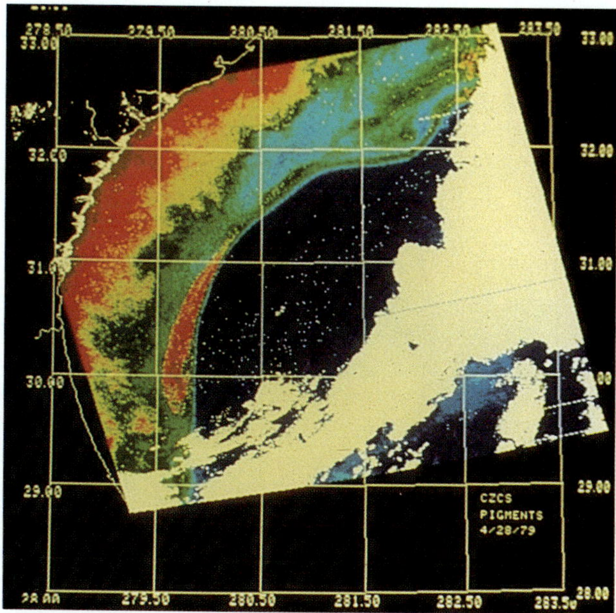

Fig. 8 from p. 175

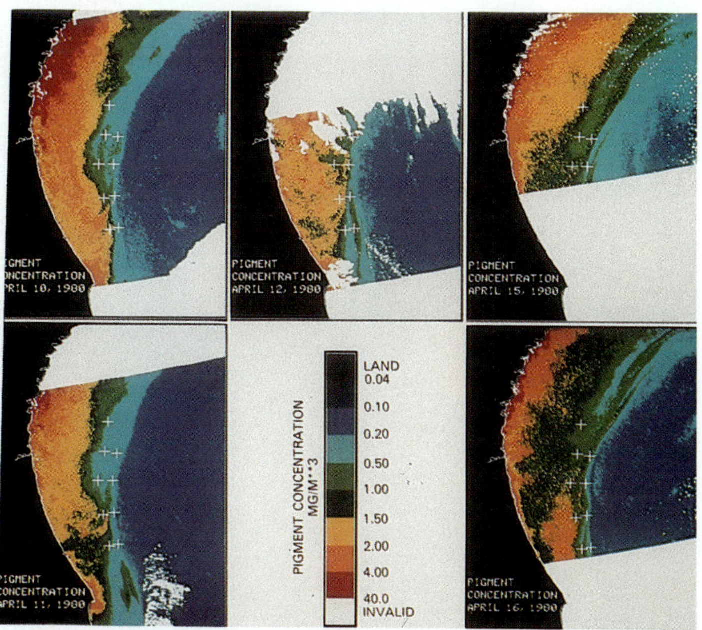

Fig. 10 from p. 177

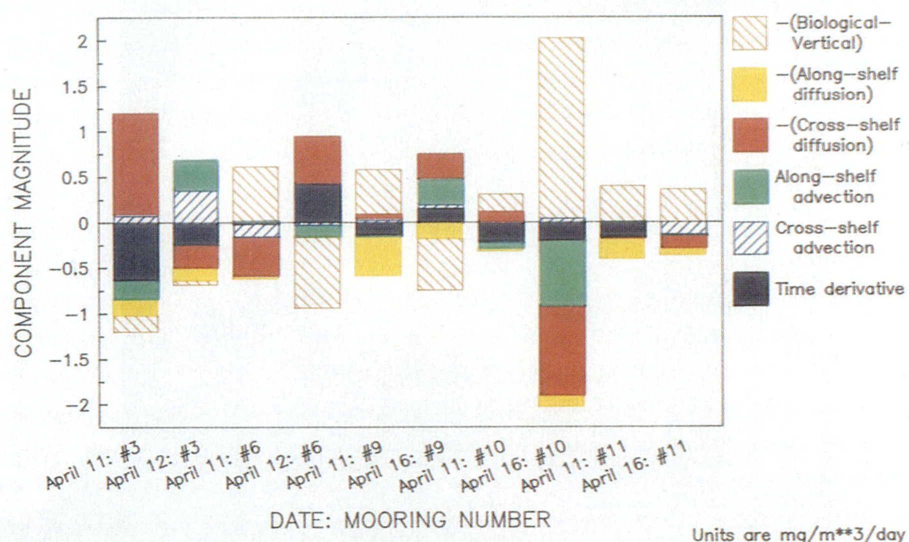

Fig. 11 from p. 178

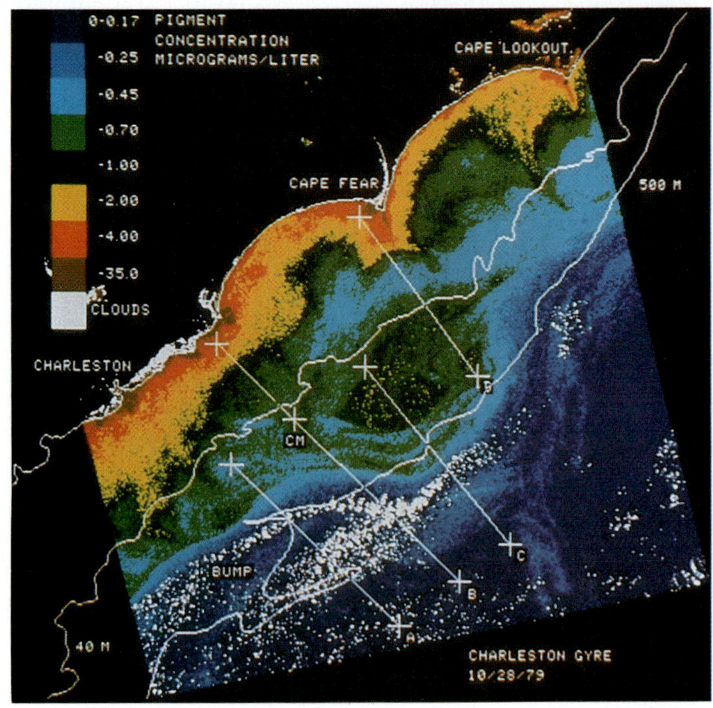

Fig. 12 from p. 179

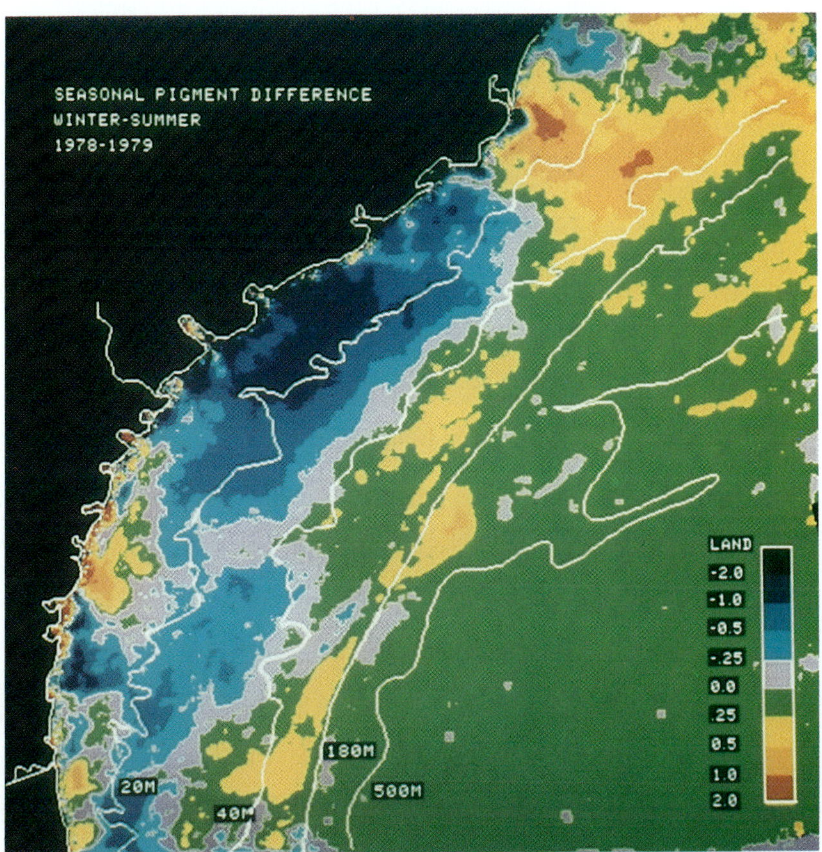

Fig. 13 from p. 180

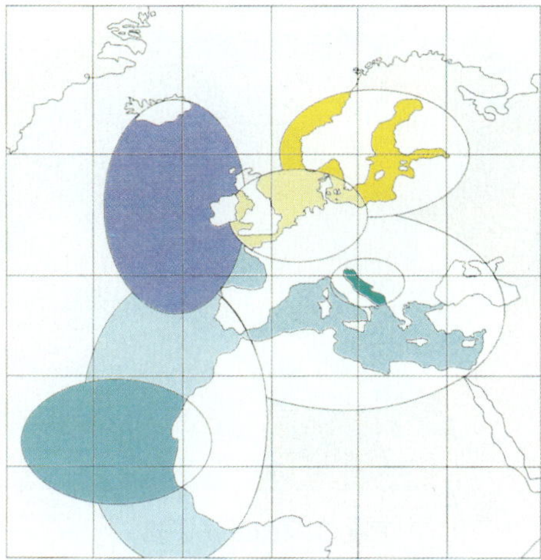

Fig. 2 from p. 197

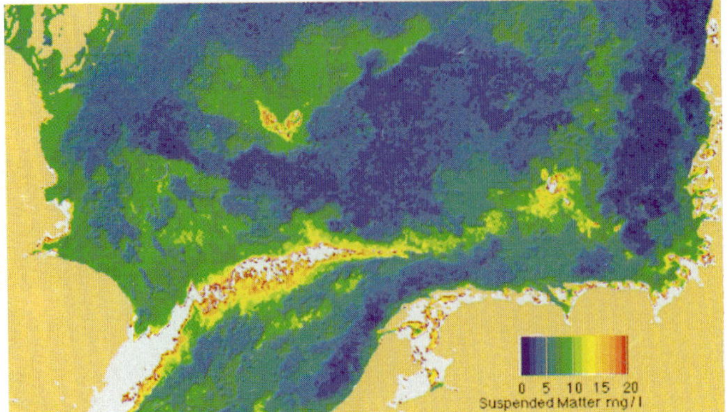

Fig. 3a from p. 198

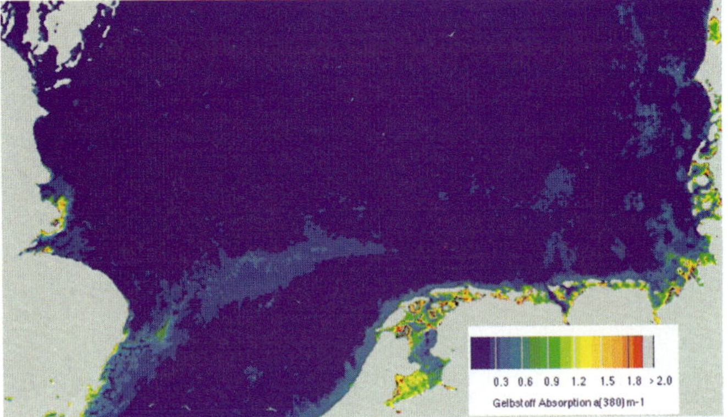

Fig. 3b from p. 198

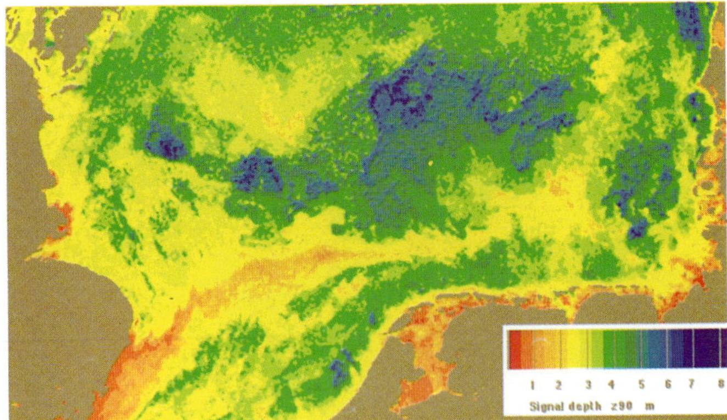

Fig. 3c from p. 198

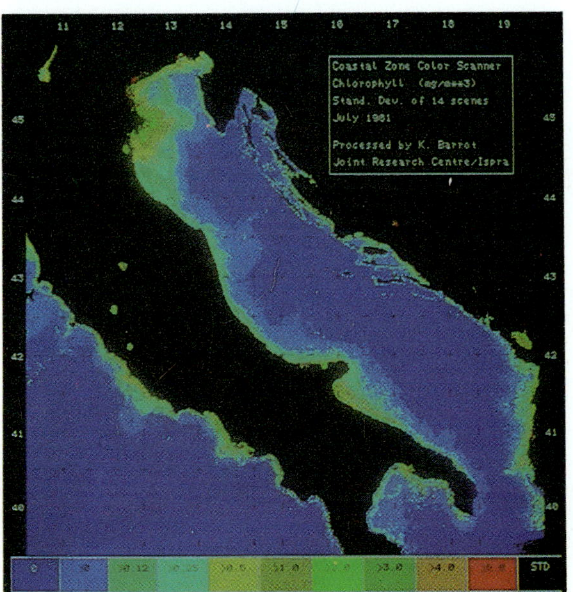

Fig. 4a from p. 199

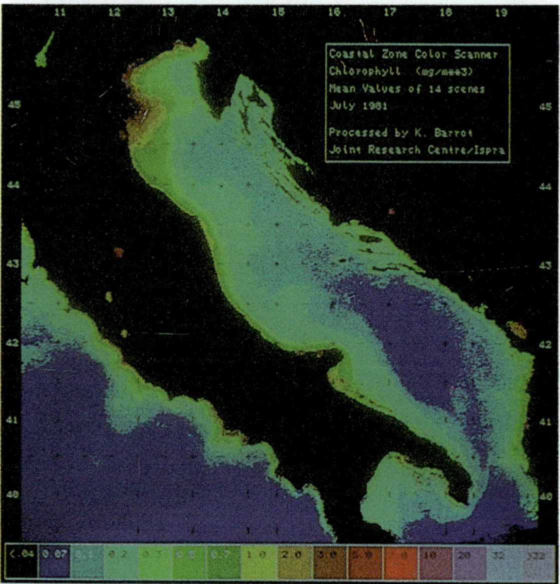

Fig. 4b from p. 199

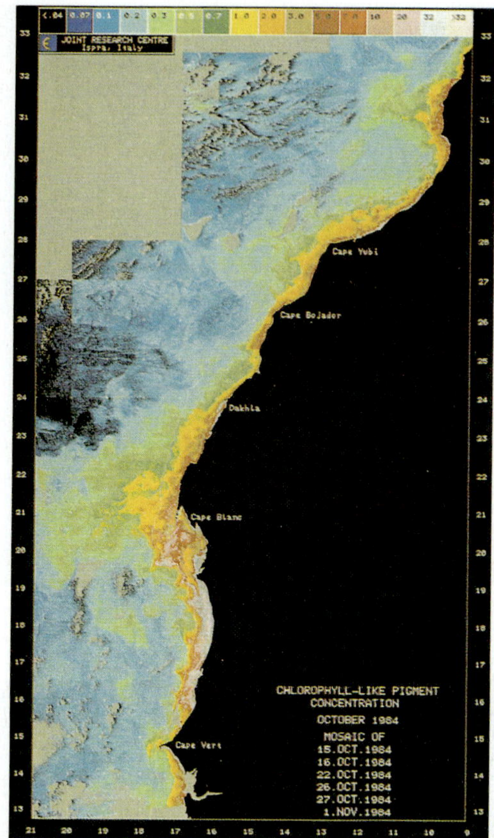

Fig. 5 from p. 201

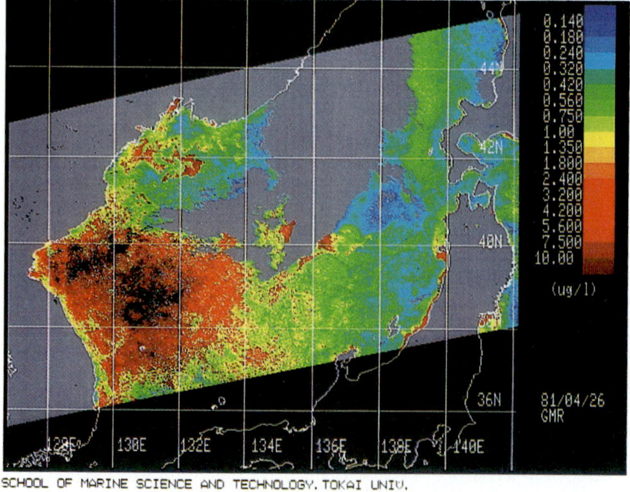

Fig. 4a from p. 218

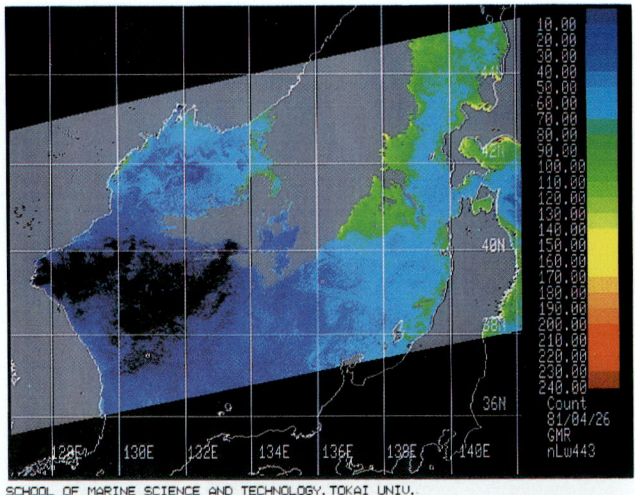

Fig. 4b
from p. 219

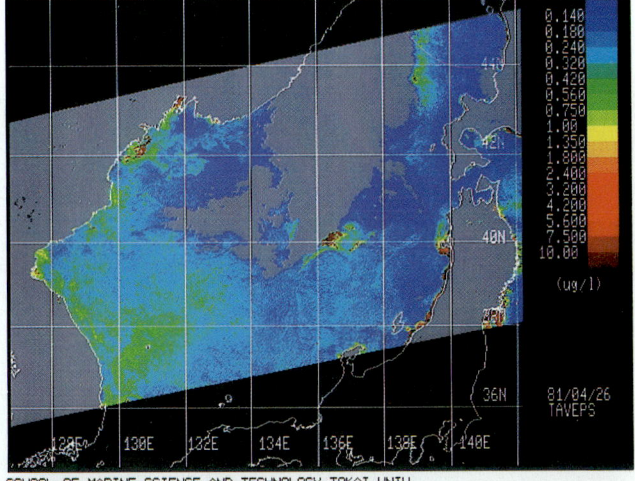

Fig. 7a
from p. 223

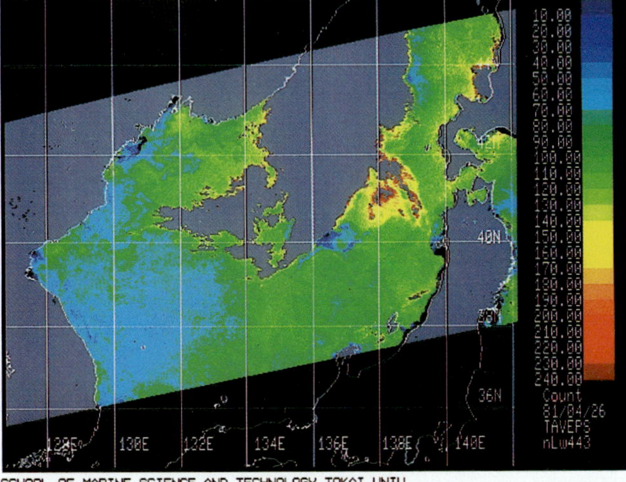

Fig. 7b
from p. 223

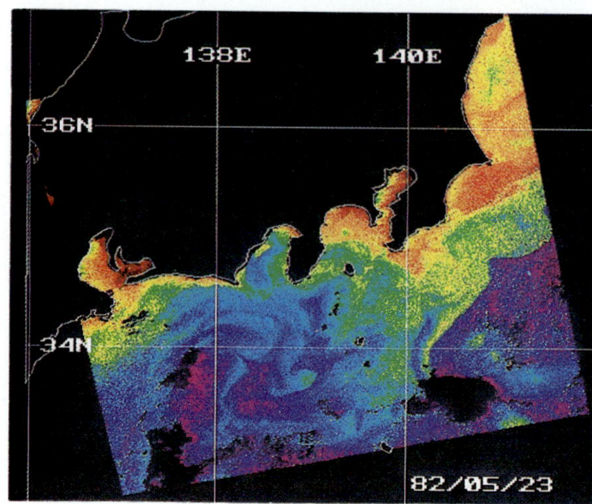

Fig. 9a from p. 226

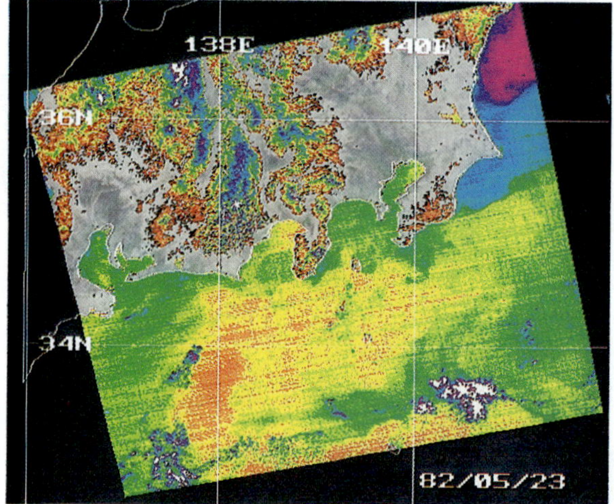

Fig. 9b from p. 226

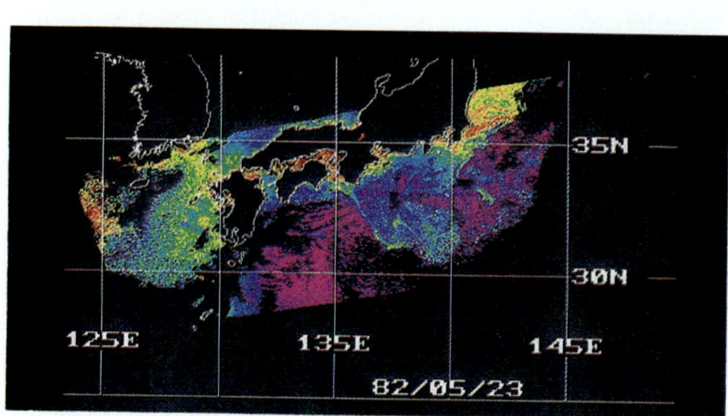
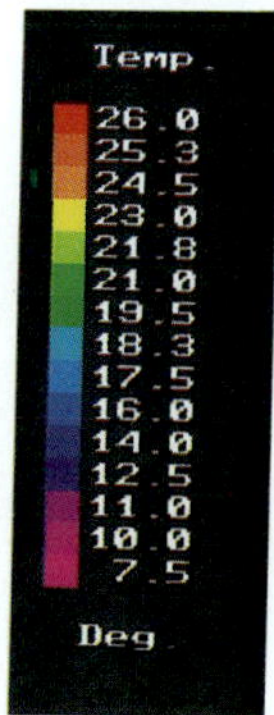

Fig. 10a from p. 227

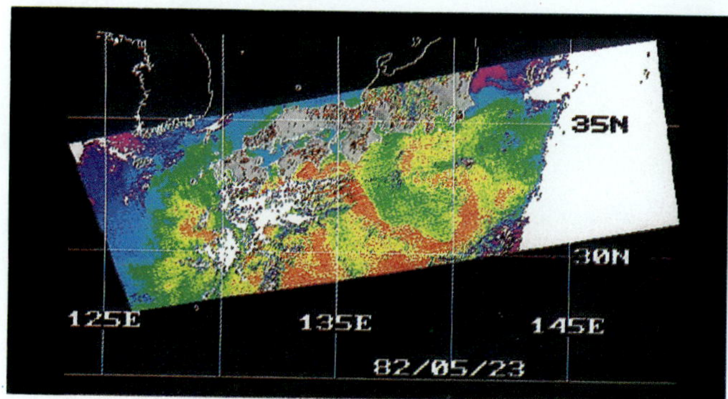

Fig. 10b from p. 227

362

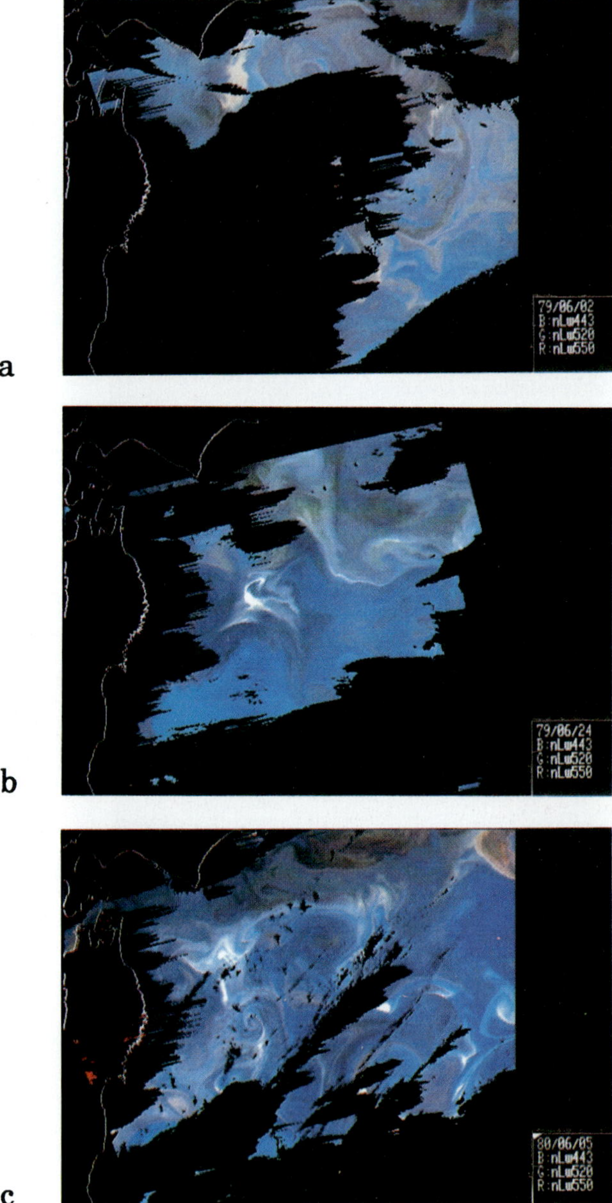

Fig. 12 from p. 230

Fig. 14 from p. 232

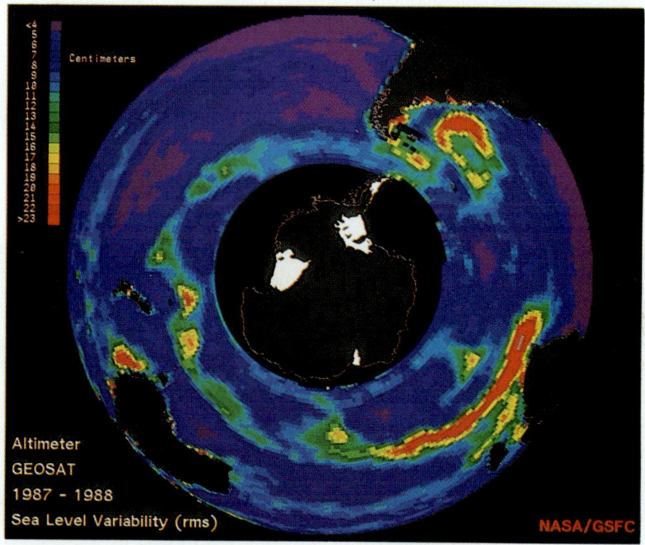

Fig. 1 from p. 241

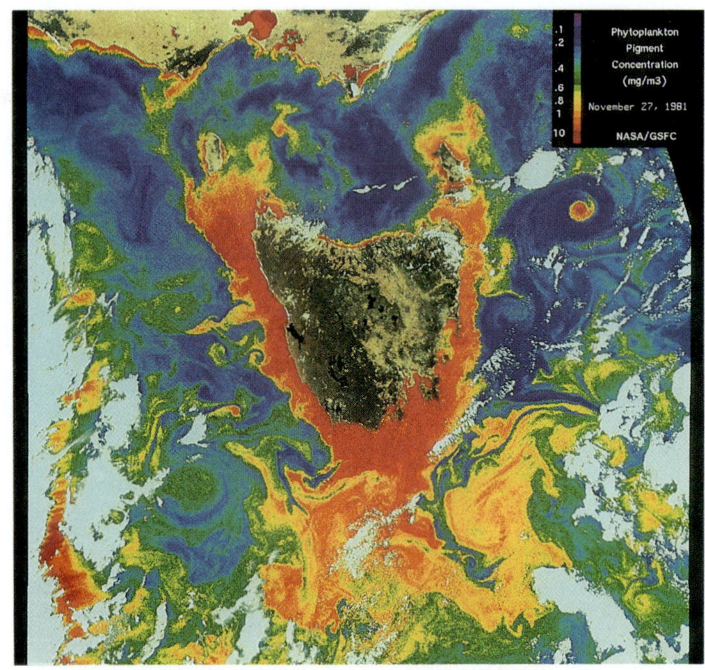

Fig. 2 from p. 244

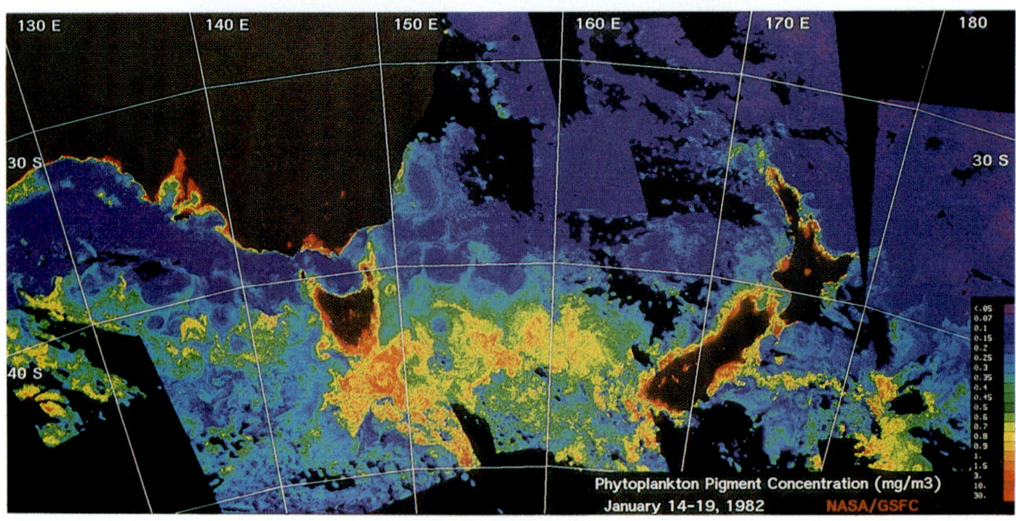

Fig. 3 from p. 244

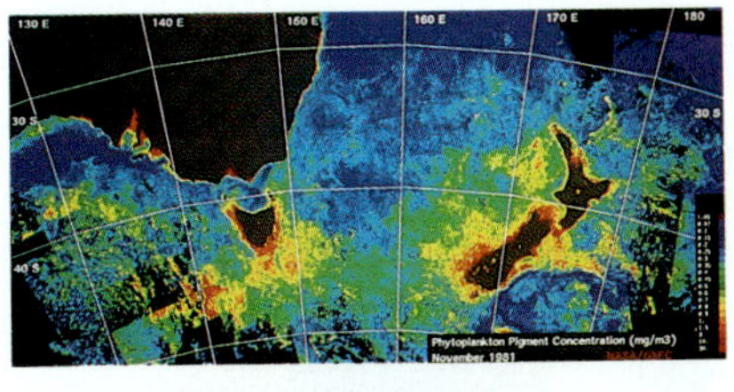

Fig. 4a from p. 245

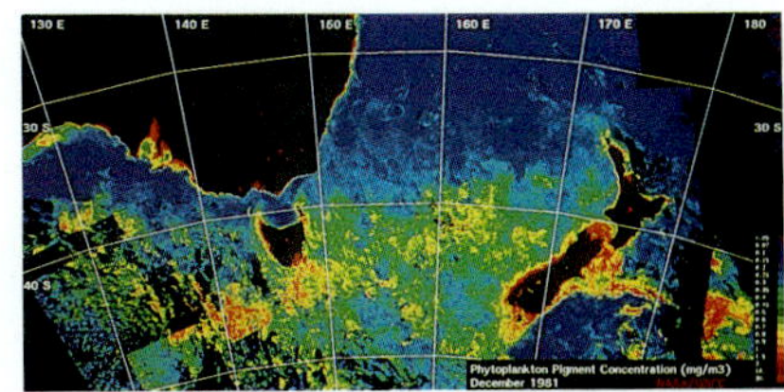

Fig. 4b from p. 245

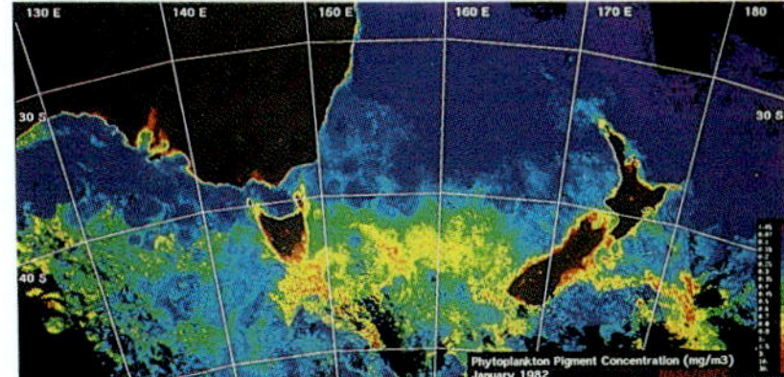

Fig. 4c from p. 245

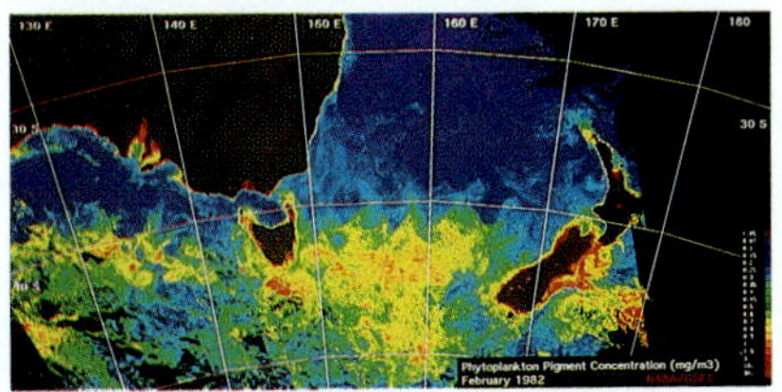

Fig. 4d from p. 245

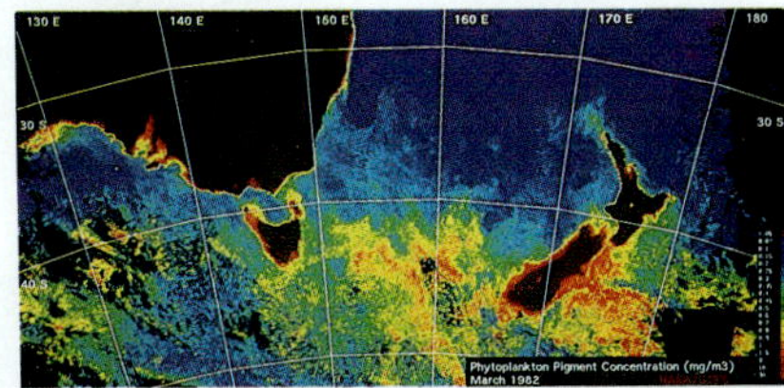

Fig. 4e from p. 245

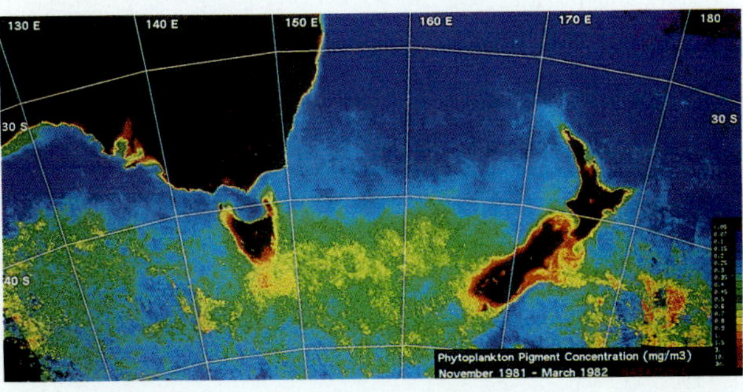

Fig. 5 from p. 246

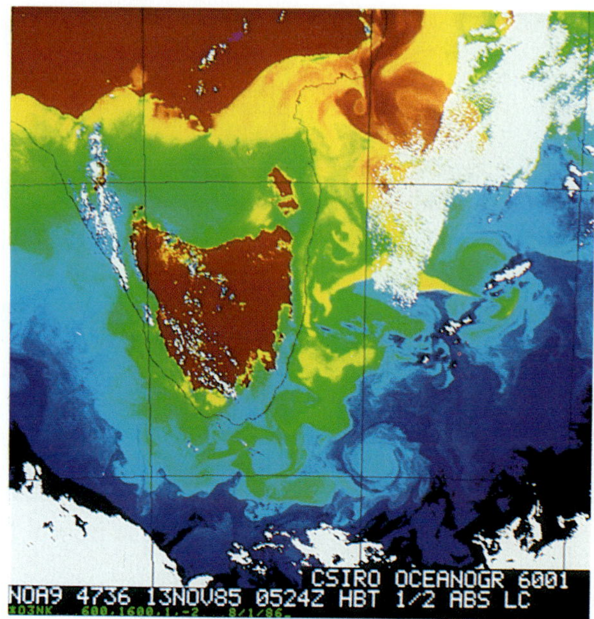

Fig. 6 from p. 247

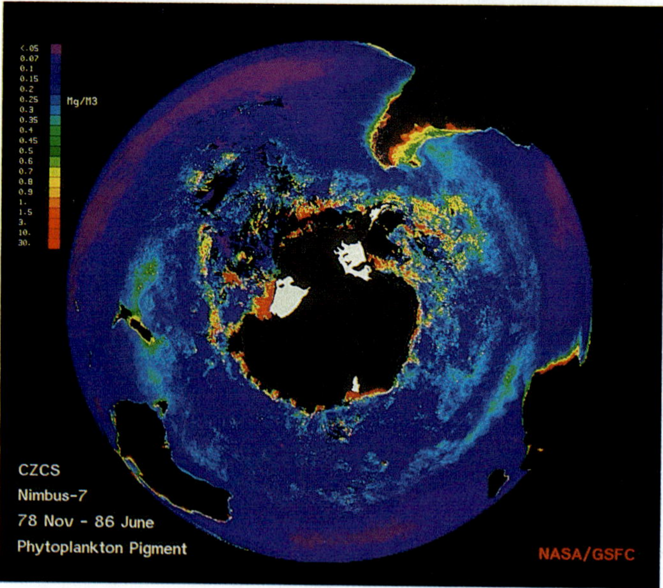

Fig. 7 from p. 248

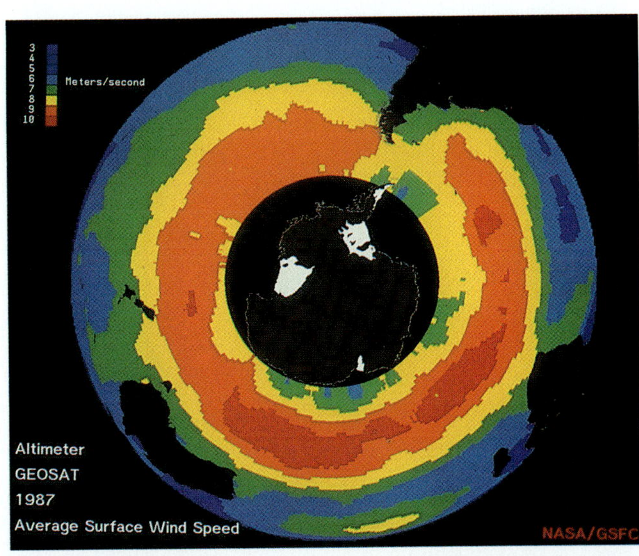

Fig. 8
from p. 249

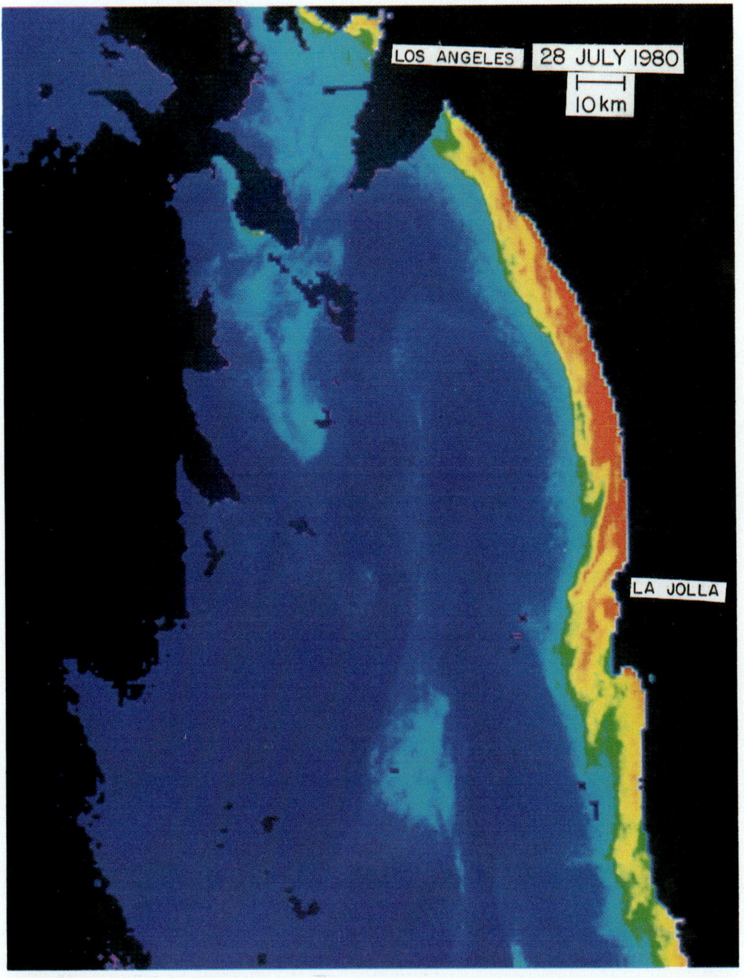

Fig. 6
from p. 305

<u>EURO</u>

C O U R S E S

REMOTE SENSING

KLUWER ACADEMIC PUBLISHERS – DORDRECHT / BOSTON / LONDON